About Island Press

■■■

ISLAND PRESS, a nonprofit organization, publishes, markets, and distributes the most advanced thinking on the conservation of our natural resources—books about soil, land, water, forests, wildlife, and hazardous and toxic wastes. These books are practical tools used by public officials, business and industry leaders, natural resource managers, and concerned citizens working to solve both local and global resource problems.

Founded in 1978, Island Press reorganized in 1984 to meet the increasing demand for substantive books on all resource-related issues. Island Press publishes and distributes under its own imprint and offers these services to other nonprofit organizations.

Support for Island Press is provided by The Geraldine R. Dodge Foundation, The Energy Foundation, The Charles Engelhard Foundation, The Ford Foundation, Glen Eagles Foundation, The George Gund Foundation, William and Flora Hewlett Foundation, The James Irvine Foundation, The John D. and Catherine T. MacArthur Foundation, The Andrew W. Mellon Foundation, The Joyce Mertz-Gilmore Foundation, The New-Land Foundation, The Pew Charitable Trusts, The Rockefeller Brothers Fund, The Tides Foundation, and individual donors.

Freedom for the Seas
in the 21st Century

Freedom for the Seas in the 21st Century

OCEAN GOVERNANCE AND ENVIRONMENTAL HARMONY

Edited by Jon M. Van Dyke,
Durwood Zaelke, and Grant Hewison

Washington, D.C. ❑ Covelo, California

Library of Congress Cataloging-in-Publication Data

Freedom for the seas in the 21st century : ocean governance and environmental harmony / edited by Jon M. Van Dyke, Durwood Zaelke, and Grant Hewison.
                                                                    p.   cm.
                        Includes bibliographical references and index.
                ISBN 1-55963-241-0 (cloth) : —ISBN 1-55963-242-9 (pbk.)
    1. Marine resources conservation—Law and legislation.   2. Marine pollution—
    Law and legislation.   3. Environmental law, International.   4. Maritime law.
    I. Van Dyke, Jon M.   II. Zaelke, Durwood.   III. Hewison, Grant.   IV. Title:
                    Freedom for the seas in the twenty-first century.
                                                        K3485.6F74   1992
            641.7'62—dc20   93-22665
                                                                CIP

Printed on recycled, acid-free paper

Manufactured in the United States of America

10 9 8 7 6 5 4 3 2 1

Contents

Preface   *Nga Kupu Timatanga* (There Is a Story That
   Needs to Be Told)   *Moana Jackson*   xiii/xv

Acknowledgments   xvii

Sponsoring Organizations   xix

Introduction: Traditions for the Future   3

I. A New Look at Ocean Governance   7

1. International Governance and Stewardship of the High Seas
   and Its Resources   *Jon M. Van Dyke*   13
     The Evolution of a New High Seas Regime
     Principles for a Comprehensive Regime of
        Ocean Governance and Stewardship

2. The Process of Creating an International Ocean Regime to
   Protect the Ocean's Resources   *Elisabeth Mann Borgese*   23
     Claiborne Pell's Model Treaty
     The Ocean Regime
     The Yugoslav Model
     Arvid Pardo's Draft Ocean Space Treaty

✓3. Perspectives on Ocean Governance   *Arvid Pardo*   38

4. Indigenous Law and the Sea   *Moana Jackson*   41
     International Law and the Sea
     International Law and Indigenous Law
     Indigenous Law and the Sea
     Revising the Law: Reviewing the Base

5. *Mare Nostrum:* A New International Law
   of the Sea   *Philip Allott*   49
     Principles of a New Law of the Sea
     The 1982 Convention

6. Changing Concepts of Freedom of the Seas:
   A Historical Perspective   *R. P. Anand*   72
      Freedom of the Seas: An Overriding Principle
      Application of the Law of the Sea
      Renewed Challenges to the Freedom of the Seas

II. Pacific Approaches toward the Ocean Environment   87

7. An Introduction to Some Hawaiian Perspectives
   on the Ocean   *Poka Laenui* (Hayden Burgess)   91
      Fundamental Philosophical Differences
      Hawaiian Philosophy and Its Role in the
         International Community
      Hawaiian Philosophy in Practice
      Conclusion

8. The Unfinished Agenda for the Pacific to
   Protect the Ocean Environment   *Mere Pulea*   103
      Background to Ocean Protection Policies
      The SPREP Action Plan
      Regional Legal Arrangements: Strategy for
         Environmental Protection
      The SPREP Convention
      Other Considerations for the Ocean Protection Agenda

9. The Convention for the Protection of the Natural
   Resources and Environment of the South Pacific Region: Its
   Strengths and Weaknesses   *A. V. S. Va'ai*   113
      Pollution
      The Real Problem
      Convention Area
      Pacific Rim Countries
      Small Island States
      Liability and Compensation
      Conclusion

10. Steps Taken by South Pacific Island States to Preserve
    and Protect Ocean Resources for Future Generations
    Florian Gubon   121
      The Pacific Island Environment and Ocean Resources
      Steps Taken by South Pacific Island States
      Conclusion

11. Governance and Stewardship of the Living Resources:
    The Work of the South Pacific Forum Fisheries
    Agency   *Gracie Fong*   131
        South Pacific Fisheries
        South Pacific Regional Fisheries Initiatives
        Conclusion

III. Controlling Ocean Pollution   143

12. The Need for a New Global Ocean Governance System
    W. Jackson Davis   147
        The Oceans as a Planetary Life Support System
        The State of the Oceans
        The Regional Approach to Ocean Governance
        The Global Approach to Ocean Governance
        Science, Politics, and Indigenous Belief Systems

13. Mending the Seas through a Global Commons Trust Fund
    Christopher D. Stone   171
        The Predicament of the Commons
        The Potential Value of Guardians
        Services a Global Commons Trust Fund Would Underwrite
        Sources of Revenues for the Fund
        A Brief History of Comparable Proposals and Their Fates
        The Lessons of This Brief History: What Are the
            Practical Prospects?

14. International Ocean Protection Agreements: What Is Needed?
    Clifton E. Curtis   187
        Existing Support Structures
        What Is Needed?
        Conclusion

15. Seaborne Movements of Hazardous Materials
    Miranda Wecker and Dolores M. Wesson   198
        What If the *Exxon Valdez* Had Carried Hazardous Materials?
        Assessing the Risks
        The Existing Legal Framework for the 1982 United Nations
            Convention on the Law of the Sea
        Prevention of Pollution
        Response to Maritime Accidents
        Emerging Legal and Political Issues

Conclusion: Evolution or Disintegration of the
International Legal Order?

16. Protected Marine Areas and Low-lying Atolls
Jon M. Van Dyke   214
Low-lying Atolls: Unique, Fragile Ecosystems That
Need Protection
Low-lying Atolls: An Inherent Part of the Marine
Environment
Global and Regional Efforts to Establish Protected
Marine Areas
Classification of Low-lying Atolls
Conclusion

IV. The Living Resources   229

17. Unregulated High Seas Fishing and Ocean Governance
William T. Burke   235
Types of Problems Arising from Unregulated Fisheries
The Interests at Stake
Previous Decisions: Principles of International Law for
Straddling Stocks
Proposed Principles Applicable to High Seas Fishing, with
Particular Reference to Straddling Stocks and Drift Nets
Conclusion

18. High Seas Fisheries, Large-Scale Drift Nets, and the
Law of the Sea   *James Carr and Matthew Gianni*   272
Is Effective Management of High Seas Drift Net
Fishing Possible?
The UN Driftnet Resolutions and the United Nations
Convention on the Law of the Sea
Global Standards for the Conduct of Fisheries on the High Seas
Conclusion

19. The International Legal Issues Concerning the Use of Drift
Nets, with Special Emphasis on Japanese Practices and
Responses   *Kazuo Sumi*   292
What Are the Real Problems?
Toward a Solution
Conclusion

20. Reconsidering Freedom of the High Seas: Protection of
     Living Marine Resources on the High Seas
     Catherine Floit   310
         High Seas Drift Nets
         Limits to High Seas Freedoms: Customary
             International Law Obligations
         Obligations under the 1982 Convention on the
             Law of the Sea
         Conclusion

21. Recent Developments in the Use of International Trade
     Restrictions as a Conservation Measure for Marine Resources
     Melinda P. Chandler   327
         Trade in Endangered Species
         Trade in Species Whose Commercial Harvest Results
             in Incidental Taking of Protected Species
         Restrictions on Trade in Commercial Species Whose
             Harvest Is Regulated or Managed
         Protection of Marine Habitats
         Issues to Consider in Utilizing Trade Restrictions
             as a Conservation Measure

22. The Legacy and Challenge of International Aid in Marine
     Resource Development   *Claudia J. Carr*   340
         The Role of Southern Marine Living Resources in the
             Global Economy
         Fisheries Modernization: Toward What and for Whom?
         The Fisheries Aid Establishment in Capture and Culture
         "Sustainable" Fisheries Development: Beyond Rhetoric

V. The 1982 Convention on the Law of the Sea and
   the Nonliving Resources of the Deep Seabed   375

23. Excision of the Deep Seabed Mining Provisions from the
     1982 Law of the Sea Convention: Reappraising the Principle
     of a Treaty's Integrity under the New Realities
     Artemy A. Saguirian   379

24. A Response to Dr. Artemy A. Saguirian
     Elisabeth Mann Borgese   388

25. Environmental Ethics, International Law, and Deep Seabed
     Mining: The Search for a New Point of Departure
     Ian Townsend-Gault and Michael D. Smith   392

VI. Military Activities and Peaceful Uses of the High Seas   405

26. Security Regimes for the Oceans: The Tragedy of the
Commons, the Security Dilemma, and Common Security
Andrew Mack   409
 The Security Dilemma
 The Security Dilemma at Sea
 Solutions

27. Denuclearizing and Demilitarizing the Seas   *Joshua Handler*   420
 Naval Nuclear Weapons
 Nonstrategic Naval Nuclear Weapons
 Strategic Naval Nuclear Weapons
 Naval Nuclear Reactors
 Reduction in Fleet Sizes and Operations
 Conclusion

28. Navies, Ocean Resources, and the Marine Environment
Joseph R. Morgan   435
 A Naval Hierarchy
 Naval Operations and the Environment
 Law of the Sea Considerations
 Comprehensive Security: Naval Operations and the Environment
 Sharing the Waters: A Source of Conflict?
 Conclusion

29. Military Exclusion and Warning Zones on the High Seas
Jon M. Van Dyke   445
 Military Warning Zones on the High Seas
 Confrontation over Warning Zones
 The Governing Legal Principles
 Applying Customary International Law to Warning Zones
 Conclusion

Summary and Conclusions—Ocean Governance: Converging
Modes of Idealism   *Douglas M. Johnston*   471
 Environmentalism
 Structuralism
 Humanism
 Nonstatism
 Transnational Bureaucratism

Perspectives on Environmental Harmony   477

Biographies   479

Index   487

Preface
Nga Kupu Timatanga

••

MOANA JACKSON

E nga mana, e nga reo, e nga iwi, tena koutou, tena koutou, tena koutou katoa.

I nga wa o mua e mohiotia nei e tatou ko te reanga o kui ma o koro ma, me ka pa mai he raruraru he awangawanga ranei ki te iwi, tere tono o ratou whakaaro ki te whiriwhiri. He aha te putake o tenei raruraru? Akuni pea he whiu, he tikanga ano mo tenei ahua. He korero ano pea hai korerotia.

No reira, i te wa i puta ai nga whakaaro o o tatou tipuna, kia whakawhiti ratou i te Moana-nui-a-Kiwa, wehi ana o ratou whatumanawa ki te taumaha o tenei kaupapa. Te tawhiti, te taitai o nga ngaru o te moana. Ka whakaaro ano ratou, me pehea ka taea e ratou tenei huarahi tipua. Ka korero ratou mo nga manu, e taea nei e ratou te whakawhiti nga moana i raro, i nga manaakitanga a Tawhirimatea, to ratou tipuna. Koia nei, te matua o nga hau. Maana e manaaki ratou i te huarahi—ma nga whetu i te po ma tama-te-ra i te awatea. Ma ratou e honohono nga ara, a, tae noa ki nga whenua o tera taha i korerotia nei e Kupe. Ko enei nga tikanga a iwi i waihotia nei e ratou ma. Mai i Tawhiti-nui, Tawhiti-roa, Tawhiti pamamao, te hono ki wairua, tau ana ki Aotearoa.

I waenganui i enei mahi, i to ratou unga mai, ki tenei whenua, ka kite ratou te maha me te tini o nga kai moana o te whenua nei. Ka moumoutia e ratou nga kai moana, me nga kai o te ngahere, engari kaore i roa ka kite ratou i o ratou i o ratou he. Ka hoki ano ratou ki nga tohutohu o o ratou tipuna, ko enei nga akonga rahui, i heke kia tatou o tenei wa. Mai i nga mahi i mahia e ratou, kua heke mai enei hekenga a iwi ki a tatou. Te mauri o nga mea katoa, nana nei i whakatangata tatou.

I roto i a tatou mahi i nga wa katoa ko nga kaupapa i heke mai, ko nga koha i tukuna mai e ratou ma. Me ka takahia e tatou enei mea ataahua ka whakataurekarekatia e tatou, tatou ano. No reira, kia mau ki nga mea a kui ma a koro ma. Tautokotia A ratou akonga, ma enei ka tu rangatira tatou.

Ka kite tatou mai i a ratou korerorero, i a ratou waiata, i heke kia tatou te rangatiratanga o to tatou iwi. I whiria e ratou i nga whiringa mai i Hawaiki, nui, i a Ranginui, i a Papatuanuko, i a Io te whakamaramatanga o nga kete or te wananga.

Te kete tuauri,
te kete tuatea
to kete aronui.
Kua whakatauria ki runga i a tatou.

Preface
There Is a Story That Needs to Be Told

MOANA JACKSON

IN THE DAYS THAT WE CALL THE PAST, and in the times when doubt or trouble confronted the Maori, wise people would seek explanation and say, "There is a story that needs to be told."

And when the ancestors looked in awe across the never-ending sea, and wondered how they could ever navigate its vast and unknown loneliness, a story was told. A tale of migratory birds guided by the moving winds of *Tawhirimatea*, and of star paths linking their flights to the far horizon. And from the story came a certainty that created a tradition of great voyages to new and distant places.

And when the ancestors turned to the ocean for food, and wondered how to maintain the bounty that was waiting for them, another story was told. A tale of careless men robbing once-rich reefs and an unforgiving *Tangaroa* (Maori for "God of the Sea") claiming back the *mauri* (life force) of his bounty. And from the story came a certainty that created a tradition of respect for the gifts that are given to humans, and a realization of the need for balance in all things.

And when, in the course of everyday events, a person abused those gifts or upset the balance, and people wondered how to restore the good order and peace of the *iwi* (tribe), a story would be told. A tale of imbalance in those who had done wrong, and of the wise acts of those who in the past had restored their place and the place of those people or places wronged. And from the story came a certainty that created a tradition of precedent and law to guide the ways of all people, and a legal tradition to protect the balance in all things.

And from the telling of all those stories came a belief that a stable sense of order, of knowing one's place in the world, gave strength and understanding. And from that understanding came the solutions to many problems.

And the stories themselves came from the voices of the people and were woven from the threads of their own existence.

Today, as the harmony of life is threatened by pressures unknown to the ancestors, and as the young seem to upset the uncertain balance of their place in the world, there is a need to seek out new stories, new certainties, new understandings.

This book seeks out such new stories, certainties, and understandings.

Acknowledgments

••

Most of the authors represented in this book met at the Waioli Tea Room in Honolulu in December 1990 for a meeting sponsored jointly by the Center for International Environmental law, the Spark M. Matsunaga Institute for Peace at the University of Hawaii, Greenpeace, and the Peace Research Centre at the Australian National University. Sebia Hawkins of Greenpeace played a leading role in convincing others that the rationale for this meeting was sound and in assembling the necessary financial support. Other important Greenpeace personnel who are not represented directly in the book but who helped with logistic and other types of support were Duncan Currie and Michael Hagler. Among the many persons from the Matsunaga Peace Institute who helped make this meeting a success were Majid Tehranian, Rhoda Miller, Terry Kessel, and Jennifer Baker.

Early drafts of the chapters in this book were presented and discussed at that meeting, and a remarkable interchange took place. This meeting was especially notable because two of the most thoughtful commentators on ocean governance issues over the decades were in attendance—Elisabeth Mann Borgese and Arvid Pardo. Their presence gave this event an aura of significance and helped all the other participants realize the complexity of the problems that needed solution and the nobility of searching for solutions. The other particularly exciting part of this meeting was the presence of so many thoughtful Pacific Islanders, including leaders of indigenous communities and many persons who had participated directly in the negotiation and implementation of the 1986 Convention for the Protection of the Natural Resources and Environment of the South Pacific Region. In addition to those represented in this book, the Pacific Islanders in attendance were Transform Agorau of the Solomon Islands, Roman Bedor of Palau, Kilifoti Eteuati of Western Samoa, and Ratu Joni Maraiwiwi of Fiji. Others at the meeting who served as commentators or rapporteurs were John Bardach of the East-West Center; John Craven of the Law of the Sea Institute; Norton Ginsburg, who was then with the East-West Center (and who also assisted with critiques of the early drafts); Paul Holthus of the South Pacific Commission; Casey Jarman of the William S. Richardson School of Law at the

University of Hawaii; Gwenda Matthews of the United Nations Law of the Sea Office; Linda Paul of Earthtrust; Carolyn Stephenson of the University of Hawaii political science department; and Mark Valencia of the East-West Center.

During the preparation of this book, the editors received enormous assistance from Margaret Spring, a law student at Duke University interning at the Center for International Environmental Law (CIEL), who drafted the introduction to part IV, among many other contributions. Additional editorial assistance was provided by Sarah Oyer, Claudia Saladin, Elizabeth Thagard, and Shannon O'Fallon, all legal interns at CIEL, as well as CIEL attorneys David Hunter, David Downes, and Chris Wold and adjunct attorney Margaret Ricker. Further assistance was provided by Marija Urlich of Auckland, New Zealand. The William S. Richardson School of Law at the University of Hawaii also supplied substantial logistic support for the preparation of the book, and Helen Shikina was instrumental in preparing the manuscript for publication, as were Nancy Benton and Nicola Ray of CIEL. Pat Harris served as final copy editor, ensuring consistency of style and grammar.

Sponsoring Organizations

Most of the papers in this book were originally presented at a meeting in Honolulu, Hawaii, in December 1990 to reexamine the increasingly outdated principles underlying the historic concept of "freedom of the seas" in light of today's understanding of the global environment. The meeting was sponsored by the Center for International Environmental Law, the Spark M. Matsunaga Institute for Peace at the University of Hawaii, Greenpeace, and the Peace Research Centre at the Australian National University.

Center for International Environmental Law was founded in 1989 to bring the energy and experience of the public interest environmental law movement in the United States to the critical task of strengthening and developing international and comparative environmental law, policy, and management throughout the world. CIEL's goals are to incorporate fundamental principles of ecology and democracy into international law, to strengthen national environmental law systems and public interest movements throughout the world, to educate and train public interest minded environmental lawyers, and to improve the effectiveness of law in solving environmental problems. CIEL works in cooperation with citizens, public interest lawyers, nongovernmental organizations, and states worldwide.

The Spark M. Matsunaga Institute for Peace at the University of Hawaii is an interdisciplinary program designed to promote peace research, peace education, and community service. It offers a Certificate in Peace Studies for undergraduates and is developing a Masters of Peace and Conflict Resolution for graduate students. It focuses on the root causes of violence, rather than the symptoms, and feels that peace and social justice must be attained through peaceful means and nonviolence. It hosted and cosponsored the conference that helped produce this volume because it views the ecological vitality of the planet as an essential element of a peaceful and just world.

Greenpeace is an independent international environmental organization with offices in 31 countries and over 4.5 million members globally. Greenpeace campaigns against environmental abuses worldwide through direct actions, lobbying, and the collection and dissemination of information. Greenpeace holds observer status in a wide range of international organizations, including many concerned with the marine environment.

The Peace Research Centre at the Australian National University was established in 1984 jointly by the Australian Department of Foreign Affairs and Trade and the Australian National University in Canberra. Its brief is to (a) carry out high quality research on topics relating to the conditions for establishing and maintaining peace on national, regional, and global scales, and (b) provide training in research in this field. The Centre publishes the periodical *Pacific Research* as well as working papers and monographs.

The views expressed are those of the authors herein and do not necessarily reflect the views of the sponsoring organizations.

Freedom for the Seas
in the 21st Century

Introduction: Traditions for the Future

··

THE IMAGE OF THE "LIMITLESS SEA" has dominated Western thought for the past millennium, and modern Western civilization has been built on the explorations, "discoveries," and conquests of the lands across the seas and the seas themselves and their resources. Because the seas have been viewed as frontier territory, an "anything goes" attitude has been accepted as appropriate when dealing with the ocean and its resources. The legal terminology that developed to justify this approach is the concept of the "freedom of the seas." As the chapters in this book explain, this concept was developed for practical political reasons by the major maritime powers during the age of imperialism, but it has maintained its force as a magical incantation to the present day, even though the colonial empires of the nineteenth century have since been dismantled.

The 1982 United Nations Convention on the Law of the Sea reduces the size of the "high seas" by recognizing that coastal states have jurisdiction over the resources in their 200-nautical-mile exclusive economic zones, but it nonetheless reaffirms that the governing regime for the remaining high seas is one of freedom of access and use. This regime guarantees the freedoms of navigation, overflight, fishing, and scientific research, and the freedoms to lay submarine cables and construct artificial islands and other installations. Although these freedoms must be exercised with "due regard" to the interests of other states, the "anything goes" ideology still prevails.

During the decade since the 1982 Convention, technological developments have dramatically demonstrated that this approach is no longer acceptable. The use of seemingly endless drift nets on the high seas in the late 1980s provided a vivid example of the possibility of overfishing of species that had been thought to be inexhaustible. The indiscriminate nature of this technology meant also that nontarget species would inevitably be taken in substantial numbers. The impact on marine mammals, turtles, sharks, and seabirds still has not been quantified in a rigorous manner, but there can be no doubt that it is significant.

"Freedom of the seas" has historically included the freedom to pollute, under the assumption that the oceans had a limitless capacity to assimilate wastes of all types and toxicity. Now we know, however, that pollutants inevitably work their way back up the food chain and that some, heavy metals for instance, present significant risks to humans. Military activities on the high seas also have historically been thought to be included under the concept of freedom of the seas, but they clearly have the capacity to pollute, to interfere with other peaceful uses of the seas, and generally to threaten the marine environment.

As food resources become more scarce, fishing nations pursue their prey in previously underfished areas, and any sense of an ocean wilderness disappears. The biodiversity of the marine environment is threatened, and the feeding habits and migratory patterns of aquatic creatures are altered.

This book brings together authors who address these concerns and offer new approaches to the challenge of protecting our marine environment. The approach of most is to challenge the prevailing view that freedom of the seas is a sacrosanct principle that promotes universally positive values. Most of the authors feel instead that the governing notion should be freedom *for* the seas, whereby the goal will be to protect the ecological vitality of the marine environment.

Instead of exploiting marine resources indiscriminately to satisfy the demands of affluent restaurant-goers in the industrialized world, it is necessary to cherish and nurture the sea creatures so that our grandchildren and succeeding generations also will have a bountiful ocean environment. Instead of viewing the oceans as the industrialized world's dump site, we must see the oceans as a fragile and finite region requiring care and protection.

One major source of difficulty is traced by many of the authors to the fundamental structural inadequacy of international law, which excludes most of international society, including indigenous peoples and nongovernmental organizations. A new and effective regime for protecting the oceans may require that these faults be addressed first.

One source of ideas on how to protect the oceans today and into the future can be found in the ancient traditions and practices of indigenous peoples, which remain strikingly relevant today. Peoples living on islands or coastal areas before the age of rapid transportation and communication understood that the ocean's resources were finite and in need of careful attention. In fact, many traditional peoples see the creatures of the sea not as different in kind but as kin and as part of the holistic life force. From this perspective of respect, they take fish and other sea creatures for their food with caution and a sense of obligation.

Today, the Pacific Islanders have begun to build a regime for the South Pacific area that builds on this tradition and introduces an approach of high environmental consciousness. The Pacific Islanders are given particular focus in this book because their experience and initiatives can help us fashion traditions for the future, traditions to guide us toward sustainable living.

Although the rest of this book is divided into traditional subject areas—

pollution, living resources, nonliving resources, and military activities—all of the chapters are connected by a common thread: Each author is searching for new ways of solving old problems. Although some of the chapters take sharply differing approaches, all share a sense that new thinking is needed and that the time is short to develop new attitudes and approaches.

Progress has been made in some of these areas even as this book has been under production. The world community has reached a consensus that high seas drift nets should not be used because of the damage they do to nontarget species and the marine environment generally. And in April 1992, the French government ment announced that it would be "suspending" its nuclear testing program in French Polynesia, a program that was universally deplored by other Pacific Islanders because of the risks it posed to the regions adjacent to the fragile volcanic atolls used for the testing. These are positive steps, but much more remains to be done.

As this book makes clear, the peoples of the world must work together to avoid destroying the marine environment, which is so vital for so many of the planet's life systems. We must adopt a perspective of respect, kinship, and environmental harmony if we want to pass on to future generations the fragile and finite ocean environment we have inherited.

I

A New Look at
Ocean Governance

THIS BOOK IS ABOUT IDEAS and their power to change reality, in this case to change the way the oceans are used and abused, to change the current freedom *of* the seas into freedom *for* the seas. It also is, implicitly, about people and their ability to conceive ideas, to encourage change, and, ultimately, to act. And it is about the pull of history and convention on our ability to think and act creatively.

The chapters in this part look at ocean governance anew in order to replace the traditional "freedom of the seas," a concept that does not have the ability to evolve to address today's complex challenges to the sea. The chapters in this part provide both history and revolutionary new theory for social change and shed light on all the chapters that follow. They also raise the question of the capacity of the 1982 Convention on the Law of the Sea to address the complex pressures on the world's oceans.

In this part, we are provided with the perspective of history, beginning with the (mercantilistic and cynical) history of the freedom of the seas concept, which prevailed only because national sovereignty could not be asserted effectively over the high seas (Anand, Jackson, Pardo). We are also provided with the (despairing) history of the self-limiting relationship between indigenous law and international law (Jackson), concluding with the recent (and hope-filled) history of efforts leading up to the 1982 Convention (Borgese, Allott). Elisabeth Mann Borgese's account of the hopes and dreams of those who prepared the way for the Third United Nations Conference on the Law of the Sea is stirring, and we can only wish to have been lucky enough to be in Santa Barbara in the late 1960s and early 1970s at the legendary—but no longer existing—Center for the Study of Democratic Institutions, working alongside Borgese, Arvid Pardo, Norton Ginsburg, and Jon Van Dyke.

Several other themes can be seen weaving in and out of these initial chapters, including the obsolescence of the state system of international society, the effects of colonialism and its collapse, the different cultural attitudes toward the sea (a theme part II returns to in greater detail), the need to move beyond conceptions of property and territory in structuring a new law to protect the sea, and the

9

need to include all of society in the benefits and management of the sea. The revolutionary concept of the common heritage of humankind echoes through the chapters, including, of course, the chapter by Ambassador Arvid Pardo.

A concise overview of the 1982 Convention is provided in Van Dyke's focused chapter, as well as in Borgese's moving historical summary. Van Dyke and Borgese, joined by Moana Jackson and Philip Allott, also offer specific principles and suggestions for a new legal regime that finally provides "freedom *for* the seas." One critical change urged by several of the authors is a change in international decision making to ensure full participation by all those who have an interest in the sea (Jackson, Allott, Borgese, Pardo). Another is the need for more sophisticated international and regional management for the oceans (Van Dyke, Borgese, Pardo, Allott). Several authors urge that the 1982 Convention be brought into force as soon as possible.

In Allott's chapter, we are also presented with the unique challenge of generating a fundamentally new international law of the sea using the 1982 Convention as a "rootstock," not through another diplomatic conference but through the much more direct process of reconceiving the theoretical basis of the law of the sea. (There are parallels in Moana Jackson's chapter.) Allott (as does Jackson) first identifies the pathological faults of an international society conceived as a system of sovereign states acting *inter se* and to the exclusion of the majority of international society, including indigenous interests, subnational interests, nongovernmental interests, and individual interests. He is optimistic that even this madness can be changed.

Maori lawyer Moana Jackson also believes that we can arrest ocean abuses and redefine freedom for the sea only by changing the fundamental basis of international law, a law that recognizes only nation-states and that denies standing to participate in international governance to indigenous peoples of the Pacific and elsewhere. According to Jackson, current Eurocentric international law is "merely the subordinate bastard child of a liaison of convenience between imperial states and the law that serves their interests." It must be fundamentally changed, Jackson asserts, to restore the rights of indigenous peoples throughout the world so that they are able to participate meaningfully in the international debate about a new law of the sea, drawing on the more nurturing values and norms of their true indigenous law.

Jackson and Allott both emphasize the need for a new guiding principle (in Maori, *te maramatanga o te ture*) for international law in general and for the oceans in particular: an international public interest. Jackson suggests that public interest guidance can be found in much of indigenous law (as that law is understood by indigenous peoples themselves, not the "arrogant legalistic nonsense" of Eurocentric law regarding indigenous peoples).

Allott elaborates on the critical importance of an international public interest. It is not a substantive concept but a form that society must fill with substance—with the social objectives of all, created with the participation of all—to direct public decision making toward the interests of society as a whole. In this conception, the public interest "is to social systems what gravity is to mechanical systems. It determines the direction of all action. . . ."

Allott's challenging chapter takes us deeper into legal theory and legal history than many readers may at first think they care to venture. But the often arduous journey offers strength for those who undertake it and, ultimately, clarity of vision for those who would protect the world's oceans. Allott's chapter also produces sparks, brilliant sparks of insight primed for action. Allott is not writing about theory; rather, he is presenting a new theory of international law and society in general and the law of the sea in particular, yet remaining ever mindful of the need to act.

Those not as well versed in the complex pathology of traditional international law may ask whether the complexity of new theory is worth it, especially when the conclusions of the new theory may perhaps appear to be obvious. International society should indeed be a participatory democracy, in which all of society is able to participate on a permanent basis in the public decision-making process rather than only occasionally through its limited effect on parliamentary ratification of treaties, a process that "puts a veneer of democratic propriety on the product of a system that is irredeemably improper."

And the role of international law should be the role of law in any society, above all asserting control over the use and abuse of government power; it should be a rich system of public law suited to the demands of an ever more complex society. And international society should be guided by an international public interest, which will involve a transfer of the center of gravity of international society from state systems to the whole earth and all of humanity. But Allott is doing more than presenting new theory. He is recruiting apostles who can explain to others the madness of the existing international system, who can urge others to question the "legitimacy of existing social reality," and who can convince others that we must all "make it our purpose . . . *to change that reality*."

In the final analysis, Allott believes that the 1982 Convention contains the genetic potential of a reconceived international law of the sea. Even by the anemic standards of the traditional international legal system, he (and Borgese) judge the 1982 Convention to be a "remarkable achievement," with nothing "in the history of multilateral diplomacy" to "equal . . . its scope, sophistication, and universality." In Allott's opinion, the only way to explain the world-changing phrase "common heritage" of humanity, and the international seabed regime, is by reference to an international public interest, guiding the negotiators beyond the mere formal resolution of the opposing interests of the state systems that they represented. In the end, it was "as if the collective consciousness of the state systems were being turned toward a previously hidden source of light, as yet only dimly seen and partly comprehended, but that they could no longer avoid and that would transform" them—and that now, perhaps, will begin to transform the reader.

1 International Governance and Stewardship of the High Seas and Its Resources

Jon M. Van Dyke

DURING THE PAST GENERATION, the nations of the world have transformed their views toward governance of the oceans. Beginning in 1945, countries began making increasingly expansive claims to the resources and waters adjacent to their coasts.[1] Four international treaties were drafted in the late 1950s,[2] but these did not resolve the controversies among nations. The drafting of these treaties was followed by extensive negotiations throughout the 1970s, leading to the 1982 United Nations Convention on the Law of the Sea.[3] The 1982 Convention has been signed by almost all of the nations of the world and is viewed as generally reflecting customary international law even by those that did not sign.[4] This agreement thus codifies internationally accepted norms that govern the sea, except for those involving deep seabed mining, which remains controversial.

Among the many important points of agreement in the 1982 Convention is the setting of limits on the jurisdiction of coastal states. In particular, it establishes that a state's territorial sea can extend to 12 nautical miles;[5] its contiguous zone, to 24 nautical miles;[6] and its exclusive economic zone, to 200 nautical miles.[7] The 1982 Convention gives the coastal state the authority to conduct fiscal and other government activities within its contiguous zone, and within its exclusive economic zone, to regulate all living and nonliving resources and to control all polluting activities.[8] The 1982 Convention also recognizes the concept of the continental shelf, which for some states extends beyond 200 miles,[9] and establishes the concept of archipelagic waters,[10] which are important to island nations like Fiji, Indonesia, Papua New Guinea, the Philippines, and the Solomon Islands. Under the 1982 Convention, the high seas beyond the 200-nautical-mile zones remains a commons, and the question now is whether keeping the high seas unregulated is appropriate given the rapidly changing technology that enables exploitation of the resources of this region.

13

The concept of freedom of the high seas has dominated Western views of the oceans since about 1800. The concept is traceable to the earlier writing of Hugo Grotius, a seventeenth-century diplomat and scholar.[11] The doctrine did not, however, prevail on its own inevitable merit. It prevailed because the imperial maritime powers of that period, namely, the Dutch and the British, had the military might to protect the right of their commercial vessels to sail unrestrained throughout the oceans. Grotius argued that the seas should be free for navigation and fishing because natural law forbids ownership of things that seem "to have been created by nature for common use."[12] Things for common use are those that "can be used without loss to anyone else."[13] From Grotius's perspective, the use of a sea-lane for transportation by one vessel did not diminish the right of any other vessel to use the same sea-lane. Likewise, the fish of the ocean seemed limitless, and thus fishing efforts by one nation's vessels did not interfere with the right of another nation's vessels to fish in the same region. More recent uses that have been recognized as freedoms of the high seas, such as submarine cable laying and scientific research,[14] also have been regarded as not diminishing the use of the seas by others.

Grotius's vision no longer holds true, however. Drift nets and other high-technology fishing methods now commonly overexploit the sea's finite living resources. The nonliving resources of the seabed are also finite and must be allocated equitably.[15] We need, therefore, to develop a new vision and a new regime to govern the high seas and its resources.

THE EVOLUTION OF A NEW HIGH SEAS REGIME

The contours of a new high seas regime are emerging, and by inspecting the recent collective decisions of coastal and ocean-using countries, we can outline the direction that a more comprehensive regime might take. Such a regime must address highly migratory species, anadromous species, marine mammals, straddling stocks, nonliving resources, pollution, military activities, and artificial islands.

Highly Migratory Species (Tuna)

Fish that travel across vast areas of the oceans during their lives have always required a special management regime. These fish—primarily the different varieties of tuna—move in and out of the exclusive economic zones of different countries and spend some of their lives on the high seas, beyond the jurisdiction of any one nation. The 1982 Convention contains a special article requiring nations that fish for these species to work with coastal nations "with a view to ensuring conservation and promoting the objective of optimum utilization of such species throughout the region."[16] Until recently, the United States argued that because the tuna are so migratory, they should not be subject to the jurisdiction of a coastal state when in its 200-mile exclusive economic zone. Increasingly isolated in asserting this position, the United States has effectively abandoned it.[17] In 1987, the United States essentially acknowledged the sovereignty of the Pacific Island nations over tuna in their 200-mile zones by agreeing to pay quite a healthy sum to those nations for fishing rights.[18]

Establishing ownership of migratory species within these boundaries leaves unresolved, however, the status of tuna outside the 200-mile zones. This problem is particularly important for coastal nations attempting to manage their stocks because large purse seining vessels now sit right outside these zones, catching large amounts of tuna and affecting the stocks within the 200-mile area. A stronger international or regional regime is needed to provide a comprehensive management approach for these important species.

Anadromous Species (Salmon)

Salmon begin their lives in streams and rivers. After a year or so, they enter the open ocean, where they live and grow for several years before returning to the stream where they were born to mate and lay eggs so new fry can repeat the cycle. Because the salmon's spawning habitat must be maintained carefully for them to reproduce successfully, the nations that pay to protect this habitat have argued that they should be able to reap the bounty of the salmon harvest. This position has been accepted in the 1982 Convention, which recognizes that "[s]tates in whose rivers anadromous stocks originate shall have the primary interest in and responsibility for such stocks."[19] This principle means that even when the salmon are beyond a nation's 200-mile zone, they cannot be caught by fishing vessels without the explicit permission of the nation of origin. Thus, one of the major high seas species already is prohibited from being freely harvested by distant-water fishing nations.

Marine Mammals

The 1982 Convention also contains separate articles for marine mammals,[20] but these provisions simply require states to work together to conserve the mammals. Earlier joint activity led to the International Whaling Convention,[21] which has established moratoriums on whaling and enabled some species to replenish themselves. Still stronger methods are needed, however, to protect dolphins, which are increasingly threatened by drift net and purse seine tuna fishing and by pollution.

Straddling Stocks

Many examples now exist in which a stock of fish is found both within a 200-nautical-mile zone and in the high seas beyond. The "donut hole" area in the Bering Sea, the rich fishing grounds off eastern Canada, and the fisheries off the western coast of South America provide some of the most dramatic examples of this situation.[22] The United States and the former Soviet Union have been sorely tempted to assert joint management of the Bering Sea "donut hole" to limit the Japanese fishing there, even though such a claim could adversely affect their interests as distant-water fishing nations elsewhere. Similarly, Canada may decide to restrict the number of European vessels fishing just outside its 200-mile limit in the Northwest Atlantic. The 1982 Convention requires only that nations cooperate in these situations;[23] a more rigorous set of principles is needed to resolve these difficult conflicts.

Nonliving Resources

In the negotiations preceding the 1982 Convention, enormous energy was devoted to creating a regime for regulating the exploitation of the seabed floor in the deep ocean, including the polymetallic nodules.[24] The result of these efforts is the International Sea-Bed Authority, which is designed both to regulate and to exploit these minerals. This cumbersome regime has yet to be implemented because many nations remain concerned about its financial implications and some are adamantly opposed to the creation of a supranational entity for resource exploitation. The high cost of harvesting the nodules relative to their current value has also dampened enthusiasm for pursuing the International Sea-Bed Authority. Some protective regime must nonetheless be established because the world's expanding population will eventually need the minerals that lie beyond national jurisdiction, and new technologies will make exploitation commercially profitable. What regime should govern these minerals to encourage their exploitation while ensuring that this global heritage is shared equitably?

Pollution

One of the historically perceived freedoms of the high seas has been the "freedom" to use the oceans as a garbage dump. According to Grotius, the oceans could be used as a dumping ground because this activity did not appear to interfere with other uses of the oceans.[25] Now that we know oceans do not have a limitless capacity to assimilate pollution, and we are seeing these limits exceeded in many locations, a comprehensive approach is needed to establish a systematic regulatory scheme to protect the oceans.

One beginning to such an effort is the London Dumping Convention.[26] This treaty classifies pollutants in categories: the most highly toxic pollutants, which cannot be dumped at all, and less toxic pollutants, which may be dumped only pursuant to careful regulation. Although this treaty has been relatively successful in addressing some of the major pollution problems,[27] it omits altogether the most important source of ocean pollution—land-based industrial activities.

Some of the "regional seas" treaties negotiated under the auspices of the United Nations Environment Program also have introduced innovative approaches to pollution issues. The Convention for the Protection of the Natural Resources and Environment of the South Pacific Region,[28] for instance, completely prohibits both the dumping of low- and high-level radioactive wastes and the disposal of any such wastes on the floor of the deep seabed.[29] This treaty, along with others, also recognizes the need to identify and protect "rare or fragile ecosystems"[30] through the establishment of marine sanctuaries. A strong argument can be made that all low-lying atolls should be recognized as fragile ecosystems that are inherent parts of the marine environment. These atolls should therefore be carefully managed and should be placed off limits to activities such as the chemical incineration plant now operating at Johnston Atoll and to military activities such as the nuclear and missile testing that has occurred at Bikini, Enewetak, Christmas Island, and Kwajalein.[31]

Military Activities

Military testing and maneuvers threaten the fragile environment of the ocean and interfere with other legitimate uses of the oceans. The 1982 Convention, however, leaves military activities essentially unregulated. Warships can therefore engage in military maneuvers throughout the high seas and also in the 200-nautical-mile zones of coastal nations.[32] This "use" of the high seas, however, can restrict other uses. When the United States and Russia launch missile tests from or into the ocean, for instance, they establish warning zones and strongly discourage any vessels from entering these areas. Yet even when ships' captains have tried to follow the instructions of the military decrees, accidents have occurred.[33]

Artificial Islands

Nations may increasingly be tempted to build artificial islands on the high seas or to expand existing uninhabitable rocks into usable islands. For example, Johnston Atoll in the Pacific was expanded from 56 to 690 acres by dredge and fill operations during the past fifty years.[34] Another example is tiny Okino-Tori-shima, one of the southernmost possessions of Japan. In its natural formation, it was the size of two king-sized beds, but Japan has spent billions of yen to expand it into a scientific station with landing facilities.[35]

Should a nation that establishes an artificial facility of this sort be entitled to claim a 200-mile exclusive economic zone around it? The 1982 Convention says that "[r]ocks which cannot sustain human habitation or economic life of their own shall have no exclusive economic zone or continental shelf."[36] Similarly, artificial islands do not generate any such zones.[37] It would thus appear that a "rock" that has been artificially expanded into a habitable island should not generate either a 200-mile zone or a continental shelf.

PRINCIPLES FOR A COMPREHENSIVE REGIME OF OCEAN GOVERNANCE AND STEWARDSHIP

All of these competing uses and threats to the oceans require the development of a comprehensive regime governing uses of the ocean and protecting its resources. Developing such a regime is not easy, but the efforts already begun provide some momentum to pursue this important goal. Some guiding principles have emerged to promote this process.

First, we must emphasize the continued ecological vitality of the oceans, giving special attention to fragile ecosystems, endangered species, and marine mammals. Our primary goal must be to perpetuate the diversity of the ocean environment for succeeding generations.

Second, all nations should act in the oceans with reasonable regard for the interests of other nations and must not act in a way that injures them. This venerable principle of international law[38] should be updated to govern shared use of this common resource.

Third, if damage or injury does result from the ocean activities of one nation or its citizens, that nation should be strictly liable for the resulting loss. This "polluter pays" principle is now widely accepted. Under this principle, any nation or entity causing a pollution-related injury to others or to a resource should accept responsibility for the loss even if the loss is not caused by negligence. This approach is the best way to internalize the costs of ocean enterprises and promote the maximum amount of care.

Fourth, regional governing bodies may be a more appropriate means to regulate the oceans than is a global international body. Because of the diversity of political systems and the different situations affecting different oceans, the world may not be ready for a single ocean governance entity. The United Nations

Environment Program has already started to divide the world's oceans into different subareas, and it has been fairly successful in identifying cohesive ocean regions. Building on these regional seas agreements may be the most productive way to proceed. The oceans are not uniform in their problems and resources, and coastal peoples adjacent to the oceans should ultimately have more control over decisions affecting their offshore areas.

Fifth, it should always be remembered that the resources of the high seas are the common heritage of humanity, to be shared equitably. The difficult regulatory decisions regarding resource exploitation should be guided by principles developed to protect the "public trust" interest in the waters of the United States.[39] Among these principles are the following: (1) a nonexclusive use should be favored over an exclusive use when choosing among competing or incompatible uses; (2) reversible commitments of resources should be favored over irreversible commitments, in recognition of the interests of ecological preservation and of future generations; (3) ocean-dependent uses should be favored over nondependent uses; (4) uses that promote and protect biodiversity should be favored over those that do not; and (5) a precautionary approach should be followed whereby a resource developer has the burden of demonstrating that the proposed activity will not unreasonably interfere with other ocean uses and will be conducted in an environmentally sound manner.

Sixth, recognizing that the ocean is a common heritage can also help in determining how to allocate its limited resources. These resources are to be shared. In many instances, the needs and industry of the adjacent coastal residents should provide the primary basis for allocating ocean resources, just as this principle justified the creation of the 200-mile exclusive economic zones. The population areas near each ocean region should have access to the resources of the oceans in relation to their needs and their interest in these resources. Persons living in landlocked and geographically disadvantaged areas who have a history of interest and investment in the oceans should also have access to these resources. Distant-water fishing nations whose vessels have worked to develop and exploit fishing resources should be given some recognition for their investments, although they cannot claim an unchanging share in these resources.

If we follow these basic principles, we should be able to fulfill our responsibilities as stewards for the creatures and resources of the oceans. We must remember that we have an obligation to them, and to the human populations who will follow us. This responsibility requires us to manage the oceans responsibly, taking what we need while ensuring the long-term viability of our marine environment for future generations.

NOTES

1. On September 28, 1945, President Harry Truman proclaimed that "the natural resources of the subsoil and sea bed of the continental shelf beneath the high seas but contiguous to the coasts of the United States" were regarded by the United States as appertaining to it and "subject to its jurisdiction and control." Proclamation No. 2667, 59 Stat. 884 (1945).

2. Convention on the Territorial Sea and the Contiguous Zone, Apr. 29, 1958, 15 U.S.T. 1606, 516 U.N.T.S. 205 (forty-five states were party to this convention as of January 1, 1987); Convention on the High Seas, Apr. 29, 1958, 13 U.S.T. 2312, 450 U.N.T.S. 82 (fifty-seven states were party to this convention as of January 1, 1987); Convention on the Continental Shelf, Apr. 29, 1958, 15 U.S.T. 471, 499 U.N.T.S. 311 (fifty-four states were party to this convention as of January 1, 1987); Convention on Fishing and Conservation of the Living Resources of the High Seas, Apr. 29, 1958, 17 U.S.T. 138, 559 U.N.T.S. 285.

3. United Nations Convention on the Law of the Sea, Dec. 10, 1982, U.N. Doc. A/CONF.62/122, 21 I.L.M. 1261 (1982) [hereinafter 1982 Convention].

4. Fifty-five countries had ratified the 1982 Convention as of March 1993. Some doubt, however, remains as to the status of Yugoslavia (which has broken apart into the republics of Serbia, Bosnia-Herzegovina, Croatia, and Slovenia) and North and South Yemen (which have reformed into Yemen). The convention will go into force when sixty countries have ratified it. The United States, the United Kingdom, Germany, Venezuela, Turkey, Israel, and a handful of other nations did not sign the treaty. The United States, the United Kingdom, and Germany have opposed part XI of the convention, relating to the deep seabed mining regime, but have agreed that the rest of the convention generally reflects customary international law.

5. 1982 Convention, *supra* note 3, art. 3.

6. *Id.* art. 33.

7. *Id.* art. 57.

8. *Id.* arts. 55-60.

9. *Id.* arts. 76-83.

10. *Id.* arts. 46-54. States with archipelagic waters can exercise greater authority in those waters than in their exclusive economic zones. Although other nations retain rights of passage in archipelagic waters, the archipelagic state has power akin to sovereignty over its archipelagic waters.

11. *See, e.g.,* Hugo Grotius, Mare Liberum (Ralph van Daman Magoffin trans. 1916).

12. *Id.* at 28.

13. *Id.* at 27.

14. *See, e.g.,* 1982 Convention, *supra* note 3, art. 87.

15. *See, e.g.,* Jon M. Van Dyke & Christopher Yuen, *"Common Heritage"* v. *"Freedom of the High Seas": Which Governs the Seabed?,* 19 SAN DIEGO L. REV. 493, 509-11 (1982).

16. 1982 Convention, *supra* note 3, art. 64.

17. *See* Jon M. Van Dyke & Carolyn Nicol, *U.S. Tuna Policy: A Reluctant Acceptance of the*

International Norm, in TUNA ISSUES AND PERSPECTIVES IN THE PACIFIC ISLANDS REGION 105-32 (D. J. Doulman ed. East-West Center, Honolulu 1987); Jon M. Van Dyke, *The United States and Japan in Relation to the Resources, the Environment, and the People of the Pacific Island Region,* 16 ECOLOGY L.Q. 217-19 (1989).

18. Treaty on Fisheries Between the Governments of Certain Pacific Island States and the Government of the United States of America, Apr. 2, 1987, 26 I.L.M. 1048 (1987). The United States agreed to pay the Pacific Island nations about $60 million during a five-year period.

19. 1982 Convention, *supra* note 3, art. 66.

20. *Id.* arts. 65, 120.

21. International Convention for the Regulation of Whaling, Dec. 2, 1946, 62 Stat. 1716, 161 U.N.T.S. 72.

22. *See, e.g.,* WILLIAM T. BURKE, MEMORANDUM ON LEGAL ISSUES IN ESTABLISHING FISHERY MANAGEMENT IN THE DONUT AREA IN THE BERING SEA (Fisheries Management Foundation, Jan. 23, 1988).

23. 1982 Convention, *supra* note 3, art. 63.

24. *Id.* arts. 133-91.

25. *See supra* notes 11-14 and accompanying text.

26. Convention on the Prevention of Marine Pollution by Dumping of Wastes and Other Matter, Dec. 29, 1972, 26 U.S.T. 2403, 11 I.L.M. 1294 (1972) (as of 1991, this treaty had been ratified by sixty-seven nations).

27. *See, e.g.,* Jon M. Van Dyke, *Ocean Disposal of Nuclear Wastes,* 12 MARINE POL'Y 82 (1988).

28. Convention for the Protection of the Natural Resources and Environment of the South Pacific Region, Nov. 25, 1986, 26 I.L.M. 38 (1987) (eleven nations, including the United States, had ratified this treaty as of 1991).

29. *Id.* art. 10(1).

30. *Id.* art. 14; *see also* 1982 Convention, *supra* note 3, art. 194(5).

31. *See generally* Jon M. Van Dyke, *Protected Marine Areas and Low-lying Atolls* (chapter 16 in this book).

32. *See* CONSENSUS AND CONFRONTATION: THE UNITED STATES AND THE LAW OF THE SEA CONVENTION 303-04 (Jon M. Van Dyke ed. 1985).

33. *See, e.g.,* Jon M. Van Dyke, *Military Exclusion and Warning Zones on the High Seas* (chapter 29 in this book); Jon M. Van Dyke, Kirk R. Smith, & Suliana Siwatibau, *Nuclear Activities and the Pacific Islanders,* 9 ENERGY 733, 734 (1984) (retelling the story of the Japanese fishing vessel *Fukuryu Maru* [*Lucky Dragon*], which was exposed to radiation from a U.S. nuclear test in 1954); Tim Ryan & Mary Adamski, *Ship Arrives with Gaping Hole,* Honolulu Star-Bull., Dec. 13, 1988, at A-8 (reporting that an Indian merchant vessel had been accidentally struck by a U.S. Navy antiship missile).

34. Jon M. Van Dyke, Ted N. Pettit, Jennifer Cook Clark, & Allen L. Clark, *The Legal Status of Johnston Atoll and Its Exclusive Economic Zone,* 10 U. HAW. L. REV. 183, 186 n.11 (1988).

35. Clyde Haberman, *Japanese Fight Invading Sea for Priceless Speck of Land,* N.Y. Times, Jan. 4, 1988, at 1.

36. 1982 Convention, *supra* note 3, art. 121(3).

37. *Id.* art. 60(8).

38. *See, e.g., id.* art. 300; Restatement (Third) of Foreign Relations § 521(3) (1986) (freedom of the high seas "must be exercised by all states with reasonable regard to the interests of other states in their exercise of the freedom of the high seas").

39. Some of the ideas that follow are drawn from Jack H. Archer and Casey Jarman, *Sovereign Rights and Responsibilities: Applying Public Trust Principles to the Management of EEZ Space and Resources*, 17 J. OCEAN & SHORELINE MGMT. 251 (1992).

2 The Process of Creating an International Ocean Regime to Protect the Ocean's Resources

● ●

Elisabeth Mann Borgese

THE MARINE REVOLUTION we are living through today did not occur overnight. Its beginnings go back to the years preceding World War II. Overfishing was already a problem at that time; offshore oil drilling was in its infancy, but its potential was already apparent. The first comprehensive effort to regulate all major uses of the sea was the first United Nations Conference on the Law of the Sea, in 1958, but its four conventions left certain gaps, which continued to widen during the subsequent decades.

States could pick and choose among the four conventions, becoming parties to any one and ignoring the others, producing a fragmented system that allowed chaotic claims to national jurisdiction, extermination of fisheries, and pollution of the marine environment. The 1958 Convention on the Fishing and Conservation of the Living Resources of the High Seas was too weak to prevent further overfishing. The outer limits of the territorial sea, as well as the limits of the continental shelf, remained uncertain; and the division on these issues was confirmed by the failure of the second United Nations Conference on the Law of the Sea, in 1960. Conservation of the marine environment and the fundamental importance of marine science and technology were not understood as yet, and, worst of all, there was no notion of the unity of ocean space and the interaction of ocean uses.

Clearly, something had to be done. Three major streams of events converged. There were conservative concerns that the limits of national jurisdiction had to be stabilized to protect the freedom of navigation of the great naval powers. There were emerging hopes of new states that had not participated in the first and second UN conferences and wanted their say in the making of a new law of the sea that would reflect their own legal traditions and economic aspirations. And there was the relentless march of technology development, which made marine resources available farther and farther out and deeper and deeper down, hastening the extinction of commercial fisheries and the pollution of the marine environment.

There was a whirlpool of excitement at the confluence of these trends, and, for a generation that had lived through fascism and nazism and World War II, a generation that was yearning for security and peace, for economic justice and the end to racial discrimination, here was a new horizon. Here was movement; here was change unconstrained by terrestrial rigidities. Here, our dreams could be taken down from their lofty heights and connected to the ground, to the seabed; here, we could step out of academia and bring innovation into the political arena. The law of the sea seemed to unite humanism, the attempt to build a human law and order, and romanticism, the love of nature and of the oceans as part of nature. It offered a starting point for a new philosophy—an "ecological worldview"—and a new economic theory—sustainable development, the economics of the common heritage.

Each one of the pioneers of the new law of the sea came to it from a different angle and a different background, but they all shared a pervasive enthusiasm and a hopefulness for the future that was hard to come by in other sectors of national or international politics.

CLAIBORNE PELL'S MODEL TREATY

Claiborne Pell, at that time already senator from Rhode Island, was driven by a lifelong association with, a profound knowledge of, and a deep love for the sea. In his 1966 book *The Challenge of the Seven Seas*, he gives a fascinating picture of what ocean development would look like thirty years later. This vision includes largely expanded and developed aquaculture; subsea systems for offshore oil production transported by nuclear-powered submarine tankers; underwater recreational facilities and resorts; fast hydrofoils; container traffic moving in "sausage trains"; "fish people" swimming freely in the deep, breathing through artificial gills ("book lungs") and communicating with the dolphins through an electronically transmitted commonly developed "language"; and an information and communication technology symbolized by a chess game played by the captain of an orbiting sea-scan satellite and the mate of a submersible hidden under the polar ice: one move each time the satellite appeared on the sub's horizon.

He was not far off the mark: All of this is possible today.

There are interesting legal and institutional developments as well in Pell's vision of the future of the oceans. He predicted that the United States would coordinate its comprehensive consolidated ocean development plan through a huge new agency, the National Ocean Agency Headquarters—NOAH for short. Globally, there would be an International Sea Patrol for surveillance and monitoring, for search and rescue, for disaster relief, for enforcement of fisheries management, and for conservation measures and protection of the great whales. New sovereign states would also emerge in the midst of international waters and be recognized by the international community, such as "Sidonia," built by King Sid on a guyot, a flat-topped seamount, within 200 feet of the surface.

After almost thirty years, the book retains an astonishing actuality in spite of, or perhaps because of, the dramatic developments that have taken place during these past three decades. On the scientific side, there is not yet any inkling of tectonic plate theory and continental drift in Pell's book. The deep-sea minerals are there, but they play a far less dominant role than they were to play through the Third United Nations Conference on the Law of the Sea (Third UN Conference). There is a certain naivete concerning institutional questions. The sea-guard function is envisaged to be exercised by the U.S. Coast Guard, under United Nations auspices. The book does not contain any awareness of how the exercise of an important function such as monitoring and surveillance of the ocean environment with a paramilitary force in reality requires an executive power and of how that executive power, unless it is to be exercised in an authoritarian manner, needs legislative control, with decision-making processes that have the confidence of the international community.

Pell's Treaty on Principles Governing the Activities of States in the Exploration and Exploitation of Ocean Space, published in 1968, shares this institutional weakness. This model treaty establishes a licensing agency with powers that considerably exceed those of the International Sea-Bed Authority under the United Nations Convention on the Law of the Sea (1982 Convention). For instance, they include the issuing of regulations concerning pollution and the disposal of radioactive waste material in ocean space as well as the establishment and the command of an Ocean Guard by the agency under the responsibility of the UN Security Council. But the treaty does not tell us who is to exercise these vast powers: a superbureaucracy—controlled by whom? Responsible to whom? It was to take fourteen years of negotiation and hard bargaining by the whole international community through the Third UN Conference to begin to sort out these questions.

Senator Pell's Treaty on Principles remains, nevertheless, a pioneering document of historic importance. His treatment of the interaction between space law and the law of the sea is interesting and apt to enhance both. His treatment of nongovernmental organizations, almost "as if they were states," is in line with the evolution of international law. Most important, the document draws attention to the need for new international "machinery" to ensure "the most efficient exploitation of the resources consistent with the conservation and prevention of waste of the natural resources of the sea-bed and subsoil of ocean space," a theme that has remained before the international community.

THE OCEAN REGIME

The next initiative that should be singled out for consideration is that taken by the Center for the Study of Democratic Institutions in Santa Barbara, California, in 1967. It led to the publication of another draft ocean treaty, The Ocean Regime (Santa Barbara Draft), in 1968; the initiation of a series of major international conferences, *Pacem in Maribus*; and the establishment of the International Ocean Institute in Malta, one of the more influential policy-research think tanks and training institutions in the world today.

The Santa Barbara Draft is indebted to Senator Pell's draft treaty, and especially to his constructive borrowing from space law. It is also deeply indebted to Ambassador Arvid Pardo of Malta, who, on November 1, 1967, delivered his historic address at the United Nations General Assembly calling for a resolution declaring the seabed and its resources to be the common heritage of humanity and for the convening of the Third United Nations Conference on the Law of the Sea to embody this principle in a universal treaty. Ambassador Pardo joined the Santa Barbara initiative from the very beginning. The deliberations at Santa Barbara, in turn, undoubtedly exercised a major influence on his later proposals in the United Nations, to which we shall return later.

The Santa Barbara Draft Convention was based on a number of basic considerations, which remain valid today:

1. The creation of the regime will be a political and constitutional task rather than an economic or technological one:

   The question of the immediate economic profitability of the oceans seems secondary. In setting out to establish an Ocean Regime, mankind is not just building a business or organizing an industry, the task is far more comprehensive. It is political in the widest sense, a new politics that must harness technology and science, that must constitutionalize science and the economy.

2. The main purpose of the regime will be to create a new form of cooperation in the international community that may set a pattern for the future activities of humanity:

   The objectives of the Regime must be based on the fact that ocean space is an indivisible ecological whole. They must be structured in such a way as to comprise the entire array of activities concerned with the oceans, at the national and international, the governmental and the nongovernmental levels. Basically, the objectives are three: development (scientific, economic, and legal), conservation (including anti-pollution), and security.

3. In view of the rapid changes in technological development and the many unknown quantities pointed out by the scientists at the Center, the emphasis must be on creating an institution to deal with these as they emerge rather than one establishing a code that might freeze development.

4. This approach may imply a de-emphasis of the usual administrative organs as we know them in the specialized agencies and a new emphasis on the deliberative organs, which would constitute, so to speak, a permanent conference on the law of the sea.

5. The organization embodying the regime must be *sui generis*, as different from other existing international agencies as its functions will be different from those of other agencies.

6. It will be apparent that all of the major problems that have to be faced are interconnected and must be solved together:

> Planning the industrialization of the sea-bed must be accompanied by planning of anti-pollution measures and the conservation of marine life. Industry, fishery, communications, the military complex are linked. Jurisdiction over any aspect of ocean activities tends to entail jurisdiction over all other aspects—or to be empty. This is the functional interdependence.

In the same way, the draft stresses "horizontal interdependence," since fish do not stop at political boundaries, nor does pollution. "If it is to be effective, the Regime must deal, therefore, with ocean space as an indivisible ecological whole." The Santa Barbara Draft thus recognizes the "permeability" or "porousness" of the boundaries that used to separate competencies and responsibilities intranationally, in government departments, as well as internationally, in the specialized agencies of the United Nations, which mirror national structural organization. Santa Barbara equally recognized that the boundaries between national, regional, and global jurisdictions had become "porous" and "permeable." All of this was to be stated authoritatively by the Brundtland Commission twenty years later in its report, *Our Common Future*.

These "interdependencies" in the Santa Barbara Draft have three major sets of institutional implications. The first two follow from the "porousness" of the boundaries between what used to be separate levels of governance—national, regional, and international—and the continuity of jurisdictions (vertical interdependence).

First, this continuity does not imply an invasion of national sovereignty. What it implies is an enlargement of the concept of "legislation"—an ever-wider range of "laws" or "norms," "regulations," "directives," "recommendations," and "opinions." This expansion is a general phenomenon, also operating in federal and even unitary states. It is connected with the role of planning, which transforms and enlarges the concept of law.

Planning is a function distinct from lawmaking. It adds a fourth dimension to government. Western constitutional theory is as deeply imbued with the conviction that government can have only three branches as people used to be with the conviction that space had only three dimensions. Then came Einstein, who proved that there was a fourth dimension, time. With planning, a fourth dimension is added to government. It may even turn out that government has more branches or dimensions than four; Chinese constitutional theory recognizes five. Riemann space is multidimensional. We must shed our Western prejudices as our interests curve around the globe, into outer space, into ocean space.

Plans do not have the character of "laws" in the technical sense. It is not of decisive importance whether plans are "enforceable"; it is far more relevant whether they benefit those who comply with them and exclude from such benefits those who do not comply. In other words, they are based on cooperative rather than coercive law, and they respect the sovereignty of nations. Twenty years later, Ambassador Chris Pinto of Sri Lanka observed that one of the innovating features of the new law of the sea is indeed that it gives rise to a "new international law of cooperation."

The second institutional implication of "vertical interdependence" is that we need a framework that provides proper linkages among national, regional, and global institutions to articulate, and make it possible to manage, this interdependence. The Santa Barbara Draft provides for a network of regional organizations, with proper backward linkages to national governments through "regional committees" and proper forward linkages to the global ocean regime and its secretariats. This notion, in a way, anticipated the Regional Seas Programme. I shall return to this later.

Finally, the "porousness" of the boundaries separating the competencies of government departments at the national level and specialized agencies of the United Nations system at the international level, caused by the interdisciplinary character of almost all issues facing the "ocean manager" and the recognition twenty years later in the preamble of the 1982 Convention "that the problems of ocean space are closely interrelated and need to be considered as a whole," has its own set of institutional implications. We need an institutional framework at national, regional, and global levels in which governments and the international community can indeed consider ocean problems in their interrelatedness, in an interdisciplinary, ecosystem fashion. We need new forms of decision making.

THE YUGOSLAV MODEL

The Santa Barbara Draft provided a rather unusual approach to this problem, inspired by Yugoslav constitutional law in the 1950s and 1960s, which came to us through the person of a very unusual Yugoslav jurist and judge in the constitutional court of Yugoslavia. Professor Jovan Djordjevic, one of the great theorists of neo-Marxism, or Marxist humanism, was involved in our work in Santa Barbara during those early years. The basic principle is simple, and it is as valid today as it was then: If the issues under consideration are interdisciplinary, the decision-making process must be interdisciplinary. It cannot be implemented by just one discipline, for example, the lawyers and the politicians (who generally are lawyers). It must involve scientists, economists, industrial managers, and all others whose disciplines are involved. The ocean regime, we reasoned, must create a new synthesis between politics, science, and economics. That implies that decisions have to be made by politicians, scientists, and economists.

The Yugoslav model provides for what could be called a "rotating bicameral system" for decision making. The fulcrum of this system is the political chamber, composed, as usual, of politicians. Their consensus is needed for any decision. But if a decision involves science or education, the consensus of the chamber of scientists is needed as well. If it involves public health, the consensus of the chamber of doctors and public health officials is needed, and so on. The Yugoslav Parliament had five chambers, reflecting the interdisciplinary nature of the issues to be dealt with and providing a structure for interdisciplinary decision making.

The assembly of our ocean regime had four chambers. The political chamber was the fulcrum, based on regional rather than national representation to keep its size within manageable limits. It provided for a chamber representing the fishing industry, a chamber representing the oil- and mineral-mining industry, and a chamber of scientists. I do not know why we forgot the shipping industry; obviously, it should have been provided for as well.

This assembly then elected a commission, which was the executive body. Together, they elected a planning board and a secretariat with various departments. The competence of the regime was comprehensive, covering all uses of the ocean, with its flexible and continuous range of enforceable to unenforceable laws, regulations and plans, its global-regional-national institutional framework, and its goal of enhancing development, environment, and security in the oceans.

That was twenty to twenty-five years ago.

Where does it rate today, on the scale between utopianism and realism? Was it an idle dream? Does it have anything to offer to the world of today? Was it just ahead of its time?

ARVID PARDO'S DRAFT OCEAN SPACE TREATY

We shall return to these questions after a brief examination of the third major effort to build an international ocean regime—and that is Arvid Pardo's monumental Draft Ocean Space Treaty (Pardo Draft), a working paper submitted by Malta to the Sea-Bed Committee in 1971.

The Pardo Draft begins by updating and modernizing conventional and customary international sea law, in particular the four conventions adopted by the First UN Conference in 1958—a task that later was to be undertaken by the Second Committee of the Third UN Conference. In some respects, Pardo's attempt was more successful than the Third UN Conference: It was more rational and less distorted by political compromise. The Santa Barbara Draft, incidentally, failed to provide this part, which is, of course, essential.

The Pardo Draft then proceeds to define coastal state jurisdiction in ocean space, which is divided into National Ocean Space and International Ocean Space, the dividing line being a single one at a distance of 200 miles from the coasts, measured from the baselines from which the territorial sea is measured. The Pardo Draft created one boundary only, for the sea floor and the superjacent waters, rather than the intricate array of boundaries that was to be provided by the Third UN Conference.

The Pardo Draft then defines the International Ocean Space Institutions: an Assembly, a Council, an International Maritime Court, and a Secretariat. Major subsidiary organs are an Ocean Management and Development Commission, a Scientific and Technological Commission, and a Legal Commission. Additional major subsidiary organs may be established by the Assembly.

To ensure balanced decision making, representation in all of the organs is weighted on the basis of population, length of coastline, gross tons of merchant shipping, possession of research and rescue vessels, the amount spent annually on marine scientific research, the amount of fish harvested annually, the amount of offshore hydrocarbons produced annually, the possession of submarine pipelines or cables in international ocean space, and, finally, the amount paid to the Institutions (which is based on revenue obtained from the exploitation of natural resources in national ocean space—a kind of ocean development tax).

States meeting the standards set up by these weighting factors belong to category A; coastal states not meeting these standards belong to category B; and landlocked states make up category C. Decisions on most issues require a majority of votes of states belonging to category A plus a majority of votes of states belonging to one of the other two categories. Some crucial decisions require a majority of votes of all three categories.

The system is undoubtedly ingenious, but in spite of all intentions, it seems almost inevitable that it would ensure a preponderance of decision-making power for the richer, more developed, industrialized maritime and coastal states.

The Assembly functions as a kind of permanent conference on the law of the sea. It has the responsibility of drafting and adopting a number of important conventions on matters of detail, which this framework convention wisely abstains from spelling out.

The Executive Council consists of all members belonging to category A, an equal number belonging to category B, and five members belonging to category C. Decisions of the Council require the affirmative vote of a majority of its members as well as of a majority of members of category A and one of the other two categories.

The commissions are composed on the basis of the same principles as the council, although the decision-making process differs slightly. The Ocean Management and Development Commission is responsible for regulation and licensing of the exploitation of living and nonliving resources in International Ocean Space, and it has to prepare and submit to the council for consideration rules relating to navigation; communications; maritime safety; seabed installations and devices; and conservation, management, and exploitation of the natural resources in International Ocean Space.

The Scientific and Technological Commission makes recommendations concerning measures to safeguard the quality of the marine environment and prepares draft regulations or conventions thereon. It also advises the Council on the proclamation of a regional or world ecological emergency in ocean space. If so requested, it may advise states on measures required to avoid pollution of national ocean space, and, most important, it advises the Ocean Management and Development Commission on the scientific, ecological, and technological aspects of licensing the exploitation of the natural resources of International Ocean Space and the exploration of its nonliving resources. This is another way of ensuring interdisciplinary decision making and the integration of development and environment, or sustainable development, as it is called today.

The most important functions of the Legal Commission are to promote the harmonization of national maritime laws and the development of international law relating to ocean space and to prepare a number of conventions on matters of detail not spelled out in the Pardo Draft. Pardo provided for a system of binding dispute settlement and for the standing of legal persons, not only states, before the International Maritime Court.

Without going into the many other details, I will recall some fundamental points that Pardo stressed in his introduction to the Draft Ocean Space Treaty:

1. The draft treaty is based on the conviction that *laissez-faire* freedom beyond national jurisdiction has become dysfunctional and that the unfettered sovereignty of the state within national jurisdiction has become equally dysfunctional.

"In contemporary conditions both must yield to the supreme interests of mankind if we are to survive and to expand our beneficial use of the oceans."

2. No state can legitimately use its technological capability, whether within or outside national jurisdiction, in a manner that may cause extensive change in the natural state of the marine environment, without the consent of the international community; the coastal state has a legal obligation to take and enforce within its jurisdiction reasonable measures to control pollution of the oceans that might cause substantial injury to the interests of other states.

3. Going far beyond his 1967 proposal, which political wisdom constrained him to restrict to the seabed, Pardo now declares that ocean space beyond national jurisdiction is a common heritage of humankind.

4. Pardo stresses the fundamental importance of marine science and technology for the development and rational management of marine resources.

5. The draft encourages the gradual establishment of a world network of parks and nature preserves (whether for recreational, scientific, or other community purposes) and of scientific stations. It also provides for measures to deal with the possible necessity of proclaiming ecological emergencies.

6. Arms control and disarmament in ocean space are mentioned, but without going into details. "If the Institutions envisaged function effectively and act wisely, it is probable that they will be requested in due course to undertake important functions with regard to arms control and disarmament in ocean space."

7. The same goes for resource management in international space. It should be made clear in this connection that the concept of resource management in ocean space beyond national jurisdiction having been firmly established, it was thought preferable to lay down only general guidelines on the manner in which the management powers of the institutions should be exercised rather than to attempt a detailed regulation of exploitation without knowledge of the conditions under which exploitation will be undertaken in practice.

The Pardo Draft consists of thirty-one chapters and 205 articles. It was far ahead of its time in its recognition that the problems of ocean space are closely interrelated and need to be considered as a whole. Sectoral approaches and absolute national sovereignty were still the order of the day. A draft convention declaring ocean space to be the common heritage of humankind and providing for international ocean institutions to regulate and manage not only the mineral resources of the international seabed but also all resources in international ocean space could not be given any attention in 1971. And yet the Pardo Draft is the real prototype of the 1982 Convention.

It was in 1980, when *Pacem in Maribus* was held in the Hofburg in Vienna, that President Amerasinghe—shortly before his untimely death—told me, "Had we looked at Arvid's draft convention in 1971, we could have spared ourselves ten years of work."

This, then, leads me to the next question: How much of our dream has survived in the 1982 Convention? Should we despair about the greediness of states, the shortsightedness of politicians, and the inadequacy of compromises; about the slowness of real positive developments, the continued degradation of the marine environment, and the exhaustion of many of its living resources? About the continued arms race in and nuclearization of the oceans? About the continued exploitation of the poor and the weak by the rich and the strong?

All this, undoubtedly, continues to give cause for concern. But we knew all along that ideas are changed in the realm of politics and that the synthesis we were seeking had to include a synthesis, as well, between long-term and short-term national and supranational interests.

The outcome, the 1982 Convention, can be described as a glass that can be seen to be half empty or half full. I prefer to see it as half full. I am, in fact, amazed not at how much of our dream has disappeared but at how much has been preserved and is now enshrined in international law.

First, the 1982 Convention contains the first comprehensive, binding, enforceable, international environmental law. This is necessarily generic: It is a framework that needs to be filled, at regional and national levels, but it is there for us to build on.

Second, the convention deals with environment in the context of development—development of natural living and nonliving resources, development of human resources, development of marine science and technology. The provisions are clear, but again, they need to be translated into institutional as well as economic and financial terms, at global, regional, and national levels.

Third, the convention deals with the advancement of peace and security by reserving the high seas (including the economic zones), marine scientific research, and the international seabed and its resources "for exclusively peaceful purposes"—a great and novel concept that needs to be interpreted and developed in legal and institutional terms.

Fourth, the convention combines, in a most creative and original way, developmental, environmental, and peace-enhancing aspects in the concept of the common heritage of humankind, which cannot be "owned" or appropriated by anybody; which must be developed equitably, for the benefit of humanity as a whole, regardless of the economic or technological stage of development of a country; which must be conserved for future generations; and which must be reserved for exclusively peaceful purposes. The common heritage concept indeed contains the seed of a new economic order, of a new economic philosophy, and of a new relationship among people and between people and nature.

Fifth, the convention goes so far as to try to embody this novel concept of the common heritage in an institutional framework, and while this, quite naturally, is not yet altogether successful, the Preparatory Commission has now succeeded in putting into place what may turn out to be a universally acceptable interim regime, not yet for commercial exploitation of the common heritage—the mineral resources of the international seabed area—but for exploration, development of human resources, and development of technology.

Sixth, the convention has peacefully achieved the most important redistribution and reorganization of ocean space ever attained in history.

Seventh, the convention has replaced a system of *laissez-faire* in the oceans with a system of management through a combination and interaction of national, regional, and international institutions, which now must be more fully developed.

Eighth, the convention is based on the recognition that "the problems of ocean space are closely interlinked and need to be considered as a whole," a simple statement fraught with the most complex institutional implications, which we now must spell out.

Ninth, the convention fosters the concept of regional cooperation, which, in a number of instances, it makes mandatory to the point that some experts have seen in it the origin of an emerging new international law of cooperation.

And tenth, the 1982 Convention contains the first comprehensive and flexible but binding system for peaceful settlement of disputes: a breakthrough in international law that eventually might well be taken over by the United Nations system as a whole.

It is in the evident interest of the world community that this convention should come into force. Only then can we build on it, be it for the development of environmental law; the progressive development of the law of the sea; cooperation in science, technology, and industry; or the building of a new world order in general.

I have already indicated several levels of action that should be pursued if we are to advance the new order in the world ocean as a model for, and part of, a new world order: if we want to advance, we should go back to our old dreams, which were dreams because they were ahead of their time.

1. An absolute priority, in this broader context, is to bring the 1982 Convention into force. What would be the point of embarking on another huge, complex enterprise to reach new agreements on the environment and development when the international community is demonstrating, through its inaction, that it cannot implement the agreements already reached, codified, and signed? New endeavors should draw on the 1982 Convention and utilize the unique experience of the Third UN Conference. The new world order must be built

solidly, one stone upon the other. The 1982 Convention on the Law of the Sea is a cornerstone. It must be solidly in place before we continue the building.

2. The next priority, once the convention comes into force, is to create a forum in which states can discuss ocean policy in an integrated manner. I mentioned that a basic recognition of the convention is that the problems of ocean space are closely interrelated and need to be considered as a whole; but since the end of the Third UN Conference, no organ in the United Nations system is considering ocean problems as a whole. The UN system still reflects a tightly closed, sectoral approach to the big interdisciplinary problems—oceans, environment, trade, energy, food, and the like—inherited from bygone times. It was indeed the Delegation of Portugal—led by my great friend Mario Ruivo—that, at the end of the Third UN Conference, pointed to the need for creating such a forum, which might take any of several forms: It might be an annual, or biannual, special session of the General Assembly on Ocean Affairs; it might be a permanent institution, such as, for instance, the Disarmament Committee in Geneva or the United Nations Conference on Trade and Development, but it must be a forum in which states can consider ocean problems in their interaction and as a whole.

3. Perestroika has put before us the concept of comprehensive security, that is, the recognition that security today does not have only a military dimension. Security can no longer be ensured through superiority in an arms race. To be secure, states and the international community need economic as well as environmental security. The concept of comprehensive security and the concept of the common heritage are complementary in this sense; the concept of the common heritage, like that of comprehensive security, has a developmental, an environmental, and a peace-enhancing dimension. In this sense, comprehensive security must be based on the principle of the common heritage, and the implementation and progressive development of the law of the sea can make major contributions toward this end.

4. The Regional Seas Program of the United Nations Environment Program, a breakthrough in the 1970s, still reflects the sectoral approach of Stockholm 1972. It focused on protection of the marine environment. It soon realized that to do so effectively, one has to deal with all sea uses as well as a number of land uses; but the institutional framework established for the Regional Seas Program still reflects a sectoral approach. This institutional framework now has to be broadened to match the broadened mandate.

Institutional innovation thus is needed at the national, regional, and global levels, and they are all interlinked. These links do not mean, however, that everything has to happen at the same time. It will take time to implement the new system.

One could start—and this would be my recommendation—with one or two pilot experiments: I would suggest the Arctic, where regional cooperation is at its very beginning and might be developed in accordance with the new comprehensive principles indicated by the Brundtland Report and put forward in quite specific terms by perestroika; and, second, the Mediterranean, where the Regional Seas Program is most advanced in institutional infrastructure and experience. For the first time since the end of World War II, the political preconditions for such an initiative now exist. Let us try there to create the institutional framework needed to implement comprehensive security, with its environmental, developmental, and disarmament dimensions. Let us declare these regional seas to be zones of peace. In both cases, institutional implementation of the concept of comprehensive security would strengthen world peace, enhance global economic development, and contribute to the conservation of the human environment. To make the Mediterranean a zone of peace could be part of, and would strengthen, an overall Middle East peace settlement.

5. Much work is to be done in the Preparatory Commission for the International Sea-Bed Authority and for the International Tribunal for the Law of the Sea, which has been meeting in Jamaica and New York since 1983 to prepare for implementation of the 1982 Convention once it comes into force. There, an interim regime is emerging for management of the common heritage, in the form of an agreement between the so-called pioneer investors—countries that have already made large investments in seabed mining—and the Preparatory Commission as a whole. (The pioneer investors, at present, are France, India, Japan, the former Soviet Union, and, most recently, China.) This interim regime will be of crucial importance not only for the future of the law of the sea but also for international cooperation in science and technology in general. Here, too, a piece of machinery is being generated that must integrate the concepts of "reservation for peaceful uses," "development," and "environment," for this is what the common heritage of humankind is all about.

6. Much work is yet to be done to give legal content to the concept of reservation for peaceful purposes. This could, perhaps, most appropriately be entrusted to the International Law Commission of the United Nations, but somebody has to take an initiative so that this can be done.

War, it has been said, begins in the minds of humans. The same can be said with regard to "integration." An institutional framework is of crucial importance for the management of "sustainable development," integrating development and environment. But behind the institutional framework there must be a conceptual framework. How many of us still think of protection of the environment as a constraint on economic development, as a cost that must be added to the cost of development? How many are capable of envisaging the environment and development as an integrated whole? Of economics, the economics of culture, as just a part of ecology, the economics of nature? How many of us really feel, deep down, that an economic system that destroys its own resource base really is not an economic system at all but is a prescription for disaster? What we need today is a new economic theory. And considering that we are transcending the age of sectoralization, that theory too will have to be broadly interdisciplinary. It will be a new philosophy, a new look at relationships among humans and between humans and nature.

Our terrestrial existence has given us the erroneous idea that we are the overlords of nature, free to treat or ill-treat her at will. The return to the sea, the penetration of the oceans with the industrial revolution, may be a triumph of our science and our technology, but it also may have its humbling effects. We are small and frail in the ocean, and nature is mighty. We may destroy it, and ourselves, but we cannot subjugate it; we must work with it, not against it, and that is what the integration of environment and development is all about. Working in the oceans, with the oceans, integrating environment and development, imposes a new paradigm, somewhat as the integration of time and space into space-time in the theory of relativity imposed a new paradigm, a new worldview. Aurelio Peccei, founder of the Club of Rome, once said that we need a new economics that is as different from classical economics as Einstein's physics is from Newton's. I like to call it the "economics of the common heritage."

This worldview—more humble than Western tradition has been for the past few hundred years, and less aggressive—that humankind is part of nature, and that if we destroy nature we destroy ourselves, we must now bring from our ocean experience to our terrestrial experience. In this sense, the oceans are our great laboratory for the making of a new world order, foreshadowed in the process of implementing, interpreting, and progressively developing the new law of the sea and the principle of the common heritage of humankind as the basis of a system of common and comprehensive security.

3 Perspectives on Ocean Governance

Arvid Pardo

I WOULD LIKE TO MAKE A FEW GENERAL OBSERVATIONS on the principles on which the governance of ocean space should be based and some observations on the principles that are at present governing our international system.

First, we live in a system of sovereign states jealous of their powers. This is quite basic, even though the system itself is obsolescent.

Second, for the past two thousand years traditional international law regarding the seas has swung between two principles: sovereignty and freedom. Two thousand years ago, we had freedom within one world, the Roman world. The political system underlying the Roman Empire collapsed, and it was followed by a period of anarchy that lasted for several hundred years. We then witnessed the rise of another principle—the principle of sovereignty. The principle of sovereignty as applied to land would also be made applicable to the seas. For about 800 years, this principle of sovereignty expanded in the marine environment until it took over the entire marine environment with the Treaty of Tordesillas. But then, after about 150 years, this principle was undermined by the growth of technology and by the collapse of the political system that sustained it, the Portuguese and Spanish empires. Then, with Grotius, we again had the advocacy of freedom. Until World War I, freedom had expanded over virtually the entire marine environment.

Third, the world needs to use ocean space ever more intensively and needs to exploit the resources of ocean space ever more extensively. At the same time, the world needs to avoid the adverse consequences of exploitation of ocean space, including the inequitable consequences deriving from inequalities in technology and access to the open seas. So the world wants to use and exploit but at the same time wants to avoid the consequences of such activities.

The principle of freedom guarantees equal possibilities for all states to use ocean space with reasonable regard for the interests of other states. But this is a negative principle. It does not avoid the adverse consequences of intensive use of ocean space. The principle of freedom is a negative freedom, it is a permissive freedom, and it ignores adverse consequences. It is not a positive freedom—in other words, it does not permit management.

The fragmentation of ocean space among more than 100 different sovereignties, although permitting management within each sovereignty and although excellent for governing the exclusive uses of the sea, cannot adequately govern inclusive uses of the sea. We must resolve the dichotomy between the need to use and exploit ocean space and the need to avoid the adverse consequences of such use. Neither freedom nor sovereignty can avoid this pitfall. In fact, it cannot be done within the framework of the existing world order, despite the progress in the law symbolized by the 1982 United Nations Convention on the Law of the Sea.

The world community needs, therefore, to establish a new legal order governing ocean space as a whole. Such a new legal order must safeguard the common interests of all users, accommodate inclusive and exclusive uses of the marine environment, and provide expanding opportunities to all countries in the use of ocean space. These goals can be accomplished only through effective management and development of ocean space resources beyond national jurisdiction for the benefit of all countries and through equitable sharing of the benefits derived therefrom.

To achieve these aims, international agreement is required on the concept of ocean space itself, that is, on the fact that the surface of the seas, the water column, the seabed, and the subsoil of the seabed form one indivisible whole.

International agreement is also required on the concept of the common heritage of humankind, which must replace both traditional freedom and traditional sovereignty as the basis of the law of the sea. Obviously, we have not reached this point yet. The implications of the concept of the common heritage of humankind are far-reaching.

The first implication is already accepted by the traditional law of the sea: that the sea can be used, but it cannot be owned. The second implication is that the use of the common heritage requires a system of management involving all of the users. Although not everybody necessarily has to share to the same extent, everybody participates in management. This view is one of the revolutionary bases of the common heritage concept and must apply not only to the seabed but also to ocean space as a whole.

The third implication of the common heritage concept is the active sharing of benefits. This means the sharing not only of financial benefits but also of the benefits derived from shared management and from the transfer of technologies.

Fourth, the concept of common heritage has environmental implications. It requires reserving the common heritage area for peaceful purposes and for future generations. (I should also note that the common heritage concept can be applied not only to ocean space but also to other environments.)

Fifth, to manage our common heritage resources, the world community must create not merely an International Sea-Bed Authority but also a balanced international system for ocean space as a whole. Despite the many constructive provisions contained in the 1982 Convention, international agreement has not been reached on any of the points that I have mentioned.

We must have an international agreement on the concept of the common heritage of humankind. We also need international agreement on the concept of regional development within the framework of global organization. We further need a clear definition of the limits of national jurisdiction in ocean space—and we do not have that yet in the 1982 Convention.

But we must not look to a fourth UN conference, which could be less successful than the Third UN Conference and which would endanger the consensus achieved in the 1982 Convention.

We must work instead for ratification of the present convention, with all its flaws, and implement and improve on the positive points contained therein for the future.

4 Indigenous Law and the Sea

Moana Jackson

THE LAND FROM WHICH I COME, Aotearoa, is part of the great ocean of Kiwa. To *pakeha*, or white people, the land is New Zealand and the ocean is the Pacific, or "peaceful sea."

But what is happening in that ocean today is far from peaceful. It provides instead a sad illustration of what is happening to oceans throughout the world in the name of progress and development. It provides a bitter commentary on the self-interest of major nations responsible for a notion of "progress" that has resulted in pollution and despoliation. But perhaps most of all, it provides continuing evidence of the inadequacies of an international law regime that is shaped by the economic and commercial interests of colonial and imperial powers.

If the continuing abuse of the Pacific and other oceans is to be arrested, there needs to be a reassessment of the existing international law regime and a redefinition of the concept of the freedom of the seas. The two are interrelated, and any review that focuses only on redefining the notion of freedom of the seas within the extant framework of Euro-American "international" law will ultimately fail.

This chapter discusses not just whether there can be freedom *for*, rather than *of*, the seas but also whether extant international law is capable of accommodating such a change. It discusses these issues within the context of the relationship between international law and indigenous peoples, especially those who are a minority in their own land.

INTERNATIONAL LAW AND THE SEA

Since the Dutch jurist Hugo Grotius wrote *Mare Liberum* more than 300 years ago, there has been a presumption that the seas should be "free." Like all legal presumptions, however, it developed in accord with Oliver Wendell Holmes's

41

dictum that the law is the product not of logic but of experience. Because none of the European maritime nations—England, Spain, Holland, and Portugal—could assert effective sovereignty over the high seas, expediency forced the recognition of free navigation for all.

That freedom, however, and the area over which it was to be exercised, remained subject to the economic, military, and mercantile interests of the European nation-states. Whether the freedom was proclaimed under the doctrine of res nullius or that of res publica, it was ultimately limitable by the ability of nations to assert claims and interests over the use of the seas.

The area that has remained free from those claims has rapidly shrunk since World War II. The creation of 200-mile exclusive economic zones in the 1982 United Nations Convention on the Law of the Sea and the declaration by major states of exclusionary zones for military purposes have drastically restricted the extent of the "free" sea. These actions illustrate two main weaknesses in the current notion of freedom of the seas.

The first arises with the concept of "reasonable use," which exists concomitantly with the concept of freedom itself. On the one hand, the freedom of the high seas has never been regarded as an immutable absolute. Thus, free use was permitted only if it was reasonable—for example, unimpeded passage for trade was a reasonable use; piracy was not. On the other hand, certain restrictions on others' freedom were permissible if the restrictions themselves constituted a reasonable use. Thus, "no-go" zones for nuclear testing were clear restrictions but were permissible as a "reasonable use" in the interests of national security.

This application of the reasonable use doctrine has been defined as a legal conclusion invoked to justify a policy preference for certain unilateral assertions as against others. As such, it illustrates that whether the test of reasonableness is that of the legendary "man [sic] on the Clapham omnibus" or the ubiquitous standard of what is "fair" between parties, it is a subjective and relativistic criterion. And in practice it has been subjective, determined solely in the economic, strategic, and sovereign interests of the Euro-American states.

The doctrine framed to protect seventeenth-century mercantilism thus continues to protect the same interests today. The interests of smaller states, or of indigenous Pacific peoples living under colonial regimes, are scorned. Their concepts of freedom of, and access to, the sea are ignored, dismissed, or marginalized. The concept of freedom, which for centuries regarded passage by slave ships as a reasonable use, now allows the passage of toxic wastes and the testing of weapons. In each instance, the nation-state beneficiaries and profiteers are the same, as are the indigenous victims who bear the social, economic, cultural, and environmental costs.

The second weakness illustrated by the doctrine of the freedom of the high seas in its present form flows from the first. If the doctrine has developed and been used to protect the interests of the major nations, its consequences reflect those interests. The idea of freedom of the seas has always been seen as a freedom to use the seas—they have been a resource to be developed, explored, and exploited for profit and economic gain. The environmental costs of that exploitation are now recognized and acknowledged. The freedom to use the seas has become a license to abuse, with a consequent depletion of the fisheries and a rise in pollution from both marine dumping and land-based sources.

Less acknowledged, certainly in international law circles, are the socio-economic costs inflicted by the doctrine, particularly on the indigenous peoples of the Pacific. It is the submission of this chapter that any attempt to remedy the environmental consequences must also seek to ease the costs borne by the people most affected and that the marine environment cannot be restored in the Pacific without restoring the rights of those indigenous peoples who have been marginalized and disrupted by its exploitation. Such reform cannot be achieved through international legal legerdemain or by a mere refinement of existing conventions. It can be done only by addressing the fundamental reasons for the exploitation and the complementary rationale of the law that permitted it.

To use a truism, one must remedy the consequences by addressing the basic cause. But the cause is not just the unthinking acquisitiveness of a throwaway and polluting society; nor is it just the proliferation of the waste-producing industries that have been established to serve that acquisitiveness. Its cause also lies in the philosophical basis of the international law that promoted and protected the acquisitiveness. Maori people call such a philosophical basis *te maramatanga o te ture*, the guiding principle of the law. And as the capitalist-mercantilist history of international law makes clear, the guiding principle behind the law of the sea has been a belief that the oceans are simply another resource to be used for economic gain: either indirectly, as a passageway, or directly, as a source of food or other wealth. The key has been that nations were to be protected in their use of the sea, without any parallel protection for the sea itself.

If this philosophical basis of the law of the sea is questioned and changed, the effects on the environment and the effects on indigenous peoples will both be alleviated. If it is not, the law of the sea will remain the ineffectual servant of those nation-state interests that have permitted the current situation to develop.

A process of reform must therefore move away from the current philosophies of the law of the sea and seek a comprehensive new basis that allows a regulated freedom of use that is subject to a regime of protection. Such a basis of nurturance and use is found in the jurisprudence of much indigenous law. There is, of course, an irony in our turning now to the philosophies of indigenous law, since the ethos of Euro-American law, which has permitted ocean despoliation, was also responsible for the attempted dismissal and suppression of indigenous law during the establishment of imperial hegemony in the seventeenth, eighteenth, and nineteenth centuries.

In most indigenous societies, the philosophies and ancient precedents of their law survived, although the notion of indigenous law itself has been redefined through the processes of legal imperialism. If one is to seek guidance from indigenous law in addressing issues of the sea, it is therefore necessary to distinguish between that law as circumscribed by Euro-American doctrines and that law as understood by indigenous people themselves. This requires a brief discussion of the historical contact between international law and indigenous law.

INTERNATIONAL LAW AND INDIGENOUS LAW

The relationship between the two systems of law in many ways reflects the relationship between the distinct societies they served. International law was a part of the destructive process of colonization that has affected indigenous peoples, and particularly Pacific indigenous peoples, for more than 300 years. Its special character was

> [d]etermined by that of the modern European state system, which was itself shaped in the ferment of the Renaissance and the Reformation. . . . That state structure became the nexus of a people's power and their nation's sovereignty . . . and the means by which their interests, economic, military and cultural, could be protected.[1]

In that sense, international law became the force that enabled the colonization of "undeveloped" non-European nations to occur. As such, it sought to impose its interests and its doctrines on the extant law of the indigenous people. This eventually led, in the common law, to the development of the precepts of aboriginal title.

At about the same time that Grotius was postulating his *Mare Liberum*, the school of Salamanca Divines, led by Vittoria and Las Casas, was debating the rights of European states in relation to the Native Americans. The issue was simply whether the Christian princes of Europe could assume an *imperium* over native peoples. Not surprisingly, the jurists concluded that they could. If the native peoples were conquered, breached the notion of *jus gentium*, or prevented the exercise of a European right to trade, *imperium* or title could be asserted. Title could also, of course, be claimed by a right of "discovery."[2]

The charters and proclamations issued by the Stuart monarchs in relation to North America built on these views and claimed an inherent right to constitute an *imperium* unrelated to the consent of the indigenous peoples. However, the work of another jurist, Emmerich von Vattel, was later to establish a presumption that consent was required when *imperium* was sought over an independent and equal state.

In *Le Droit des gens*, Vattel described international law as "the science of rights which exist between nations" and stated that a foreign nation has no inherent right of interference over another state. Indigenous societies were recognized as nation-states, although their state of savagery meant that their title could be restricted to areas for which they had a "special need."[3]

The British refinement of Vattel's approach was to distinguish by means of a standard of civilization those countries that were nation-states and those that were not. The ethnocentrism and racism of British colonial policy and law ensured that most indigenous peoples were defined as the latter and thus were viewed as less capable of claiming and protecting their rights.

Their consent was still required, however, and most often the Crown claimed *imperium* through a voluntary submission or cession by the natives in treaty. That has certainly been the basis of the Crown's assertion of sovereignty in Aotearoa–New Zealand—a claim still rejected by Maori people.[4]

With the proclamation of sovereignty, indigenous rights became subject to the authority of the Crown or the colonizing government.[5] The ways in which those rights are defined and given effect make up the doctrine of aboriginal title and the corpus of what is often called indigenous or customary law.

That Eurocentric body of law, however, is not indigenous law. It is merely the subordinate bastard child of a liaison of convenience between imperial states and the law that serves their interests. It acknowledges no general sovereign indigenous right that is not dependent on the authority of the colonizing nation, and it admits of no indigenous right to, or jurisprudence for, the use and protection of the sea.

This Eurocentric law stands in sharp contrast to indigenous law as defined by indigenous peoples. To many indigenous peoples, certainly to the Maori, proclamations of Crown *imperium* and international rejection of indigenously defined rights are arrogant legalistic nonsense.[6] It matters not, in our case, what the British declared or proclaimed to themselves. Aotearoa is still Maori land, and the seas that surround it could still be better nurtured through the philosophies that underlie our law. It is that to which we now turn.

INDIGENOUS LAW AND THE SEA

It is not, of course, possible to speak of one body of indigenous law that governed the use of the sea. However, the ethos of Maori law has commonalities with the law of other indigenous peoples. In particular, it shares the basic indigenous belief that the sea is not a property resource existing only for humans to exploit.

For the Maori people, *te tikanga o te moana*, or the law of the sea, is predicated on four basic precepts deeply rooted in Maori cultural values. First, the sea is part of a global environment in which all parts are interlinked. Second, the sea, as one of the *taonga*, or treasures of Mother Earth, must be nurtured and protected. Third, the protected sea is a *koha*, or gift, which humans may use. Fourth, that use is to be controlled in a way that will sustain its bounty.

From this cultural and divinely ordained matrix the actual *tikanga*, or law, is developed. Thus, for example, *te tikanga o nga kaitiaki* aims to stop pollution of coastal water not by seeking more effective methods of waste disposal, as is often advocated today, but by ensuring that any activity produces as little waste as possible at its source. Such laws were rigidly enforced by various sanctions and through *korero tunhonohono*, or agreements among the various tribal nations that make up the Maori polity.

It is impossible to detail here all of the various laws of *te tai ao*, the environment. It is submitted, however, that within the values and norms that shaped those laws are the seeds of understanding that could transform current international thinking on protection of the global marine environment.

Two of the many claims laid by Iwi Maori before the Waitangi Tribunal give an indication of the interrelated view of the environment encapsulated within indigenous Maori law. In the Kaituna claim, the people of Te Arawa objected to the discharge of sewage into the Kaituna River. For the Maori, the river was part of an interconnected water system involving Lake Rotorua, Lake Rotoiti, and the Maketu Estuary. Each body of water is important to the people of Te Arawa as a source of food, as the site of important historical events, and as the base of spiritual sustenance. To discharge sewage into such waterways not only would be a *hara*, or wrongful act that breached the laws protecting tangible food sources, but would also breach the laws protecting the intangible spiritual sources of the Iwi's well-being. *Nga tikanga o te tai ao*—the laws of the environment—sought to protect both the tangible and intangible, since together they strengthen *te korowai a Papatuanukau*—the cloak of Mother Earth.

In a similar claim, the Te Ati Awa people sought to prevent pollution of their seafood beds by industrial waste and sewage. In simple terms, the mixing of water contaminated by waste with water used for food gathering or spiritual purposes disturbed the balance of the environment. Maori law exists to preserve that balance.

Unfortunately, the colonial government's dismissal of those norms and the laws that grew from them has meant that the Maori have been unable to contribute in any meaningful way to the debate about a new regime for the sea.

Indeed, because Euro-American international law has accepted the Crown's claim of sovereignty by treaty, the Maori are never accorded standing to participate in international forums. In the bitter irony alluded to earlier, the international law that has essentially permitted "free" use and abuse of the seas is the same law that has denied the efficacy of an indigenous law predicated on concepts of sustainability and protection.

REVISING THE LAW: REVIEWING THE BASE

The concerns of the authors of the chapters in this book, the work of the Preparatory Committee for the United Nations Conference on Environment and Development (UNCED), and the involvement of many nongovernmental organizations indicate that there is general agreement about the need to reexamine and redefine the governing law. It is the submission of this chapter that there must be a reexamination of the global and regional goals of environmental protection as stated in existing conventions and agreements and a parallel reexamination of the general concept of ocean use.

The world community must then question whether the present law, and the present international agreement process, can effectively give effect to a new regime. Greenpeace International, in an August 1990 paper prepared for UNCED, made the following statement:

International agreements all too often reach decisions that reflect the lowest common denominator of compromise. All too often, the meetings of parties to those agreements are a pressured market-place that often elevates the semblance of achievement above actual achievement. Experts in the field of international negotiations have expressed their doubts.

[It is unclear] whether traditional treaty techniques will prove to be suitable for meeting the technical requirements of effective "eco-management" on the global or regional scale, once international action passes from the declaratory to the operational stage. Environmental problems characteristically require expeditious and flexible solutions, subject to current updating and amendments to meet rapidly changing situations and scientific/technological progress. In contrast, the classical procedures of multilateral treaty making, treaty acceptance and treaty amendment are notoriously slow and cumbersome.

One excellent example of a rapidly changing situation involves the burgeoning opportunities to employ "clean production methodologies." In order to meet the technical requirements of effective protection of that type though, it is essential that far better mechanisms be in place to provide for the exchange of technology, as well as training, data and information.[7]

The establishment of such effective mechanisms will, however, flow only from reexamination of the basic law and concepts that underpin the present treaty and lawmaking regime. At present, much effort is being expended on developing new perceptions of the freedom of the seas, but little appears to be being done in relation to international restrictions on, and nonrecognition of, the precepts of indigenous law.

Nongovernmental organizations are now frequently allowed to have input at various ocean-environmental conferences, but indigenous peoples who are minorities in their own land are excluded, thus depriving the international community of valuable and ancient insights. This exclusion also, of course, maintains the unjust exclusion and denial of indigenous law and the rights that flow from it.

The emphasis on "recognized" nation-states not only has limited the available sources of potential input but also has maintained the legalistic and political belief that only lawyers and politicians have the wisdom to make law. In fact, of course, those people who are affected by a law have an interest in it and should be involved in its framing. This was the basis on which the Maori developed their legal processes, and it should be part of modern procedures, especially in something as fundamental as environmental law.

NOTES

1. J. L. BRIERLY, THE LAW OF NATIONS: SOME PAPERS 121 (1955).
2. See P. McHUGH, THE ABORIGINAL RIGHTS OF THE N.Z. MAORI OF COMMON LAW (1987).
3. See William J. Masters, Maori & Indigenous Rights (dissertation, University of British Columbia, 1987).
4. See J. KELSEY, A QUESTION OF HONOUR (1990).
5. See B. Slattery, *Understanding Aboriginal Rights*, 66 CAN. B. REV. 727.
6. See M. Nepia, *Te Whaka—Marama*, Maori L. Bull., Feb. 1990.
7. Greenpeace International, paper prepared for UNCED, Aug. 1990 (on file with the editors of this book).

5   *Mare Nostrum*: A New International Law of the Sea*

Philip Allott

> That princes may have an exclusive property in the soveraigntie of the severall parts of the sea, and in the navigation, fishing and shores thereof, is so evidently true by way of fact, as no man that is not desperately impudent can deny it.
>
> SIR JOHN BOROUGHS[1]

USING THE 1982 UNITED NATIONS CONVENTION on the Law of the Sea as a rootstock, it is possible to generate a fundamentally new international law of the sea. Such a regeneration will not be the product of yet another diplomatic negotiation among the representatives of the governments of states. It will be brought about by a much more direct and efficient method. It requires nothing more nor less than a reconceiving of the theoretical basis of the law of the sea, a fundamental change in the underlying conceptual structures. To talk new theory is to talk about the possibility of change. To talk new theory may be to talk social revolution.

Law conceives social reality as abstraction, then reabstracts it in the form of generalized legal relations. To do law is necessarily to do theory. Law is the application of ideas to material reality, with a view to re-forming human consciousness, that is to say, with a view to being a cause of conforming behavior. To do law is, inevitably, to act philosophy.

Ideas meet material reality to produce law, but the reality itself is a product of many other meetings between human being and human being, between human individual and society, between society and society, between humanity and the natural world, and between all these things and their conceptualizing as ideas.[2] It

---

* A version of this article appears in 86 AM. J. INT'L L. 764 (1992).

follows that international law in general, and the law of the sea in particular, has only the most superficial appearance of innocent necessity.

It also follows that those who wish to effect a change in international law must choose their point of entry. They may operate on the material reality (by some form of direct social or economic action), or they may operate through the value-actualizing political process (in international matters, ultimately through the diplomatic process, given current conceptions of international social organization), or they may operate through the social-philosophical process (by working on ideas directly), as this chapter does.

The 1982 Convention is the product of a total international social process extending back, philosophically and historically, to the sixteenth century and far beyond. And in its turn, it has itself entered into the material reality of international society, becoming an active element in the generation of further practical and philosophical effects. From the point of view of international social philosophy, the 1982 Convention is unusually interesting. It seems to be an actualizing of well-known conceptual structures, but it contains within itself the potential negation of those structures and hence the potentiality of a structurally new law of the sea.

In the 1982 Convention, a particular structure of international social organization is re-forming itself, undergoing a process of structural metamorphosis. In its half-formed new structural uniqueness, it is full of painful ambiguities and exciting possibilities, full of the "inharmonious harmony [that] is fitted to the growth of life."[3]

To understand the unique genetic and historical phenomenon that is the 1982 Convention, we must look at it in a new light, a light of another wavelength, the light of another structural philosophy, other than the now-traditional conceptual structure of the law of the sea. To uncover the deep-structure potentiality of this particular manifestation of the law of the sea, we must reimagine the philosophy of law of the sea and the philosophy of international law in general.

PRINCIPLES OF A NEW LAW OF THE SEA

The following is a set of primary axiomatic principles of a newly conceived international law of the sea.[4]

Principle 1. The international law of the sea respects the natural integration of land, sea, and air and the natural integration of all social phenomena.

Principle 2. The systematic relationship of humanity to the sea is not merely a legal relationship of property but rather is a social relationship of participation.

Principle 3. The international law of the sea uses legal relations as a method of actualizing the international public interest and of implementing international social objectives.

Principle 4. The management of the relationship of humanity to the sea is socially and legally accountable.

Principle 1: Integration of Natural and Social Phenomena

Law is necessarily a taxonomy. That is to say, law must classify material reality for its own purposes. It must determine kinds of persons, kinds of events, kinds of behavior, and kinds of places. But in the law, classifying is also prescribing. If the law recognizes the so-called state as a legal person, a subject of legal relations, then it has conferred a form of existence on that notional entity that it would not otherwise have. If international law fails to recognize the human individual as a legal person, a subject of legal relations, then it affects the possibilities, and hence the nature of the social existence, of the human individual. If the law uses the notion of "the sea" as a primary classificatory concept, then it sets the sea apart for legal purposes, thereby giving rise to the possibility of assimilation and discrimination on the basis of that isolating identification.

The obvious distinction between land and sea and air might have offered an obvious basis for distinguishing between a law of the land, a law of the sea, and a law of the air. Historically, however, the so-called law of the sea did not develop as part of such a rationalizing legal taxonomy. There was, and is, no distinct international law of the land governing the acquisition, use, and disposition of land territory. There is only a miscellany of rules that in one way or another relate to such matters. There was no need for a law of airspace until airspace came to be economically significant, and even then, so-called air-law came to be essentially the law of air transport rather than a general legal regime of the use and abuse of the air medium.

So far as the law of the sea is concerned, the great theoretical-political struggle of the seventeenth century was a struggle with the evidently anomalous character of the sea itself. It was outside land-territorial sovereignty, but it was an area of intense material interest to the new sovereigns as moat, highway, and fish pond. As the state-sovereignties closed in on themselves politically and territorially, the sea space between them became practically conflictual and theoretically problematic. Economic interests, imperializing interests, and defense interests conspired to produce a theoretical structure that matched a land-territory omnipotence (all-power) with a sea-space omnicompetence (all-freedom). This then created a mixed area of interaction between the two—the territorial sea—an area whose limiting concept was formed by negation as an area neither of all-power nor of all-freedom.[5]

It is now necessary to negate and surpass the time-honored conceptual separation of land and sea, for three reasons in particular:

1. Since the conclusion of the 1982 Convention, there is generally thought to be a much more complex legal interface of land and sea, with many areas of mediation between all-power and all-freedom—territorial sea, contiguous zone, international strait, archipelagic sea, exclusive economic zone, fisheries zone, continental shelf, safety zone, exclusion zone, deep seabed and ocean floor, and superjacent high seas. One may say that the shared zone has now become the rule rather than the exception. It is the general high seas that would have come to be anomalous if it were still to be regarded as essentially an area whose limiting concept was all-freedom.

2. The nature of a state's so-called sovereignty over its land territory has profoundly changed in recent times. A state's territory is no longer regarded as a fortress, protected by the gates of "political independence and territorial integrity" and by the bailey of "domestic jurisdiction." International society, through international law and through nonlegal means, now has a direct interest in all that happens within any state system anywhere. The naturally communal character of the sea space is no longer so clearly differentiated from the no longer exclusive character of land territory. Exclusive political control of land territory is tending to become a residual phenomenon. Preconceptions of exclusive political control over naturally communal sea areas must tend to become anomalous to the same extent.

3. From both the scientific and the economic points of view, it has become much more obvious in recent times that land, sea, and air form a single physical and economic system. Land events cause sea effects. Sea events cause land effects. Land and sea interact within the single envelope of air. From world climate change to mass production fishing, from deforestation to acidification, from nuclear weapons testing to sewage disposal systems, from crude-oil pollution to sea-lane congestion, from warm ocean currents to highly migratory species of fish, there is a natural continuum between land-based activity and sea-based activity, directly or through airborne effects. International law is required to respond with a corresponding natural monism.[6]

The present challenge is how to move the sea space out of its anachronistic social and legal isolation, to integrate it into the total social process of international society, in which all of humanity participates. In seeking to meet that challenge, we might even find that a reconceived sea space can play a leading part in the progressive reconstruction of international society as a whole.

Principle 2: Participation in the Use and Governance of Our Sea

Society's natural relationship to the sea is more than a relationship of property. It is a relationship of participation, in the forming of social objectives for *our* sea and in the sharing of its benefits and its governance.

In the history of particular national societies, there has been a continuous development over time of conceptions of property. Political struggle, including revolutionary struggle, has often, and in many societies all over the world, turned on the issue of property.[7] An important part of the dominant ideology of human societies, from the earliest known societies to the most complex societies of the present day, has been devoted to structures of ideas concerning the distribution of social and individual control over things. With the emergence of the social technique of legislation—formalized prospective lawmaking—the equilibrium between the individual and society in the control of things could be established with unlimited flexibility and precision. In complex modern societies, a vast mass of lawmaking and law applying has turned property into one of the most intricate phenomena of a social system, with property power shared, in different ways for different kinds of property, between the property owner and a society that intervenes powerfully in all property relations.

To understand how property came to play so significant a role in the making of international law and, in particular, in the international law of the sea, it is necessary to refer to a particular historical role played by property in the social development of the European societies that had a predominant influence on formation of the relevant international law. In the period when feudalism was a dominant form of social system (say, from the tenth to the thirteenth centuries of the modern era), government and property were not clearly differentiated. Institutions of property played a virtually constitutional role. The king was, in some sense, the principal landholder, even if he might also be conceived, or conceive himself, as the sovereign, and hence the source of ultimate social and legal authority, in some more mystical sense.

The story of the development of constitutional monarchy and then of liberal democracy was, among many other things, the story of the isolating of *government* as a distinct systematic category, so that it came to be seen as the communal ascertaining of social objectives and their realization through law and administration. Society was then left with two fundamental problems that became the substance of much subsequent social, and even revolutionary, struggle—the problem of how to govern the government and the problem of the relationship of government to property.

In the succeeding centuries, the problem of property came to be conceived as the problem of whether government was the servant of property or whether property should be the servant of government. Theories of social contract and then theories of capitalism and socialism were episodes in a struggle to fix the society-property relationship at a theoretical level and so make that fixed relationship into the more or less unspoken premise of all lawmaking and law enforcing.[8]

In addition to the cloudy confusion of government and property, two other features of the medieval feudalist consciousness played a fateful part in the development of international law in general and the law of the sea in particular: the subjection of the sovereign to law and the intrinsic reciprocity of social relations. There was never a single generally accepted answer to the question of how a sovereign could be both sovereign and under the law, but the notion was, perhaps, more effective for being notoriously obscure.

A hierarchical view of social relationships, which seemed to be an intrinsic feature of feudalism and, perhaps, of the Christian medieval worldview in general, was likewise accompanied by a broad sense of social reciprocity. Social relationships seemed to be, in some sense, a form of exchange, of service in return for protection, of obedience in return for social responsibility, of payments in return for social services.

These three features of medieval social consciousness—the government-property nexus, the sovereignty-higher law paradox, and social reciprocity—turned out to be very useful features in the primitive structuring of the external relationship among the new sovereigns. The new *states* could claim to have a sort of inherent property in their national territories, without the need to spend much time on the problem of how there can be property relations without a social-legal system capable of creating property relations.

In other words, the new states could be said to have formed their new international society on the basis of a sort of social contract, a hypothetical agreement among themselves as to the formation of society. The terms of the contract would have been on these lines: "We, the new nation-states, hereby agree to form a society under law, but a society that recognizes no source of governing authority other than ourselves and that will recognize the reciprocal right to exist of all recognized members of the society, including the exclusive right of each to govern its own national territory as sole landowner."[9]

The new nation-states had thus formed themselves into a mutual protection society of landowners. The sea space between them then fitted easily into the new framework. Property government could be extended to a security zone around the land territory (territorial sea). The waters beyond—called "high seas," presumably as an English mistranslation, possibly itself due to a French mistranslation, of the Latin *altum mare*, "deep sea"—could be conceived as a sort of "common land," which had been a familiar feature of feudal land arrangements, organizing various forms of propertylike use and propertylike control of particular land areas without needing finally to settle the precise property status of such land.

What happened next was what did not happen next: (1) The gradual isolation of property and government as separate social systems occurred within the nation-states but did not occur in the international system. Property and government remained, and still remain, in a stifling embrace throughout the international system. (2) The development, within the nation-states, of the tense dialectical relationship between property and government did not occur internationally. The socializing of international state landholding has been minimal, at least until the recent emergence of the catalytic concept known as "the environment." (3) Internationally, there occurred only a weak reflection of the internal social process, which led to the ever more effective subjection of government to law within the nation-states, through constitutionalism, through legislation, and through the development of ever more sophisticated law-applying and law-enforcing techniques.

To negate and surpass the notion of property contained like a fossilized skeleton within international law in general and the law of the sea in particular, two strategies are required—to reconceive the notion of property in international law and to re-form some of the property relations of international law into relations of government, so that all of society can participate. As discussed in part III of this book, the significance of the 1982 Convention is that it makes use, however modestly and unconsciously, of both of these strategies.

To reconceive the notion of property in international law, it is necessary first to understand how the concept of property has transformed itself in national legal systems. The following is tentatively proposed as a contemporary, post-mystical conception of legal relations of property.

Property is a particular form in which a legal system may create legal relations. Or, rather, property is a portmanteau for a bundle of legal relations whose purpose is to classify and order the legal relationships among particular classes of persons in relation to particular kinds of things. Through the manipulation of property relations, society can choose which particular powers to delegate to the property owner and which social objectives and social controls to include. Society also can delegate property power to itself, to institutions deemed to be acting directly on behalf of society as a whole. Thus, particular property relations lie along a spectrum that extends from essentially socialized property to essentially individualized property.

In this view, the philosophical priority between property and society is transcended as property comes to be seen as integral to the systematic self-ordering of society in accordance with society's objectives. Moreover, the opposition between property and government is transcended in that property has become something akin to a form of delegated government power. The property owner is seen not as an "other" whom society must tame and control but as an integral participant in the systematic social action of society. A modern conception of government would also transcend the separation of government and governed, seeing their relationship as some sort of division of labor within the total system of society.

Nowhere would such a conceptual transformation be more appropriate than in relation to the law of the sea. If one were seeking to determine the natural relationship of humanity to the sea, one would certainly not begin by supposing it to be a relationship involving a set of assorted so-called states, acting through so-called governments. Nor would one begin by supposing that the relations of such states were relations of property or quasi-property relations, let alone legal relations of property.

Other contributors to this book have drawn attention to the way in which certain cultures have conceived of the natural relationship between humanity and the sea and its resources.[10] Such conceptions seem to have a significantly numinous aspect, leading to attitudes such as humility, respect, fellow feeling, an absence of a sense of domination and exploitation, and an absence of *meum* and *tuum*. Over the past century and a half, natural science has also been leading human consciousness back to a sense of human integration in nature.[11] And in the most recent period, an idea of the integrity of the total system of the earth has been forcing itself into human consciousness throughout the world, as humanity is made aware of the mutual dependence of everything on earth, living and nonliving.

The sea is naturally neither *mare liberum* (Grotius) nor *mare clausum* (Selden) but rather is *mare nostrum* ("our sea"). Grammarians classify the word "our" as a possessive pronoun. But it is important to realize that this classification conceals different uses of such pronouns. It would be better to say that in such phrases as "our country," "our world," and "our sea," the word is, as it were, a participatory pronoun, indicating not a relationship of possession but a relationship of participation. The sea is our sea because we find ourselves to be cohabitants with the sea on the planet Earth and because all human beings naturally share its potentialities.[12]

If we nevertheless find that in reality the sea is abused through the exercise of social power constituted in legal relations of property, and if we find that all human beings do not have free and equal access to their sea, then, in the spirit of Rousseau, we must ask: *What can make legitimate humanity's social relationship to the sea?*[13] If we are not satisfied with the legitimacy of existing social reality, then we must make it our purpose, in the spirit of Marx, *to change that reality.*[14]

Principle 3: International Public Interest

The master of an oil tanker who cleans out the ship's tanks at sea after passing through the Strait of Gibraltar is not subject to traditional international law. The corporation whose chemical factory discharges toxic waste through a pipeline into the Adriatic Sea is not subject to international law. The master of a fishing vessel who releases 5 miles of drift net into the Pacific Ocean is not subject to international law. The oil-producing corporation that makes an unauthorized survey of a particular section of a continental shelf is not subject to international law.

International law is traditionally regarded as a law between states. Its effect on natural and legal persons other than states depends on social processes that are independent of the systematic structure of international law. Legal relations of international law must be implemented through the behavior of actual persons, natural and legal. But the connection between international legal relations and the actual behavior of persons can be made into a relationship of law only by the intervention of other legal systems. The constitution of a particular state must enable the relevant legal relation of international law to be embodied in a legal relation of national law, with or without the interposition of national legislation.

The international social objectives that are realized in legal relations of international law, including the law of the sea, are formed by a process of double aggregation. Each state system aggregates through its internal social process a view of its interests (subjective social objectives) in relation to a given matter. The national interests are then aggregated through the international system—negotiation, agreement, and decision making of international organs. International legal relations are then said to be formed as the product of two particular forms of aggregation—through the accumulation of lawlike behavior into legal relations of customary international law and through the conclusion of treaties.

At the heart of the international system, and of international law, is thus a hidden theory of representation. The system relies on the internal social aggregation process of each state system that enables a so-called government to represent the national interest internationally. Such a systematic structure is unlikely to incorporate adequately other important features of international reality. It excludes those subnational interests that are not adequately represented through government representation (for example, the interests of nationals that are contrary to those of their government). It must simply be assumed that the national political process successfully achieves social aggregation for external purposes, whether by fair means or foul. Such a structure also excludes cross-national interests that cannot adequately be contained within the social aggregating of one particular state system (for example, the interests of multinational corporations). It excludes, finally and above all, such common interests of humanity as are not adequately represented by the state systems collectively.[15]

There is, however, a more obscure and still more insidious distorting effect of the unreformed self-conceiving of the international system. It contains as a leading social process what may be described as a false dialectic that cuts across the natural struggle to find and actualize social objectives. The interaction of the aggregated national interests takes on a life of its own. Instead of being merely a way of aggregating individual, subnational interests into a collective, so-called international interest, the respective aggregations at the state-system level come to seem to be original interests.

The result has been that the system, although still seeming to the uninitiated to be a process for the collective forming of social objectives, turned itself into a system that gives effect neither to universal social objectives of all humanity nor to the social objectives of all human individuals collectivized through the state systems. It came to be dominated by an independent dialectic at the median level between the two, at the level of relations between so-called states. The international system came to be one whose leading social process was so-called diplomacy, and diplomacy came to be the controlled interacting of the notional national interests. Humanity had formed itself into a society whose social process was interstate relations.

At least until the 1982 Convention, the law of the sea has contained nothing other than the international aggregate of aggregated national interests. At least until very recently, there has been no possibility of forming international social objectives in relation to the use of the world's sea space, other than as by-products of international-national aggregation. And hence there has been no question of forming international law as the realization of genuinely international social objectives formed by humanity as a whole from consideration of the world sea space as a whole.

It has taken time, and it will take an intellectual effort, to make a change of perspective in relation to the social organization of the world. It involves transferring the notional center of gravity of international society from the level of the state systems to the level of the totality of humanity, from the level of the separate national territories to the level of the whole earth. The notional center of gravity of a society is to be found in the idea of the public interest. The public interest is to social systems what gravity is to mechanical systems. It determines the direction of action of all social forces. It conditions the application of effort by participants in the society. It even provides a formal basis for evaluating the performance of the society. It is an integral systematic element in all lawmaking and law applying.

Public interest is not itself a substantive concept. It is a categorical form into which each society puts substance. But the formative effect of public interest is that it causes public decision making to be directed at the interest of the society as a whole, to be directed at the benefit of the totality. Society generates social objectives endlessly as a leading product of its social process. Those social objectives are formed by the interaction between subsocietal interests (of individuals and subordinate societies) and the public interest of the society as a whole. This is the systematic expression of that feature of *participation* which has been considered in relation to Principle 2 discussed earlier, the reconciling of property and government in a modern society. Participation in a modern society means participation in the forming of social objectives in such a way that the social objectives of society can become not merely objectives *for* all participants but objectives *of* all participants.

So long as international society lacks any conception of the public interest of international society as a whole, its social process will remain vestigial and primitive. Nowhere is this more true than in relation to the law of the sea. The natural integrity of the sea, considered in relation to Principle 1, means that the sea space of the world is not only an integral part of the world's social and economic system but also a single, undivided feature of the material reality of the world, a continuous structure extending over the whole planet. If we did not have the experience of four centuries to contradict us, we might have supposed that it would be impossible for humanity to organize its relationship to the sea other than by respecting its natural integrity through the application to it of a conception of the world public interest.

As part of the process of forming its social objectives, society chooses the appropriate means for realizing those objectives. One of those means is law, by which the public interest is made actual in the behavior of individual human beings. Law is thus a systematic mechanism for connecting ideas to behavior. Law is a product of society that also produces society, in the sense that law organizes the behavior of the members of society in accordance with the objectives society has formed. Society thus makes itself in its own image. As a society gives substance to its public interest, so it conceives its potentiality.

A society, such as international society, that has little or no sense of the public interest of society beyond the aggregating of the self-interest of some of its members cannot have a legal system that enables it to form itself effectively as a society. Society uses law in particular cases, rather than other social mechanisms (education, religion, morality, economic incentives), because law can efficiently perform certain specific functions.[16] Law is a visible intervention by society in the general social process, a purposive intrusion into the working of other, seemingly more natural and diffuse, social forces.

Law can thus acquire a sort of magical social authority by reason of its being seen as a direct expression, the authentic voice, of society's public interest. To conform to the law is, at least in principle, to universalize the particularity of one's behavior, to act not only for oneself but also for society, which has realized (disaggregated) the public interest in the law to which one's behavior is then found to conform. It is for this reason, among others, no doubt, that international law has failed to acquire the dominating authority that law can have. If it is the law of a society that is not systematically organized to realize the public interest through lawmaking and law applying, then it cannot be heard as the authentic voice of the whole of society. If it is the law of a society that cannot generate social objectives in the universal public interest, then it is not the law of a society that is forming itself through law.

A practical consequence of these anomalies of international law is that it has had to make do with a rather rudimentary range of forms of law. Forms of law reminiscent of national laws of property, contract, and delict have been the forms of substantive law available to the international law process. These are forms of law structured around bilateral relationships: the owner and others; contracting party and contracting party; tort-feasor and injured party. In international law, distributive justice, such as it is, lies hidden in the interstices of corrective justice, like gold in dross.

It is, therefore, not surprising that international law has had difficulty in articulating legal relations among individual human beings or between substate groups and international society as a whole. So-called international crimes and human rights are still regarded as systematic anomalies. The promotion of social justice through law, especially economic justice, has foundered at the level of legal principles and rules and has had to be pursued, if at all, through the medium of bilateral international relations or through the administrative decision making of international agencies. And, as the most general consequence of international law's structural disabilities, international law has not been able to make use of the great elaboration that has occurred in the role of public law within national legal systems.

When the explanatory focus of international society is shifted to the level of the public interest of international society as a whole, the status and role of the state systems redefine themselves. The state systems are constitutional organs of international society, the focus of delegated government power, which is participatory, rather than of property power, which is exclusionary. The purpose of the delegation of power is to serve the public interest of international society by realizing its social objectives. And because international society is a participatory democracy, the state systems are simply privileged participants in the forming of international social objectives.

The role of international law in such a restructured international society is then the role of law in any society. And the forms of law available to the international legislator are at least as many and various as the forms of law available to the national legislator. In particular, and above all, the control exercised by law over the behavior of the state systems is primarily a control over *the use and abuse of delegated power* rather than merely a control over the rights and duties of a property owner or a control over the *inter se* relations and *inter se* delicts of property owners. It is, first and foremost, a system of public law.

In this way, the international legal system can at last begin to realize the extraordinary potentiality of a customary law system. In such a system, the subjects of the law, by willing and acting for the sake of obeying the law, are privileged to become part of the legislator because their behavior, in being intentionally law conforming, is behavior for the sake of the public interest of the whole society.

The control by international law over the behavior of other persons then becomes a matter for international law to determine. It may, as it very largely does at present, delegate the law control to the state systems to exercise through national law. And national law may make use of property relations, however individualized or socialized, as one among many other institutional structures and always subject to overriding international social objectives, as a means of subdelegating power to substate persons and societies.

In this way, the social objectives of international society can be written into the very existence of the state systems, can be transmitted by them into the internal forum of the state, and can be made to reach and affect the behavior of all human beings, including the behavior of public officials and the decision makers of corporations. These international social objectives, including environmental objectives, can be achieved through public–law control of the use of power by the governments of the state systems, through public–law goal-determined duties on governments to use their intrasocietal power to achieve those objectives, and through rules and principles applying directly to nonstate actors whose decisions affect the sea space and whose behavior may be legally controlled either directly by international law or, as appropriate, through powers delegated to the state systems.

By means of a newly conceived law of the sea made within the social processes of a newly conceived international society, humanity's participation in and with the sea space of the world can be a participation of unlimited sophistication and sensitivity, not stifled by the artificial bonds of *meum* and *tuum*, not condemned to the rule of the past over the future, but free to create a social future of material and moral progress.

Principle 4: Accountability for Abuse of Government Power

In the newly conceived international democracy, as in any democracy, social decision making is a process, not an event. The past and the future flow into the present because society is what it has been and because society, in its decision making, is creating its own future. Because today is tomorrow's yesterday, future generations haunt the decision making of the present generation. In democratic theory, accountability is the power of the future over the present.

In democratic state systems, the hold of the future on the present is ritualized in the form of elections. Elections are not, however, the only or the most efficient form of social accountability within a democracy.[17] In complex modern democracies, accountability through elections has come to be supplemented by accountability through public participation. Through maximization of the information available to the public concerning matters of public interest and through the increasingly sophisticated organization of public input into the decision-making process, democracy becomes a system of permanent participation rather than a system of occasional validation. In such a system, law becomes a particular social subsystem by which society may create the future through particular forms of social action, conditioning behavior through legislative formulas and through procedures for implementation, application, and enforcement of legislation. Legal accountability, a special case of social accountability, is designed to ensure that decision making under the law makes a future for society that lies somewhere within the limited range of possibilities proposed by the law.

Of social and legal accountability in the broader sense there is very little in the international system. The adoption of texts of one kind or another through diplomatic processes has come to rival war as the leading form of international obsessional behavior. Each text is seen as some sort of victory of the present over the future. A legal text, complete with seals and signatures, is seen as the biggest victory of all.

The obsession with texts is a clear sign of the impoverishment of the international system as a political system and of the rudimentary nature of the international system as a democracy. Text-dominated diplomacy tends toward monopolization of power in the text makers. Text making then becomes a form of sacramental behavior, in which only the initiated may participate.[18]

After the text is formalized, implementation occurs either through national processes or through the international mechanisms created or enlisted by the terms of the text. Accountability, the true hold of the future on the present in a democratic system, will then amount to no more than the sum of the elements of accountability provided for in or under the text or available in the national systems.

The net result is a perfectly familiar fact. Most of humanity is disenfranchised in the current international system, participating only notionally in the making and implementing of international social decisions. Countless individual human beings; countless nations and peoples not specifically organized as state systems; and countless corporations, nongovernmental organizations, and subordinate social systems of all kinds have no specific effect on international decision making unless they have some specific and actual effect on a recognized international political actor.

Once again, the international law of the sea offers eloquent testimony in favor of fundamental change. It follows from Principle 1 that the world's sea space should be seen not only as integrated in the total physical and economic system of the world. The management of the world's sea space is also integrated into the social system of the world and into all its social subsystems. It follows from Principle 2 that all those who have an interest in the future of the sea space must be seen as natural and inevitable participants in the world's social decision making, participating directly and decisively. It follows from Principle 3 that the dynamic focus and origin of the law of the sea must be found at the level of international society and not at the median level of the false dialectic of the international relations of the state systems. A worldview that is geocentric (relating international society to the earth as a total system); zoocentric (relating international society to the society of all living things); and anthropocentric (relating international society to the whole of humanity, in all its teeming particularity) will shape the form of all social objectives concerning the world sea space and of all legal relations of the international law of the sea in the direction of a truly universal public interest of the world.

It follows from Principle 4 that such a worldview generates an international system that naturally contains social and legal accountability of all kinds and naturally engenders an international politics in the widest sense. In such a new system, law is reconceived as a *process* rather than an event. Legal texts are not merely to be used as if they were truces or treaties of peace in the social struggle but rather are seen as episodes in an entirely continuous social process, embedded in the flowing development of social reality, as international society creates itself not only through lawmaking but also through every other form of social activity.

In particular, international control of the sea space through agencies acting specifically in the public interest will come to be seen as a task of government management rather than simply a matter of legal control. Just as the task of government within some complex modern democracies (including the European Community as a democracy of democracies) has taken on more and more of the character of a system of public management of a physical and social system, so the international task will come to be seen also as an inextricably complex systemic interaction of science, technology, economics, and human values that cannot be organized simply by occasional text-centered lawmaking.

This means that what international society expects of all the persons and subordinate societies whose behavior affects the world sea space has to be conceived as something more than merely an expectation of conformity with law. The social development of modern democracies has been led by demand in the sense that the scale and complexity of the social task, including especially the economic management task, came to exceed the capacity of a social system in which the hold of the future on the present had to be expressed in the form of more or less primitive legal relations and intermittent civil war. In managing a system so vast and complex as the world sea space, international society will also have to go beyond the old familiar international social process (diplomacy and war) by transforming all international actors into subjective participants in a social process of unlimited complexity.

The hold of the future on the present will be expressed in a form of accountability that will go far beyond a mere calling to diffuse political account or to bilateral legal account. It will become a permanent system of social auditing, self-auditing and other auditing, monitoring, foreseeing, preventing, responding, rethinking, and remaking. And all those disenfranchised in the present international system—individual human beings, nongovernmental organizations, corporations, nonstate nations, and peoples—must come to be seen not as anomalies and irritants in the international social process but as fully participating members of a universal international society.

THE 1982 CONVENTION

The potentiality of a reconceived international law of the sea can be found in the *discors concordia* of the 1982 Convention. In the convention, the international law of the sea and, perhaps, international law in general have set off on a journey of self-discovery and self-regeneration.

The genetic inheritance of traditional international law is found in the property relations ethos of the 1982 Convention as it creates the framework for the law by assigning specific sea areas among the various state systems of the world. It is found in the way in which the convention is focused at the level of the *inter se* relations of the state systems. It is found in the assumption that accountability, if it means anything, means legal accountability; and legal accountability is focused on the resolution of so-called disputes, that is to say, on the adjustment of the bilateral relationship of the state systems on the basis of claims made by one against the other. The essentially bilateral relationship is made trilateral only by the consensual invoking of a third party, in accordance with part XV of the convention, that third party being conceived as a representative of the law-order of the convention itself rather than as a representative of the general public interest of international society.

Above all, the traditional genetic inheritance is present in the overall assumption of the 1982 Convention that its legal relations are not the expression of international social objectives but are the product of aggregated national interest. The social objectives of the convention are thus presumed to be identical with the content of the legal relations. In short, the 1982 Convention is the authentic voice of a society of state systems, represented by their respective governments, behaving characteristically within the social process of diplomacy. And judged as such, it must be admitted that the convention is a remarkable achievement of that society and that process. In the history of multilateral diplomacy, there has been nothing to equal the 1982 Convention in its scope, sophistication, and universality.

The judgment of history may come to be that the 1982 Convention was made possible only by an underlying change of consciousness that had already begun to affect the participants in the social process, although very few of them may have been conscious of the significance, or even of the existence, of any such thing. It is as if the collective consciousness of the state systems were being turned toward a previously hidden source of light, as yet only dimly seen and partly comprehended, but that they could no longer avoid and that would transform the human social world.[19]

The clues to the subtext of the 1982 Convention, the convention-in-waiting, may be tenuous, but they are not negligible.

1. The international seabed regime in part XI of the 1982 Convention is explicable only on the basis that the government representatives who took part in formulating it were influenced by a consciousness, however latent and unacknowledged, that traditional property relations could no longer meet the needs of a new kind of international social objective, an objective that obviously went far beyond the competing claims of the state systems. Part XI provides that the International Sea-Bed Area and its resources are the *common heritage of humankind and that no state shall claim or exercise sovereignty over any part of the Area or its resources; nor shall any state or natural person or juridical person appropriate any part thereof.* The convention provides that *activities in the Area shall be carried out for the benefit of humanity as a whole.*

Whatever conventional social theories may actually have determined their behavior, government representatives at least knew that they were doing something unusual, perhaps too unusual, in drawing up part XI. But in drawing up part V of the convention, on the exclusive economic zone, they may well have seen themselves as players in the great, age-old game of diplomacy. Yet part V is perhaps the best illustration of the convention as a half-completed metamorphosis, enacting the painful and elusive struggle of the past with the future. Although article 56 is articulated as a traditional distribution of property-type rights—sovereign rights, jurisdiction, and powers—article 55 and part V as a whole create an intricate network of legally constituted social interactions, which can easily be seen as something more than a mere accumulation of essentially bilateral relationships. In a more creative perspective, it might just as well be seen as essentially and naturally communitarian, a system of social management organized through a distribution of legally constituted social power.

In short, both part XI and part V use traditional property relations, by negation in part XI and by affirmation in part V; but, one may choose to believe, they use them to transcend them.

2. The articulation of many provisions in terms of classes of states (coastal, landlocked, archipelagic, geographically disadvantaged, developing, and transit, and states bordering international straits) creates a new dialectical level, intermediate between the interstate and international society levels, causing a sharing of consciousness based not merely on the purely formal category of statehood but also on actual substantive characteristics. A substantive by-product of this new structural consciousness is a new conception of the distributivist values at issue. The new identification is created not merely for formal or institutional reasons. It reflects new kinds of claims as to actual needs and potentialities, new kinds of claims on international social justice.[20]

3. The 1982 Convention contains many legal relations that are articulated in terms of social objectives rather than contingencies, suggesting that the convention is meant to be inserted into a social process, a process of progressive social development and not merely the progressive development of international law. The contracting parties are required to achieve, in their future interactive social behavior, *equal treatment, equitable sharing, regard for legitimate interests, effective protection of human life, effective protection for the marine environment, cooperation in the conservation and management of the living resources of the high seas, and optimum utilization of the living resources of the exclusive economic zone.* There are provisions that list specific governing policies in particular areas: for example, article 59 (which includes a reference to "the international community as a whole") and articles 123 and 150. And there are provisions on revenue sharing regarding the continental shelf and the international seabed area and on sharing regarding the exclusive economic zone.

In the case of part XI of the convention and part XII (on the protection and preservation of the marine environment), international social objectives and international social process are apparent. It requires an act of willful incomprehension to read them as anything other than the actualization through legal forms and processes of an international public interest.

The keystone provisions in parts XI and XII contain the two most elegant and eloquent instances of the intersection of text and subtext in the convention, the new wine of communitarianism spilling over from the old bottle of legal formalism. Article 136 provides that "the Area and its resources are the common heritage of mankind"—transmuting a property metaphor into a universal ideal. Article 192 provides that "States have the obligation to protect and preserve the marine environment"—transmuting a banal legal relation into a universal social objective.

4. When the convention is seen as a whole, its gestalt seems to be much more that of a public law system than that of a contractual arrangement.

It is striking that every sea area, whatever its conceptual articulation in terms of property relations, is conceived in the convention as being, not incidentally but inherently, an area of power and interest shared by two or more state systems.[21] The exercise of the supposed property right is, in all cases, actually a process of decision making within procedural and substantive constraints.

The 1982 Convention relies greatly on the rule making and standard setting of intergovernmental organizations, most notably in relation to protection of the marine environment. The convention also provides for various forms of convention-based decision making—on a grand scale in relation to the international seabed, but also in relation to the outer limit of the continental shelf, fish stock surpluses, and sea boundary delimitation, not to mention dispute settlement under part XV.

Also, perhaps in distant recognition of the principle of integration considered earlier, the legislative focus of the convention flows freely through the outer membrane of the state systems into their internal processes—the making of laws and regulations, the exercise of criminal and civil jurisdiction, the decision making of executive authorities. And it uses as its legislative reference points every kind of actor and activity associated with the sea.

The conventional view may be that the 1982 Convention is a characteristic and distinguished example of a treaty codifying and developing general international law. A variation of the conventional view would see it as a so-called lawmaking treaty, a category designed to appropriate some of the charisma of socialization without undermining an essentially contractual social phenomenon. But it is also possible to see the convention in the light of quite a different ideal type of society-constituting activity: the routine social legislation of a modern democracy.

It would be possible to legislate internationally for the integral phenomenon of the sea from either direction: either piecing together the legal rules necessary to produce the law-conforming behavior required to control the interaction of the state systems or setting social objectives and distributing public-law powers (legislative, executive, and judicial) with a view to actualizing those objectives. The same kind of choice became available in national legislation. Within the framework proposed in the present study, the first approach is what might be called a property approach; the second is a government approach.

It was suggested earlier that as societies become very much more complex, with greatly increased levels of social (especially economic) energy and of potential social conflict, legislation must more and more mean the creation of decision-making systems, set within an interlocking structure of ever more general social expectations, so that the most particular of human desires can be integrated into the most general aspirations of society as a whole. Public law has developed as a way of organizing this intensely dynamic and subtle form of social self-ordering. The intensity, the subtlety, and the dynamic character of the 1982 Convention may be read as the outward signs of an international society that is now undergoing the same kind of fundamental development.

National experience shows that a price has to be paid for such social development. Traditional power holders naturally resist the redistribution of their personal power. Individual members of society naturally fear and resent the impersonal power of social systems. It is clear that modern democracies have not yet solved the psychological problem of the two new forms of social alienation, of those who manage the new systems and those who are managed. A democratically conceived international society will share in the progressive development of democracy.

Whatever else it may be, the 1982 Convention is an education, for governments and citizens alike, in the new demands of the new world in which we live. The governments of the state systems may make their judgments of the convention on the basis of their traditional conceptions of their own functions and interests. For all those persons and societies, subnational and supranational, that are not themselves members of governments, there is another kind of judgment to be made. And if we conclude that it is important in the public interest of international society as a whole that the convention, in its entirety (including, above all, part XI), be brought into force soon, we should do what we can, as participants in the international democracy, to see that this event occurs.

Then, whatever the imperfections and limitations of the 1982 Convention, it can enter into the reality of international society as a powerful creative force, preparing the minds of all to manage a world in which global social problems call for solutions that far exceed the potentialities of traditional diplomacy and traditional international law. There is no better place than the universal social phenomenon of the sea to begin learning to integrate universal social phenomena into the self-socializing of the human species.

NOTES

1. SIR JOHN BOROUGHS, THE SOVERAIGNTY OF THE BRITISH SEAS, PROVED BY RECORDS, HISTORY AND THE MUNICIPAL LAWS OF THE KINGDOME (written 1633, published 1651).
2. New conjunctions of material reality (for example, new capacities to exploit natural resources or new techniques of production), new ideological conjunctions (for example, the rationalizing of capitalist accumulation or the social organization of

labor), new social value clusters (for example, popular sovereignty or nationalism or self-determination), new forms of social self-consciousness (for example, the identifying of a Third World or the reidentifying of indigenous peoples)—such disparate things enter into the total social process and meet existing social phenomena, including actualized social-philosophical conceptual structures (constitutional institutions, law, property, the family, religion), and, through the wonderful dialectic of the social process, new social reality is formed, including new law and new ideas. And so the social process goes on, forever feeding on itself.

3. OVID, METAPHORPHOSES, bk. I, l. 433, at 33 (Miller trans. Loeb Classical Library 3d ed. 1984).

4. These principles, although proposed here as self-explanatory axioms, are also intended to be derivations at the level of practical theory from the pure theory of international society presented in PHILIP ALLOTT, EUNOMIA—NEW ORDER FOR A NEW WORLD (1990).

5. For a new perspective on these historical developments, see R. P. Anand, *Changing Concepts of Freedom of the Seas: A Historical Perspective* (chapter 6 in this book).

6. For further consideration of the integration of the sea space into the rest of the physical world, see W. Jackson Davis, *The Need for a New Global Ocean Governance System* (chapter 12 in this book).

7. "In the opinion of some, the regulation of property is the chief point of all [in framing the constitution of a state], that being the question on which all revolutions turn." ARISTOTLE, POLITICS II.7.2, at 72 (Jowett trans. Oxford University Press 1909).

8. In social systems structured on the basis of centralism—that is, on the basis of the ultimate authority of the will of one person or one group (the party or a priesthood)—social reality can contain means of resolving decisively even social-theoretical problems. In societies that conceive of themselves as liberal democracies, social-theoretical disputes are never fully or finally resolved. They are ritualized in the process of politics. An endless succession of partial and temporary resolutions is mediated and enacted through the legal process (lawmaking and law applying).

9. Kant has offered the best explanation of the utility of the social contract hypothesis as a device of social theory. "It is in fact merely an idea of reason, which nonetheless has undoubted practical reality; for it can oblige every legislator to frame his laws in such a way that they could have been produced by the united will of a whole nation, and to regard each subject, in so far as he can claim citizenship, as if he had consented within the general will." IMMANUEL KANT, *On the Common Saying: "This may be true in theory, but it does not apply in practice,"* in KANT'S SELECTED POLITICAL WRITINGS 79 (Reiss ed. 1970).

10. *See* chapters 4 and 7, by Moana Jackson and Poka Laenui, in this book.

11. In the theory of evolution, biological science suggests that humanity has blood ties, as it were, with nonhuman animals ("genetic continuity," in the antiseptic language of science). Modern cosmological theory suggests that all life, and even all matter, may have a common origin. In this way, natural science, which works on the basis of an epistemology of empiricism, a metaphysic of materialism, and an ethic of

pragmatism, seems to be leading human consciousness on a path that is at least parallel to that of those many religious traditions that have taught the metaphysical and moral integration of the human being in the universe of all-that-is.

Still more surprisingly, evolutionary biology suggests that human beings may, in some sense, be descendants of animals that inhabited the sea, so the profound feelings that humanity has always had in relation to the sea, revealed in art, myth, and analytical psychology, may have some profound source other than the awe that the power and the wealth of the sea naturally inspire.

12. In the Roman use of the phrase *mare nostrum* in relation to the Mediterranean Sea, the possessive pronoun may have included some sense of Roman cosmopolitanism, however provincial that may now seem to have been. In the unfortunate use of the phrase by a later Italian government in relation to the same sea, the pronoun was certainly intended to be possessive, even if it was meant as a claim rather than as a description of established fact.

13. "Man is born free; and everywhere he is in chains. One thinks himself the master of others, and still remains a greater slave than they. How did this change come about? I do not know. What can make it legitimate? That question I think I can answer." JEAN-JACQUES ROUSSEAU, *The Social Contract; or, Principles of Political Right*, in THE SOCIAL CONTRACT AND OTHER DISCOURSES 165 (Cole trans. rev. ed. 1973).

14. "The philosophers have only *interpreted* the world, in various ways; the point, however, is to change it." KARL MARX, *Theses on Feurbach, XI*, in MARX & ENGELS, I SELECTED WORKS 15 (1969). *Cf.* "Without revolutionary theory there can be no revolutionary movement." V. I. LENIN, WHAT IS TO BE DONE? 25 (1947).

15. It is possible for such a theory of representation to remain hidden on account of the antiquated philosophical basis of the international system that has been considered in relation to Principle 2. The government-property nexus simply gives rise to an assumption that the self-sufficiency of each individual state system must be taken as axiomatic fact.

16. In criminal law, law can determine behavior most directly by the imposition of appropriate penalty costs. In public law, law can distribute decision-making power so that social objectives are realized by discretionary choices made within limits set by the law. In social legislation, law can create the material conditions for a "good life" as conceived by social objectives, if necessary through reallocation of society's scarce resources. In civil law, in general, law can create equal conditions for the conduct of social relations, including economic relations (the "level playing field" function of law).

17. Elections give rise to two kinds of social distortion—false horizons and false validation. The horizon of an election may come to be a decisive factor in public decision making, distorting the perspectives of all public decisions, especially those with much longer-term natural horizons. And there is a temptation to use elections as a form of vindication for past, and a form of authorization for future, behavior.

The phenomena of false horizons and false validation are not confined to public office. Corporate decision making is also conditioned by artificial horizons—price, a share issue, a merger or acquisition, the annual accounts, the annual general meeting of shareholders—which may be significantly inappropriate in relation to other, more natural horizons (such as the time required to develop a product or process, long-term investment in safer or cleaner manufacturing practices, or the handling of long-term social and physical effects of corporate decisions). And false validation of otherwise antisocial behavior of a corporation may come from profitability, the share price, credit rating, the absence of shareholder dissent, or the awarding of government contracts.

There is yet another form of social distortion, which consists in the bureaucratization of the system because of the relative permanence of officials. The bureaucratization of the international system means, in particular, that what has been called, in the foregoing consideration of Principle 3, the false dialectic of international relations takes on a dual aspect, consisting of a continuing interaction of permanent officials accompanied by a staccato interaction of politicians, as they pass across the international scene.

18. Texts in the form of treaties may be submitted to national parliaments for retrospective validation, and, in some national systems, national parliaments may be involved in some way in formulating (validating in advance) the "policy" that the adepts will seek to sacralize in the mystical text. The institution of treaty ratification is, perhaps, the leading international example of false validation. With its origins in predemocratic political structures, it puts a veneer of democratic propriety on the product of a system that is irredeemably improper.

19.
> The storm we long expect
> Shall whirl the vessels round upon their route,
> Setting the fleet to sail a course direct;
> And from the blossom shall come forth true fruit.

DANTE, THE DIVINE COMEDY: PARADISE, Canto XXVII, II, 145-48, at 295 (Dorothy L. Sayers & Barbara Reynolds trans. Harmondsworth 1962).

20. It is more than sociologically interesting that some of the delegations to the Third United Nations Conference on the Law of the Sea, most notably that of the United States, were unusually diverse in their composition, including representatives of nongovernmental interests and specialists of various kinds. There was even evidence of fractional interaction among delegations, a sort of communion of specialist interests and expertise.

21. This point is developed further in Philip Allott, *Power Sharing in the Law of the Sea*, 77 AM. J. INT'L L. 1 (1983).

6 Changing Concepts of Freedom of the Seas: A Historical Perspective

R. P. Anand

FREEDOM OF THE SEAS: AN OVERRIDING PRINCIPLE

The history of the law of the sea is to a large extent the story of the development of the "freedom of the seas" doctrine and the vicissitudes through which it has passed over the years. For nearly the past two centuries, this doctrine has been an undisputed principle of international law. All other rules relating to interstate conduct on the sea more or less revolved around this doctrine. Thus, even though a coastal state's jurisdiction over its territorial sea was recognized as essential for the protection of its security and other interests, the limits of this jurisdiction were kept as narrow as possible to maintain this freedom over the widest possible area. Even limited jurisdiction beyond the territorial sea for protection of coastal fisheries was totally denied until the end of World War II. A contiguous zone for the protection of coastal economic, health, and financial interests was either refused or merely tolerated in the name of the freedom of the seas by Great Britain, the biggest maritime power for more than 200 years.[1]

Origin of the Principle

It is generally assumed that it was the seventeenth-century Dutch jurist Hugo Grotius who first propounded the doctrine of freedom of the seas in the modern period. Although the principle was accepted under Roman law and had been reduced to a legal formula according to which the sea was *commune omnium*, that is, common property of all, this view was lost and forgotten after the disintegration of the Roman Empire.[2] "The reawakening of the principle," it is believed, "was brought about by Hugo Grotius."[3]

Grotius propounded his thesis relating to the freedom of the seas in his famous book *Mare Liberum*, or *The Free Seas*, published anonymously in 1609. He wrote this remarkable book to defend his country's right to navigate in the Indian Ocean and other Eastern seas and to trade with India and the East Indies (Southeast Asian islands), over which Spain and Portugal asserted a commercial monopoly as well as political domination.[4] In fact, *Mare Liberum* was merely one chapter (chapter 12) of a bigger work, *De Jure Praedae* (*On the Law of Spoils*), which Grotius as advocate of the Dutch East India Company had prepared in 1604 and 1605 as a legal brief but had refrained from publishing.[5]

Whether Grotius was indeed the originator of the doctrine of freedom of the seas in the modern period or whether he merely "plucked the ripe fruit" of the Spanish theologians and publicists of the sixteenth century who had previously argued for the freedom of the seas might be debatable.[6] But modern Eurocentric writers on international law have no doubt that the doctrine of freedom of the seas originated in Europe, was based on European beliefs and concepts, and was derived from European state practice.[7]

Asian Traditions Ignored

It is submitted, however, that freedom of the seas existed long before Grotius was ever heard of and before Europe emerged as a formidable force on the international stage. It was actually being practiced in the sixteenth century by the Asian countries in the so-called East Indies. Thanks to the Asians' liberal traditions of freedoms of peaceful navigation and international maritime trade, and their willingness to allow foreign merchants to establish themselves and apply their own personal laws in their personal affairs, the Europeans obtained an easy foothold in Asia.[8] Whether expressed in the form of a doctrine or not, the unobstructed freedoms of navigation and commercial shipping were accepted by all countries in the Indian Ocean and other Asian seas for centuries before history was ever recorded. Besides historical records, numerous travelers' memoirs testify to this state of affairs.[9] Freedom of the seas was also a recognized rule in the Rhodian Maritime Code and was unequivocally adopted in Roman law. From the first century A.D., regular maritime commercial relations were established between Rome and several states in India and the Indian Ocean region, and they continued for nearly 300 years.[10]

On the eve of European penetration into the Indian Ocean, not only was the principle of freedom of the seas and trade well recognized in customary law of Asia, but also in some states this principle was codified and well publicized. Examples include the maritime codes of Macassar and Malacca, which were compiled at the end of the thirteenth century, based on customary practices.[11] Resisting the Dutch attempts to monopolize the trade of the Spice Islands, the ruler of Macassar is reported to have said in 1615 that the sea was common to all and that "it is a thing unheard of that anyone should be forbidden to sail the seas."[12]

Freedom of the Seas: A Casualty in Europe

While the salutary practices of freedom of navigation and unobstructed maritime trade continued to prevail and prosper in Asia, in Europe the Rhodian and Roman tradition of freedom of the seas foundered in the turbulent waters of the disputes and conflicts of the numerous smaller states that emerged from the ruins of Rome, each vying with the other. Maritime commerce died in a "state of wild anarchy" in Europe, and even the memory of Rhodian law did not last beyond the thirteenth century. By this time, all European seas came to be more or less appropriated by European states, leading to numerous disputes and almost continuous warfare. Thus, in addition to the wide claims of Spain and Portugal, Venice claimed sovereignty over the Adriatic Sea, Genoa occupied the Ligurian Sea, England dominated the undefined British seas, and Denmark closed the Baltic by closing The Sound and extended its control over the northern seas.[13]

Portugal Disturbs Peaceful Navigation in the Indian Ocean

When the Portuguese arrived in India at the end of the fifteenth century, they found no maritime powers, no warships, and no arms in the sea. The absence of armed shipping in the Indian Ocean helped tiny Portugal to control vast areas of the ocean. The Europeans were sea powers trained in the rough waters of the Atlantic and the North Sea, whose challenges hardened them into expert navigators and naval warriors. Portugal sought to apply European custom to control the vast Indian Ocean and enforce its control by its armed carracks and galleons against the unarmed Indian Ocean ships engaged in peaceful trade. Although Portugal was fairly successful in gaining a share of the Asian spice market and in disturbing peaceful navigation in the Indian Ocean, it could not wipe out the Asian maritime trade.[14] But the Portuguese monopoly of the Eastern spice trade and its huge profits aroused the jealousy of other European powers, which began to challenge Portugal's authority late in the sixteenth century.

Contest of Wits and Arms in Europe

As noted earlier, Grotius, taking his cue from the Asian maritime practices of free navigation and trade, propounded his doctrine in a brief for the Dutch East India Company to contest the Portuguese monopoly. The company asked Grotius, who was associated with it as a lawyer, to defend the company's capture of a Portuguese vessel in the Straits of Malacca in 1604. After learning as much as he could about India and the East Indies, their traditions of free trade and commerce throughout history, and the Portuguese attempts to stultify the traditional freedom of navigation to these countries, Grotius wrote *De Jure Pradae* in 1605 to defend the action. He tried to "show that war might rightly be waged against, and prize taken from the Portuguese, who had wrongfully tried to exclude the Dutch (and others) from [trade with eastern countries]."[15] His greatness lies in his keenly observing the maritime customs of Asian countries; presenting them in the form of a doctrine supported by logical arguments, Christian theology, and the authority of venerable Roman law; and recommending these views to the European countries, which had forgotten these traditions. This fact of history has been generally ignored by historians of international law.[16]

Besides Asian traditions, Grotius relied on logic. He tried to establish two propositions: first, "that which cannot be occupied, or which never has been occupied cannot be the property of anyone, because all property has arisen from occupation"; and second, "that which has been so constituted by nature that although serving some one person it still suffices for the use of all other persons, is today and ought in perpetuity to remain in the same condition as when it was first created by nature."[17] The air belongs to this class of things, and so does the sea. Therefore, argued Grotius with the disarming logic of the time, "[t]he sea is common to all, because it is so limitless that it cannot become a possession of one, and because it is adapted for the use of all, whether we consider it from the point of view of navigation or of fisheries."[18]

It must be pointed out, however, that in spite of all this learning and logic, neither Grotius nor Holland was in favor of freedom of the seas as a principle. Grotius conveniently forgot the "freedom of the sea" principle he had propounded in 1609 with such fervor and went to England in 1613 with a Dutch delegation to argue in favor of a Dutch monopoly of trade with the Spice Islands. In fact, he was surprised to find that his own book, published anonymously, was being quoted by the British against him.[19] Successive attempts by each European state to demand freedom of the seas for the lucrative spice trade of the East Indies, and later attempts by each state to try to create a monopoly for itself, led to a spate of books by numerous scholars in Europe. In this battle of books and wits, it was not Grotius, as is generally assumed, who won. The real victor was John Selden, a brilliant British scholar and statesman, whose *Mare Clausum, sen de Domino Maris Libri Duo* (*The Closed Sea; or, Two Books Concerning the Rule Over the Sea*), written at the behest of the English Crown, remained the most authoritative work on maritime law in Europe for the next 200 years.[20] Although several other publicists countered Selden's arguments, all of the European countries continued to follow his prescription in controlling as much ocean as their power would permit. Selden won this protracted battle not by the brilliance of his arguments but by the "louder language" of the powerful British navy.[21]

Freedom of the Seas Becomes the Rule

In the nineteenth century, after the Napoleonic Wars, freedom of the seas came to be revived under the patronage of Great Britain. Commercial exploitation, the riches of the Asian trade, and the vast colonial empires in America led to the industrial revolution in Europe. The needs of the industrial revolution—larger markets and raw materials—and the surplus capital that could not be invested in Europe led to huge colonial empires in Asia and Africa. As Europeans became more interested in commercial prosperity and free trade, and ever more Europeans needed to travel to Asia and Africa, Selden's *Mare Clausum* became an anachronism. Great Britain, as the greatest naval and industrial power, became the strongest champion of freedom of the seas and its police officer.[22] Grotius, a false prophet for 200 years, was proclaimed as a great hero, and his arguments, illogical in several respects, came to be chanted as holy mantras.

APPLICATION OF THE LAW OF THE SEA

Apart from a few general principles, much of the maritime law in the nineteenth century and the first part of the twentieth century was nothing more than a panorama of conflicting claims. Freedom of fisheries, which England had come to accept only after three wars with Holland and other conflicts with its neighbors, continued to be a subject of serious dispute among Europeans. There was no agreement about a uniform limit on the territorial sea or about freedom of navigation through it or through straits, especially for warships. Nations also disagreed about the contiguous zone, and England, ever since the repeal of its own Hovering Act in 1876, had continued to question the legality of such jurisdiction exercised by other states. Moreover, a large part of the law of the sea relating to war, contraband, blockade, and the rights of neutrals was always at the mercy of belligerents who stretched their rights according to their power and the contingencies of war. During World War I and World War II, the belligerents, led by Germany and Great Britain, expanded their authority over the sea on the basis of controversial doctrines such as "ultimate enemy destination" and "long-distance blockage" and enforced the doctrines over the strong protests of neutrals.[23]

Law Helps the Powerful

Freedom of the seas meant essentially nonregulation and *laissez-faire*, which was in the interests of the big maritime powers. This lack of law under the freedom of the seas doctrine was often used in the nineteenth century by European powers to threaten small states and obtain concessions from them or simply to subjugate them. There is no dearth of cases of trigger-happy Western naval commanders using naval ordnance against the "backward" peoples of Asia and Africa on the smallest excuse, or no excuse at all.[24] Even later, it gave the big powers a license to use their "freedom" in furtherance of their immediate interests—whether for navigation, fisheries, or military maneuvers— irrespective of the rights of others. The situation became even more serious during and after World War II, when the maritime powers took the liberty of expanding this "freedom" further and enclosing even wider areas of the sea, either for defeating the enemy or for conducting nuclear and missile tests, threatening the life and liberty of all peaceful users of the sea.[25] Protests by smaller countries over such uses of the sea were almost always rejected on the ground that what was not prohibited in law was permitted and that these were "reasonable" measures of security and self-defense.[26]

This situation was tolerated not only because of the overbearing influence of the maritime powers but also because the sea was of only limited importance and use. Nations did not have a strong need to provide an elaborate law for use of the sea for limited purposes, such as navigation, fishing, and, occasionally, fighting. Eventually, however, the imprecision of the law began to be seen as a problem. An attempt was made to codify the law under the auspices of the League of Nations in 1930, but it failed, mainly because the big maritime powers, especially Great Britain, insisted on a narrow, 3-mile territorial sea, and the smaller coastal states were deeply concerned about protecting their fisheries and other interests in wider zones.[27]

The Post-1945 Era: A New World

By the end of World War II, the whole balance of forces had changed. The Western European powers, which had dominated the world scene for nearly 300 years, were no longer at the center of the world stage. Out of the ruins of the world holocaust emerged the United States and the Soviet Union, with enough strength to dominate the world and to challenge each other seriously. The world divided into two power groups and plunged into a bitter cold war that affected all aspects of international relations and law.

With the weakening of Europe, colonialism collapsed and numerous Asian and African countries emerged, but for a long time they had no status and played no role in the formulation of international law. In postwar society, these states, along with the disgruntled Latin American states—the so-called Third World—acquired a new influence. Not aligned with either of the power groups, these countries aligned together to play an important role in international legal and political structures in pursuit of their interests.

There was another development. The tremendous advances in marine technology after World War II revealed a new world, with nine times as much vegetation available in the sea as was cultivated on land. Even more important, natural resources and minerals in quantities beyond anyone's imagination were present not only in the water of the sea but also on the ocean floor and in the underlying subsoil. By 1945, geologists had confirmed that huge quantities of sorely needed oil and gas resources lay buried under the seabed off the shores of various countries, outside the territorial sea, and that technology existed to exploit these resources. These invaluable hydrocarbon resources could not be left untapped, or be allowed to be exploited by other states, as had been the case with fisheries for centuries.

The development of new technology also revolutionized fishing mechanics. Significant technological breakthroughs in the ability to detect, concentrate, and harvest fish in the high seas increased the capacity of a few technologically advanced countries to indulge in overfishing, threatening entire fishery resources near the coasts of other states. The need to protect coastal resources—both living and nonliving—had become all the more evident.

Truman Proclamations: Serious Challenge to the Freedom of the Seas

The first and most important challenge to the traditional freedom of the seas doctrine in the period following World War II came from the United States, which had emerged as the strongest maritime power after the war. The twin proclamations by President Harry Truman on September 28, 1945, referred to developments in technology as necessitating the extension of U.S. coastal jurisdiction to establish conservation zones in contiguous high seas areas to protect fisheries and the right to exclusive exploitation of the mineral resources of the continental shelf.[28] In both proclamations, the littoral state extended its limited jurisdictional powers to areas of the high seas close to its coasts, without any claim to an extension of territorial waters, and specifically declared unaffected the high seas character of the areas and the right to free and unimpeded navigation in those waters. In spite of this disclaimer, the Truman Proclamations were certainly novel claims that modified, if not grossly violated, the freedom of the seas doctrine.

The United States's proclamations led to numerous claims by other states not only for continental shelf jurisdiction but also for protection of their fisheries. By 1958, nearly a score of countries had made such continental shelf claims. Some Latin American countries went even further. Argentina, Chile, Peru, Ecuador, Costa Rica, El Salvador, and Honduras all extended their jurisdiction or sovereignty to 200 miles to protect their fisheries from depredations by outsiders. Practically every proclamation claiming special rights to the continental shelf or fisheries contained the statement that freedom of the high seas was fully recognized and maintained. But the 1950 United Nations Memorandum on the Regime of the High Seas suggested that these disclaimers should not be taken seriously.[29]

The author of the memorandum suggests that concern over such claims should not be reason "for rejecting the continental shelf theory, any more than acceptance of the theory should cause the principle of freedom of the high seas to be consigned to the lumber room."[30] He recommended that "the principle of the freedom of the seas must be made more flexible so as to allow for the theory of the continental shelf, just as it was adapted to make room for sedentary fisheries in the high seas."[31]

Conflicting and Diverse Claims

Confusion during this period over the legal validity of these claims of continental shelf and fisheries jurisdiction was compounded by widening territorial sea claims. By 1958, at least twenty-seven of the seventy-three independent coastal states had claimed specific breadths of territorial sea in excess of the so-called traditional 3-mile limit. These claims ranged between 5, 6, 12, and 200 miles. Six others, although rejecting the 3-mile rule, did not specify their limits.[32]

Some countries sought to achieve the same purpose without extending their territorial waters or fisheries jurisdiction by adopting straight baselines for measuring the territorial sea joining outermost islands, islets, or rocks off their coasts. Thus, Norway essentially extended its territorial sea by redrawing its baselines and enclosing vast bodies of waters, large and small bays, and countless arms of the sea and making them internal waters subject to the absolute sovereignty of Norway. This method for protection of coastal fisheries from outsiders was upheld by the International Court of Justice in the *Anglo-Norwegian Fisheries* case in 1951.[33]

UN Efforts to Codify the Law

The divergent standpoints adopted by different states since World War II on the territorial sea, fisheries jurisdiction, continental shelf, and other issues of the law of the sea made the already ambiguous and uncertain situation "a confused medley of conflicting solutions."[34] To bring order to this confusing situation, the United Nations organized two conferences in 1958 and 1960 to develop and codify the law in a systematic manner. Four conventions[35] were concluded in 1958 that, on the whole, reasserted the traditional freedoms of the sea and also accepted coastal states' sovereign and exclusive jurisdiction over their continental shelves. Although coastal states were permitted to extend maritime zones and adopt fish conservation measures over adjacent waters, no agreement could be reached about the extent of the territorial sea or fisheries jurisdiction, and the agreement on the definition of continental shelf was vague and controversial. Another attempt was made in 1960 to reach agreement on the territorial sea, but it also failed.

Many coastal states still wished, and some claimed, wider territorial seas, but they were unsuccessful. During the 1958 and 1960 conferences, there was a continuous struggle between two groups—the numerically strong but poor newly independent Asian and African nations and their allies in Latin America, supported by the Soviet group, on the one hand, and the politically dominating, rich, satisfied, Western maritime powers and some other small Asian-African countries under their influence on the other.[36] While the maritime powers recounted and reasserted the virtues of the freedom of the seas as a "time-honored" principle, the dissatisfied states of the Third World thought it was a "timeworn" old doctrine that could still be useful but only if modified and adapted according to the changed needs of the changed international society.[37]

Rejecting the 3-mile rule for the territorial sea as a "fallen idol," the new members of the international community said that "agreement among maritime powers alone was not law"[38] and that "rules should be based on general State practice, not on that of a handful of States that had repeatedly been challenged and [was] now finally rejected."[39] But the Western maritime powers were still strong enough to enforce the traditional law of *laissez-faire* that favored them. The developing countries did not like this law but could not change it.

RENEWED CHALLENGES TO THE FREEDOM OF THE SEAS

The accelerating pace of technological, economic, social, and political change has altered our relation to the sea. The sea is no longer vast and inexhaustible and is not to be used merely for navigation and fishing by only a few maritime powers. Even in navigation, the old rule of each ship charting its own route has disappeared, and new, complicated navigational rules and routes have been devised. Ships cannot be permitted to navigate unregulated in high-traffic areas like the narrow Straits of Dover or in busy ports like New York. Moreover, navigation, especially crude-oil transport by huge tankers, has evoked the danger of steadily growing seawater pollution that could destroy biological balance and exterminate many useful species in the marine environment. The *Torrey Canyon*, the *Exxon Valdez*, and other catastrophes are constant reminders of the dangers posed by transporting crude oil by ship.

New uses and interests have come forward to compete with old uses, and these must be accommodated on the basis of "equitable apportionment." More authority and law are needed in what was hitherto largely a legal vacuum. Exclusive national authority in the ocean would be disastrous unless constrained by internationally agreed-on provisions and protected by the international institutions. Moreover, the new majority of the worldwide community of states has been generally critical of the traditional law, codified in the 1958 Convention on the High Seas, and of the freedom of the seas, which, they believe, has been inimical to their interests. They want to overhaul the old maritime law and develop a new, more balanced and equitable regime under which they would be equal partners in sharing newfound riches of the sea and deep seabed. They hope that the new regime for the sea might help them in augmenting their meager economic resources.

There is no doubt that the law is changing. In 1950, the French jurist Gidel had already said that in "fisheries and mineral resources the Grotian tradition of freedom of the high seas is losing its paramountcy which, generally speaking, had survived fairly well down to the present day."[40]

The trend to curb the freedom of the seas by extending coastal state jurisdictions for protection of security and economic interests of the coastal states increased after 1960. By the end of 1973, nearly 35 percent of the ocean, an area equal to the land mass of the planet, was claimed by the coastal states. Deploring this trend, some well-meaning jurists regretfully felt that the era of *mare liberum* "may now be drawing to a close."[41] Others, like Hersch Lauterpacht, felt that "in so far as the original conception of the freedom of the seas, as it came to full fruition in the nineteenth century, acquired a rigidity impervious to needs of the international community and to a regime of an effective order on the high seas, 'the loss of paramountcy' provides no occasion for anxiety."[42]

Demand for a New Law

In 1967, a perceptive representative of a very small state, Arvid Pardo of Malta, informed the United Nations General Assembly about the inadequacies of the current international law and freedom of the seas, which could and would encourage appropriation by technologically advanced nations of the vast areas of the sea that suddenly had been found to contain untold wealth. To avoid a potentially disastrous scramble for sovereign rights over the seabed, he suggested the creation of an effective international regime for the seabed and ocean floor beyond a clearly defined national jurisdiction, and acceptance of that area as a "common heritage" of humankind that would not be "subject to national appropriation in any manner whatsoever, to be used and exploited for the exclusive benefit" of humanity as a whole.[43]

Pardo's essentially internationalist approach was heralded by many as an idea whose time had come. The General Assembly, in a resolution adopted on December 18, 1967, recognized the "common interest" of humankind "in the seabed and ocean floor" and declared that "the exploration and use of the seabed and ocean floor . . . should be conducted . . . in the interest of maintaining international peace and security" and for the benefit of humanity as a whole. It also established an ad hoc Sea-bed Committee to prepare for the Third United Nations Conference on the Law of the Sea. On December 17, 1970, it unanimously adopted a Declaration of Principles Governing the Sea-bed and the Ocean Floor. The General Assembly declared, *inter alia*, that the seabed beyond the limits of national jurisdiction was not subject to national appropriation or sovereignty but was "the common heritage" of humankind and must be exploited for the benefit of humanity as a whole, "taking into particular consideration the interests and needs of the developing countries."[44]

Although the maritime powers sometimes denied the legal force of these declarations of the General Assembly, there was clear indication that the new majority had started asserting itself. At the Third UN Conference, organized to regulate new uses of the sea for the vastly extended international society, the new states were determined to play a more vigorous role. Over the objections of the "old guards" and defenders of the traditional law, who preferred a conference only for formulation of law for the exploitation of the seabed beyond national jurisdiction, these states wanted a comprehensive conference to review the whole international law of the sea. To the overwhelming majority of states, the status quo was unsatisfactory. They wanted to be able "to analyze, question and remold, destroy if need be, and create a new, equitable, and rational regime for the world's oceans and deep ocean."[45]

Third UN Conference

At the Third UN Conference, which met at its first substantive session in 1974 in Caracas, Venezuela, the new majority of the developing countries made it clear that it was only the strong countries "that profited from these unlimited and undefined freedoms" of the traditional law.[46] The continuing *laissez-faire* on the high seas had ceased to serve the interests of international justice.[47] In seeking to establish a new legal order, the developing countries said, they would be "seeking not charity but justice based on the equality of rights of sovereign countries with respect to the sea."[48] Only a new international law could establish this new order, because "between the strong and the weak, it is freedom which oppresses and law which protects."[49] The developing countries, in short, were determined, as the president of Venezuela said in opening the conference, that the sea could not be permitted to "be used in such a way that a few countries benefitted from it while the rest lived in poverty, as had been done with the riches of the land."[50]

In an attempt to reconcile the freedom of the seas with the wider, inclusive interests of the enlarged and yet increasingly interdependent international society, the conference achieved agreement on a wide range of issues. Besides a general consensus in favor of a 12-mile territorial sea, a 200-mile exclusive economic zone, and a continental shelf extending to the end of the continental margin, the seabed beyond the limits of the national jurisdiction came to be reaffirmed and accepted as the common heritage of humanity as a whole. Although the exact meaning and content of the term "common heritage" may be somewhat vague, like numerous other concepts of international law, an international machinery for exploitation of the ocean's resources has come to be devised and accepted by an overwhelming majority of states in the wide-ranging 1982 United Nations Convention on the Law of the Sea.[51]

Although some of the Western maritime powers have refused to sign the 1982 Convention, and although it has yet to come into force, the basic premise of the consensus reached at the Third UN Conference is clear and beyond doubt, namely, that in the future the sea must be used for the benefit of all and not merely for the interests of a few great powers.

Although navigation is vitally important, the sea is not merely a navigational route, as it has been for centuries, but is a new area of wealth, still largely unexplored, which will be the scene of the next adventure and expansion of humanity. It is generally recognized that the sea offers the greatest promise and poses the gravest threat to the world of tomorrow. It can no longer be a largely "lawless" area or a legal vacuum.

Freedom of the seas will still be a relevant concept, but this freedom will not be unlimited. It will be the same kind of freedom that individuals enjoy in a national society, namely, freedom under agreed-on legal principles.

NOTES

1. Britain itself claimed contiguous zone jurisdiction under its Hovering Acts, passed in 1736. But in 1876, by then a great champion of the freedom of the seas, it repealed those laws and would no longer accept contiguous zone jurisdiction. Indeed, the 1930 codification conference failed to reach agreement on the territorial sea, partly

for this reason. *See* THOMAS W. FULTON, THE SOVEREIGNTY OF THE SEA 693–703 (1911) (reprinted 1976); J. L. BRIERLY, THE LAW OF NATIONS 205–06 (6th ed. 1963).

2. *See* CHRISTIAN MEURER, THE PROGRAM OF THE FREEDOM OF THE SEA 4–7 (Leo Franchtenberg trans. 1919).

3. *Id.*

4. They claimed legal title to half of the non-Christian world under a Papal Bull of May 4, 1493, by which Pope Alexander VI divided the world between the two and defined a line of demarcation running 100 leagues west of the Azores and Cape Verdes islands and granted to Spain all lands to its west and to Portugal all lands to its east. By a bilateral treaty of 1494, the two powers fortified their title. *See* K. M. PANNIKKAR, ASIA AND THE WESTERN DOMINANCE 31–32 (1954).

5. This book was discovered in 1864 and published in 1868. *See* W. S. M. KNIGHT, THE LIFE AND WORKS OF HUGO GROTIUS 79ff (1925).

6. James Brown Scott believes that the modern law of nations, including the law of the sea, rests on Spanish foundations and was first formulated by Spanish theologians, to which Grotius made only "trifling additions." *See* JAMES BROWN SCOTT, *Introduction*, HUGO GROTIUS, DE JURE BELLI AC PACIS LIBRI TRES (Classics of International Law, 1925); THE SPANISH ORIGIN OF INTERNATIONAL LAW (1928); THE CATHOLIC CONCEPTION OF INTERNATIONAL LAW (1934).

7. *See* J. H. W. VERZIJL, *Western European Influence on the Foundations of International Law*, 1 INTERNATIONAL LAW IN HISTORICAL PERSPECTIVE 435–36, 445 (1968). *See also* ADDA B. BOZEMAN, THE FUTURE OF LAW IN A MULTICULTURAL WORLD 169ff (1971).

8. *See* C. H. ALEXANDROWICZ, AN INTRODUCTION TO THE HISTORY OF THE LAW OF NATIONS IN THE EAST INDIES (sixteenth, seventeenth, and eighteenth centuries) 224 (1967).

9. *See* THE TRAVELS OF MARCO POLO (William Marsden ed. & trans. 1948); IBN BATUTTA, TRAVELS IN ASIA AND AFRICA (1325-54) (H. A. R. Gibb trans.); *Narrative of the Journey of Abd-er-Razak, A Persian Traveller and Ambassador of Shah Rukh (1442)*, in INDIA IN THE FIFTEENTH CENTURY (R. H. Major ed.).

10. *See* H. G. RAWLINSON, INTERCOURSE BETWEEN INDIA AND THE WESTERN WORLD: FROM THE EARLIEST TIMES TO THE FALL OF ROME 9–12 (1926); *see also* E. H. WARMINGTON, THE COMMERCE BETWEEN THE ROMAN EMPIRE AND INDIA 35ff (1974).

11. For a translation of both codes, *see* J. M. PARDESSUS, 6 COLLECTION DE LOIS MARITIMES (1895); Sir Stanford Raffles, *The Maritime Code of the Malays*, 4 J. ROYAL ASIATIC SOC'Y (Straits Branch), Dec. 1879, at 1–20.

12. *Quoted in* G. J. RESINK, INDONESIA: HISTORY BETWEEN THE MYTHS 45 (1968).

13. *See* FULTON, *supra* note 1, at 3–5; PITMAN B. POTTER, THE FREEDOM OF THE SEAS IN HISTORY, LAW AND POLITICS 36–38 (1924).

14. *See* MARIE A. P. MEILINK-ROELOFSZ, ASIAN TRADE AND EUROPEAN INFLUENCE 136–72 (1962).

15. W. S. M. KNIGHT, *supra* note 5, at 80.

16. *See* ALEXANDROWICZ, *supra* note 8, at 44.

17. HUGO GROTIUS, THE FREEDOM OF THE SEAS OR THE RIGHT WHICH BELONGS TO THE DUTCH TO TAKE PART IN THE EAST INDIES TRADE 28ff (James Brown Scott ed. & Ralph Van Daman Magoffin trans. 1916).

18. *Id.*

19. *See* G. N. Clark, *Grotius' East India Mission to England*, 20 TRANSACTIONS GROTIUS SOC'Y 79 (1934); *see also* Knight, *supra* note 5, at 136-43.

20. In England, "*Mare Clausum* became in a sense a law book." FULTON, *supra* note 1, at 374.

21. *See* POTTER, *supra* note 13, at 61.

22. *See* SIR GEOFFREY BUTLER AND SIMON MACCOBY, THE DEVELOPMENT OF INTERNATIONAL LAW 53 (1928). *See also* MARK W. JANIS, SEA POWER AND THE LAW OF THE SEA 15 (1976).

23. *See* JOHN C. COLOMBOS, INTERNATIONAL LAW OF THE SEA 62, 748-52 (6th ed. 1967); JULIUS STONE, LEGAL CONTROLS OF INTERNATIONAL CONFLICTS 484ff, 500ff (1959).

24. For details of numerous cases, *see* R. R. PALMER AND JOEL COLTON, A HISTORY OF THE MODERN WORLD 548ff, 615ff (3d ed. 1965).

25. *See* MARJORIE WHITEMAN, 4 DIGEST OF INTERNATIONAL LAW 554-59, 604-05 (1965). For a discussion about the legality of these tests, *see* Emmanuel Margolis, *The Hydrogen Bomb Experiments and International Law*, 64 YALE L.J. 629 (1955); but *cf.* Myres S. McDougal & N. A. Schlei, *The Hydrogen Bomb Test in Perspective: Lawful Measures of Security*, 64 YALE L.J. 648 (1955). *See also* Jon M. Van Dyke, *Military Exclusion and Warning Zones on the High Seas* (chapter 29 in this book).

26. For such defenses by both the U.S. and British governments, *see* 4 WHITEMAN, *id.* at 548ff, 585ff, 600.

27. *See* Jesse S. Reeves, *The Codification of the Law of Territorial Waters*, 24 AM. J. INT'L L. 493 (1930).

28. *See, e.g.*, Proclamation No. 2667, 10 Fed. Reg. 12,303 (1945).

29. *See United Nations Memorandum on the Regime of the High Seas* (reputed to have been prepared by the French jurist Gidel), U.N. GAOR, U.N. Doc. A/CN.4/38, July 14, 1950, at 2-3 [hereinafter *UN Memorandum*].

30. *Id.* at 86-87.

31. *Id.* at 86-87. *See also* Hersch Lauterpacht, *Sovereignty over Submarine Areas*, 27 BRIT. Y.B. INT'L L. 398, 399 (1950). Freedom of the seas could not "be treated as a rigid dogma incapable of adaptation to situations which were outside the realm of practical possibilities in the period when that principle first became part of international law."

32. *See Draft Synoptical Table Prepared by the U.N. Secretariat in Pursuance of the Resolution of the First Committee (Territorial Sea and Contiguous Zone) at its 14th Meeting (Mar. 13, 1958)*, U.N. Doc. A/CONF.13/C.1/L.11, Mar. 20, 1958; *id.* Rev. 1, Apr. 3, 1958; *id.* Rev. 1, Corr. 2, Apr. 22, 1958.

33. 1951 I.C.J. 132.

34. *UN Memorandum, supra* note 29, at 112.

35. Convention on the Territorial Sea and Contiguous Zone, Apr. 29, 1958, 516 U.N.T.S. 205, 15 U.S.T. 1606; Convention on the High Seas, Apr. 29, 1958, 13 U.S.T. 2312, T.I.A.S. No. 5200, 450 U.N.T.S. 82; Convention on Fishing and Conservation of Living Resources of the High Seas, Apr. 29, 1958, 17 U.S.T. 138, 599 U.N.T.S. 285; and Convention on the Continental Shelf, Apr. 29, 1958, 499 U.N.T.S. 311, 15 U.S.T. 471.

36. See Arthur H. Dean, The Second Conference on the Law of the Sea: Fight for Freedom of the Seas, 54 AM. J. INT'L L. 752 (1960); Robert L. Friedheim, The Satisfied and Dissatisfied States Negotiate International Law, 18 WORLD POL. 20-41.

37. U Mya Sein (Burma), Second U.N. Conference on Law of the Sea: Official Records, Summary Records of Plenary Meetings and Meetings of the Committee of the Whole, U.N. Doc. A/CONF.1918, Geneva, Mar. 17 to Apr. 26, 1960, at 58.

38. Shukairy (Saudi Arabia), id. at 74.

39. Hassan (United Arab Republic), id. at 102.

40. UN Memorandum, supra note 29, at 55.

41. Wolfgang Friedman, Selden Redivivus: Towards a Partition of the Seas, 65 AM. J. INT'L L. 763 (1971).

42. Lauterpacht, supra note 31, at 378.

43. Arvid Pardo, U.N. Doc. A/C.1/PV.1515, Nov. 1, 1967, at 6.

44. G.A. Res. 2749 (XXV), 25 U.N. GAOR Supp. (No. 28) 24, U.N. Doc. A/8028 (1970).

45. C. W. Pinto (Sri Lanka), Problems of Developing States and Their Effects on the Law of the Sea, in NEEDS AND INTERESTS OF THE DEVELOPING COUNTRIES 4 (Lewis M. Alexander ed. Kingston, R.I. 1973); see also Lusaka Declaration of the Third Conference of Heads of State or Government of Non-Aligned Countries, Sept. 1970, U.N. Doc. A/AC.138/34, Apr. 30, 1971, at 5.

46. See Vratusa (Yugoslavia), 1 THIRD U.N. CONFERENCE ON THE LAW OF THE SEA, OFFICIAL RECORDS, at 92, U.N. Sales No. E.75.V.3 (1975).

47. Warioba (United Republic of Tanzania), id. at 92.

48. H. S. Amersinghe, id. at 218.

49. Raharijaona (Madagascar), id. at 106.

50. Carlos Andres Perez, id. at 36.

51. United Nations Convention on the Law of the Sea, Dec. 10, 1982, U.N. Doc. A/CONF.62/122, 21 I.L.M. 1261 (1982).

II

Pacific Approaches toward the Ocean Environment

TO PEOPLES OF THE PACIFIC ISLANDS, the ocean is more than merely a resource; it is a living being and a home for other living beings. The Pacific Islands and their peoples were born of the ocean. Its waves carried their ancestors across time and space from forgotten places. To them, the land and the sea are inseparable, both dependent on and connected with each other. Their life, health, spirituality, and consciousness are linked inexorably with the sea.

In the first chapter of this part, Poka Laenui (Hayden Burgess) contrasts the Hawaiian perspective of kinship and sharing with the Western attitude that "sees people as distinct, separate, and above the rest of the creatures of the world." The Western presumption is that all of the earth's bounty, including the ocean, is merely a resource for humankind to dominate and control.

Laenui sets forth several fundamental philosophical differences underlying the differing perspectives toward the ocean, such as stewardship versus kinship, with "stewardship" implying that humans are "protecting the oceans as they would their manor"; and resource versus *Ke Kumu*, with "resource" relating to a purely economic perspective and *Ke Kumu*, or "the Source," seeing the ocean as "the source for cleansing, healing, and nourishment of the spirit," a source with many other values beyond economics.

The opposing Western perspective, of dominance and control over the ocean, reached its apotheosis in the nuclear testing and other military activities that have been conducted in the Pacific by France, the United States, the former Soviet Union, and others, as well as in ocean dumping of radioactive and other hazardous wastes and overfishing with indiscriminate drift nets. These and other assaults from the large industrialized countries demonstrate the vulnerability of the South Pacific's ocean environment. A. V. S. Va'ai asserts in his chapter that the Pacific Island states and territories "will cease to exist in the future if the status quo continues."

The response of the South Pacific states and territories has been one of regional solidarity and the recent creation of an ambitious regional environmental program. The centerpiece of this program, the Convention for the Protection of the Natural Resources and Environment of the South Pacific Region and its protocols, is addressed in chapters by Mere Pulea, A. V. S. Va'ai, and Florian Gubon. Gracie Fong examines the South Pacific Forum Fisheries Agency, acclaimed as one of the most effective in the world.

As impressive as the regional response has been, many of the chapters also express a wariness that the greater cooperation needed to implement and enforce the regional agreements successfully will not be forthcoming from the United States, Japan, and other Pacific Rim countries. Yet without the transfer of financial resources, expertise, and technology, the initial success of the South Pacific regional approach may never be fully realized.

7 An Introduction to Some Hawaiian Perspectives on the Ocean*

Poka Laenui (Hayden Burgess)

THE MOST APPROPRIATE WAY TO BEGIN an examination of Hawaiian perspectives on the ocean is to discuss the Hawaiian creation chant, the *Kumulipo*.[1] This chant illustrates the deep and enduring differences between Western and traditional Hawaiian ways of relating to and respecting the ocean. More than just an "environment" or a "resource," to Hawaiians the ocean is a living being—a home for other living beings and a home of living gods.

By contrast, even the most enlightened Western institutions and organizations, such as the World Commission on Environment and Development, the International Union for Conservation of Nature and Natural Resources, the World Wildlife Fund, and the United Nations Environmental Program, all presume the superiority of humankind above all of nature's other components. This presumption is based essentially on a "divine cause" concept, from which many of the environmental and property laws are based. The core of this presumption is the belief that the environment has no other purpose but to support and sustain humankind—to be dominated, mastered, and controlled for that singular purpose. This presumption sees people as distinct, separate, and above the rest of the creatures of the world.

It was not the intention of the Hawaiian priests who composed the *Kumulipo* for their chief, Ka 'I-i-mamao, at the turn of the eighteenth century, to explain

* Perspectives in this paper were contributed by many people, some through specific instructions on Hawaiian philosophy and others through their conduct, some through their writings and still others through *kukakuka* (discussions) over the years. Special mention should be made of some contributors. They are Pilahi Paki, Carl Imiola Young, Julian Hoffsnieder, Eric Enos, Roman Bedor, Walter Keliiokekai Paulo, Eddie Kaanana, Daniel Hanakahi, Rubelitte Kawena Johnson, Martha Beckwith, Queen Liliuokalani, and, especially, Puanani Burgess and our ancestors who composed the *Kumulipo*.

the universe in scientific terms. Dictated by centuries of established Polynesian custom, their intent was simply to relate a newborn chief of high social rank to his ultimate origins in the earth's very beginnings, at the point where all prehuman forms of nature and human life are but common kindred.[2] The *Kumulipo* sets forth the order of evolutionary progression, proceeding from the invertebrates of the first age to marine invertebrates in the second. Thereafter, the egg-bearing vertebrates branch off into insects, birds, and reptiles. As the cold-blooded species emerge from the marine environment to live freely out of water, the succeeding chapters of the *Kumulipo* move through the warm-blooded mammals; the breaking of light; and the appearance of La'ila'i, the woman; Kane, the god; and Ki'i, the man; followed by more than a thousand lines of genealogical husband-and-wife pairs.[3]

This creation chant does not treat the mystical notion of divine cause as the source of all life on earth. Rather, the *Kumulipo* sets forth that all life, including human life, evolved on the same creational plane and that the spiritual forces giving impetus to evolution are intricately intertwined in the evolution process. Thus, in the *Kumulipo*, we are connected both biologically and spiritually to all of the creatures and plants of the earth. Our relationship with ocean life, for example, is not merely the sharing of the same physical makeup. It is the sharing of a spiritual interrelationship as well.

FUNDAMENTAL PHILOSOPHICAL DIFFERENCES

Thus, while many international organizations may ask: How do we protect our ocean as a most valuable resource?, our Hawaiian practitioners would ask: How do we protect this womb, this sacred place of creation, this spiritual core of the earth? Clearly, differing Hawaiian and Western perspectives on the world strongly influence the relationship of each group to the ocean. The following sections highlight such conflicting viewpoints.

Stewardship versus Kinship

The word "stewardship," which is much used in today's environmental protection parlance, suggests that the relationship of humankind to the ocean is that of benevolent despot. Stewardship implies that humans are charged with the duty of protecting the ocean as they would their manor, their forest, their kingdom. But what that means is that while humans are in charge, they are separate from and superior to that of which they are the steward.

In the Hawaiian way, as celebrated in the *Kumulipo*, we are born on the same genealogical line as the sea cucumber, the *limu* (seaweed), the starfish, the slug, the shark, the dolphin, and the whale. We are part of, and kin to, the ocean and all of its living partners. Therefore, this relationship requires the same kind of protection and respect that human relations require.

Scarcity versus Abundance

The predominant Western view is that humans are surrounded by a scarcity of resources. Based on that view, all resources must be claimed and form the basis of an economic model. Out of this scarcity, either in fact or by controlling availability of goods in the marketplace, a margin of profit is maintained between production cost and sale price to drive the economy.

Hawaiian belief is built around a framework of abundance, in which it is presumed that no matter how much or how little is available, in the sharing there will be enough for all. Abundance, however, does carry with it strict responsibilities of respecting, caring for, and feeding the food supply and of recognizing the connectedness between everything. It does not imply carelessness, wastefulness, or disrespect. Hoarding is one of the great social evils: To make a profit is to take advantage of another's misfortune—a breach of proper conduct.

Aquatic Continent versus Ocean and Land

Former president of the United States George Bush, in a speech at the East-West Center in Honolulu, Hawaii, on October 27, 1990, referred to the Pacific Ocean as an "aquatic continent." If that concept is accepted, the next step in relation to the ocean is to apply continental approaches to division of territory and of resources. Lines of demarcation would thus be appropriate, since they have worked so well in Europe, the Americas, and elsewhere—or so the logic goes.

Hawaiians see the Pacific Ocean as having its own characteristic. The ocean differs from continents in that it represents another element in the Hawaiian creational framework. It represents fluidity in life, the ever-changing nature of the world—quite a contrast to land, which represents the element of stability. Specific rules, different from those for land, must be observed for the ocean. To pollute one part of the ocean is to pollute distant places touched by its waters. To destroy a species in one part of the ocean is to starve a people dependent on that species in an opposite part of that ocean. The ocean is a great connector of countries and of peoples, the common amniotic fluid from which we have all come and that we continue to share.

Although land and water are distinct elements in the Hawaiian creational framework, the ocean is inseparable from the land. To conceive of one without the other is to have night without day, a body without spirit, man without woman. Another Pacific Islander, Roman Bedor of Belau (Palau), aptly described Pacific Islanders as persons standing with one leg on an island and the other in the ocean. Both are essential to their good health and happiness. To deprive them of their island is to cut off one leg; to deprive them of the ocean is to cut off the other.

Resource versus *Ke Kumu*

Black's Law Dictionary defines resources as "[m]oney or any property that can be converted into supplies; means of raising money or supplies; capabilities of raising wealth or to supply necessary wants; available means or capability of any kind." Taking this definition, the Western view is that a resource forms the first stage of a purely economic model of the global system.

Ke Kumu, the Source, is the more traditional view held by the Hawaiian people of the ocean and all of its living beings. The ocean is the source for a multitude of things beyond economics, security, or transport. It is the source of food to island peoples and the source of health, providing a variety of medicines for physical and emotional well-being. It is also the source for cleansing, healing, and nourishment of the spirit and a place to learn the ways of nature.

HAWAIIAN PHILOSOPHY AND ITS ROLE IN THE INTERNATIONAL COMMUNITY

Having reviewed some fundamental differences between philosophical approaches to the ocean environment, let us turn to the avenues by which these Hawaiian perspectives may contribute to ongoing international environmental discussions.

Hawaii has been kept from participating fully with other Pacific Island nations in regional and international development of environmental laws relating to the ocean. This regional and international isolation is not a result of geography, even though Hawaii is the most distant land mass across a body of water in the world. Rather, Hawaii's seclusion stems from a history of political connivance and military aggression. The foundation of this isolation began with the initial invasion by the United States of the independent nation of Hawaii in January 1893 and the subsequent and continuous occupation of this Pacific nation ever since, which has prevented Hawaii from fully participating at the international level as a Pacific Island nation.

Prior to the United States's invasion in 1893, Hawaii was an active participant in international affairs, with almost a hundred diplomatic and consular posts around the world; a member of one of the first modern international organizations, the Universal Postal Union; and a nation that could boast of having been the first to have its head of state, King Kalakaua, circle the world in a voyage of friendship, commerce, and peace. Hawaii had treaties and conventions with a multitude of states—Belgium, Bremen, Denmark, France, the German Empire, Great Britain, Hamburg, Hong Kong, Italy, Japan, the Netherlands, New South Wales, Portugal, Russia, Samoa, Spain, the Swiss Confederation, Sweden and Norway, Tahiti, and the United States.

When Hawaii was annexed to the United States in 1898,[4] it became known as the "Territory of Hawaii." The United States soon built Hawaii into its Pacific military fortress and command post, taking over all control of immigration and "foreign" intercourse. The Missionary Party, the businessmen who had precipitated the U.S. invasion, having achieved their objective, secured a steady sugar market in the United States and soon developed Hawaii into their sugar-coated empire, in which anything "sugar" touched, "sugar" controlled. Thus, all banking, utilities, shipping, communications, and media and every level of politics were controlled by sugar interests.

Since the U.S. invasion, Hawaii has ceased to play any significant role as a member of the international community. Its current status, shrouded in the legal fiction called "statehood," is today being questioned by many people, predominantly those of the indigenous race. Interesting Hawaiian times are ahead. Hawaii is stepping into a period of self-examination as it reviews the struggles for independence occurring in Europe as well as the Pacific. The year 1993, declared by the United Nations as the International Year for the World's Indigenous Peoples, also marks 100 years since the U.S. overthrow and subsequent occupation of Hawaii.[5]

Notwithstanding almost a century of occupation and recycling of Hawaii's indigenous people into the American "way of life," the perspectives of Hawaii's indigenous people to their ocean are still alive and well. These perspectives have also found their way into international forums through activities of nongovernmental organizations (NGOs). Hawaiian perspectives have been adopted by the largest international indigenous nongovernmental organization, the World Council of Indigenous Peoples, and have been promoted before international organizations such as the World Commission on Environment and Development, the United Nations Working Group on Indigenous Populations, and the International Labor Organization's Committee of Experts on the redrafting of its Convention Concerning the Protection and Integration of Indigenous and other Tribal and Semi-tribal Populations in Independent Countries (Convention 107). The following is a relevant portion of such a submission to the UN Working Group:

The earth is not a commodity to be bartered back and forth to maximize profit or to be damaged for scientific exploration or tests. The earth is the foundation of indigenous peoples. It is the seat of spirituality, the fountain from which our cultures and languages flourish. The earth is our historian, the keeper of events and the cradle for the bones of our ancestors. It provides us food, medicine, shelter and clothing. It is the source of our independence. It is our mother. We do not dominate her, we harmonize with her.

Activism through indigenous NGOs is a welcome opportunity for asserting Hawaii's indigenous perspectives in international environmental discourse; yet these experiences can be very limited (because of either lack of political access or economic limitations) and often are given only cursory attention. In the Pacific region, however, Hawaii's participation in such discussions as a nation "state" may be more welcome because fellow Pacific Island political entities are willing to set aside strict political criteria for Pacific participants. The inclusion of Hawaii in the South Pacific Games and the South Pacific arts festivals are two examples of this attitude toward Hawaii's direct participation in regional activities. Yet, aside from such demonstrative cultural activities within the region, such "state" participation usually does not adequately encompass Hawaii's indigenous perspectives but instead reflects a U.S. colonial perspective.

It should be noted that the Pacific colonial and decolonization experience is ever present in the consciousness of many independent and emerging independent countries of the Pacific. The possibilities of support for indigenous or suppressed minorities' perspectives by some of these Pacific countries are very promising. However, it does not seem likely that these Pacific countries will go out of their way to solicit the concerns of the indigenous peoples under colonial rule. It is also unlikely that indigenous peoples still under colonization have the sophistication and technical training to be aware of or understand the language of ongoing international conferences on the issue. In this situation, unless some "translator" organization steps forward with sufficient resources to bring together such indigenous and sympathetic governments, the potential contributions of indigenous peoples, such as the Hawaiians, will be lost.

HAWAIIAN PHILOSOPHY IN PRACTICE

On the northwestern coast of the island of O'ahu is the community of Wai'anae, where the highest concentration of indigenous Hawaiians live today. This community has been known historically as a place of retreat and rejuvenation. Traditionally, whenever defenders of the island of O'ahu were defeated, they would gather in this area to rebuild in order to reclaim the island. It is said that the gods who protect and care for this place leave a very strong spirit in the people living there. Almost 50 percent of the population is composed of indigenous peoples, as opposed to 18 percent for Hawaii as a whole.

Driving through the main street of Wai'anae, it is not easy to see the day-to-day exercise of Hawaiian philosophy by those trained in the practice. But if one observes carefully, one can usually come across practices that reflect the continuing spirit of the indigenous people of Hawaii, such as the following scene:

> Keliiokekai Paulo's 'opelu fishnet has just been completed. This completion gave rise to celebration and blessing. A group of about thirty, ages stretching from great-grandparents to toddlers, has gathered for the dedication of the fishnet. Kupuna Kaanana stands at the front, leading the ceremony, recalling from as long ago as his childhood how his kupunas (elders) would give thanks to our ancestors' gods, in the same way we will this day. He says that we must take this time to remember that we are part of the same creation as everything about us, that just as we must respect and treat one another with aloha, we must always treat every other creation with aloha, and that the gods within every one of us are also in all creation. In his prayer of dedication, he calls on the fishnet to bring good fortune to its fishing folks and calls on us always to keep intact our aloha for the waters and for all of our relations found in and about them.
>
> The prayer having been completed, we take part in the sacred activity of eating, feeding our bodies and the gods within us, under the spread of the 'opelu fishnet.
>
> Now that Keliiokekai's net has been consecrated, he is able to fish with it. He watches for the right time of year, checks the moon phase, and prepares for fishing. Over the years, he has developed a Ko'a, a special place where the 'opelu gathers. He has been feeding and cultivating the 'opelu during the years, bringing pumpkin, taro, bread, or whatever other food he can find to feed the Ko'a. When he is ready to gather the fish, he will lower his large, cone-shaped net above the Ko'a, continue feeding the fish, and simply lift the net up and onto the boat. If he finds old friends also caught in the net, he will release them, tossing them back into the ocean. They will teach the others of the Ko'a.

CONCLUSION

This chapter has presented some Hawaiian perspectives on the environment. This is not to say that these perspectives are exclusive to the Hawaiian people, or to Pacific Islanders. They are shared by people in many parts of the world who follow many disciplines, people who have taken time and care in understanding themselves, their environment, and the inseparable relations among us all. Unfortunately, we continue to represent a small minority of opinion and conduct in the present world. Yet hope springs eternal!

Kumulipo

Kawa Akahi	The First Stage
O ke au i kahuli wela ka honua	When space turned around, the earth heated
O ke au i kahuli lole ka lani	When space turned over, the sky reversed
O ke au i kuka'iaka ka la	When the sun appeared standing in shadows
E ho'omalamalama i ka malama	To cause light to make bright the moon
O ke au o Makali'i ka po	When the Pleiades are small eyes in the night,
O ka walewale ho'okumu honua ia	From the source in the slime was the earth formed
O ke kumu o ka lipo, i lipo ai	From the source in the dark was darkness formed
O ke kumu o ka po, i po ai	From the source in the night was night formed
O ka lipolipo, o ka lipolipo	From the depths of the darkness, darkness so deep
O ka lipo o ka la, o ka lipo o ka po	Darkness of day, darkness of night
Po wale ho'i	Of night alone
Hanau ka po	Did night give birth
Hanau Kumulipo i ka po, he kane	Born was Kumulipo in the night, a male
Hanau Po'ele i ka po, he wahine	Born was Po'ele in the night, a female
. . .	. . .
. . .	. . .
Hanau ka 'Uku-ko'ako'a	Born the coral polyp
Hanau kana, he Ako'ako'a, puka	Born of him a coral colony emerged
Hanau ke Ko'e-enuhe eli ho'opu'u honua	Born the burrowing worm, tilling the soil
Hanau kana he Ko'e, puka	Born of him a worm emerged
Hanau ke Pe'a	Born the starfish
Ka Pe'ape'a kana keiki, puka	The small starfish his child emerged
. . .	. . .
. . .	. . .
Hanau kane ia Wai'ololi	Born male for the narrow waters
O ka wahine ia Wai'olola	Female for the broad waters
Hanau ka 'Ekaha noho i kai	Born the coralline seaweed living in the sea
Kia'i ia e ka 'Ekahakaha noho i uka	Kept by the bird's nest fern living on land
He po uhe'e i ka wawa	It is a night gliding through the passage

Kumulipo (*continued*)

Kawa Akahi	The First Stage
He nuku, he wai ka 'ai a ka la'au	Of an opening; a stream of water is the food of plants
O ke Akua ke komo, 'a'oe komo kanaka	It is the god who enters; not as a human does he enter
O kane ia Wai'ololi	Male for the narrow waters
O ka wahine ia Wai'olola	Female for the broad waters
Hanau ka 'Aki'aki noho i kai	Born the 'aki'aki seaweed living in the sea
Kia'i ia e ka Manienie-'aki'aki noho i uka	Kept by the manienie shore grass living on land
He po uhe'e i ka wawa	It is a night gliding through the passage
He nuku, he wai ka 'ai a ka la'au	Of an opening; a stream of water is the food of plants
O ke Akua ke komo, 'a'oe komo kanaka	It is the god who enters; not as a human does he enter
O kane ia Wai'ololi	Male for the narrow waters
O ka wahine ia Wai'olola	Female for the broad waters
. . .	. . .
. . .	. . .
Hanau ka Huluwaena, noho i kai	Born the hairy seaweed living in the sea
Kia'i ia e ka Huluhulu-'ei'ea noho i uka	Kept by the hairy pandanus vine living on land
He po uhe'e i ka wawa	It is a night gliding through the passage
He nuku, he wai ka 'ai a ka la'au	Of an opening, a stream of water is the food of plants
O ke Akua ke komo, 'a'oe komo kanaka	It is the god who enters; not as a human does he enter
O ke kane huawai, Akua kena	The male gourd of water, that is the god
O kalina a ka wai i ho'oulu ai	From whose flow the vines are made vigorous;
O ka huli ho'okawowo honua	The plant top sprouts from the earth made flourishing
O paia('a) i ke auau ka manawa	To frame the forest bower in the flow of time,
O he'e au loloa ka po	The flow of time gliding through the long night
O piha, o pihapiha	Filling, filling full

Kumulipo (*continued*)

Kawa Akahi	The First Stage
O piha-u, o piha-a	Filling, filling out
O piha-e, o piha-o	Filling, filling up
O ke koʻo honua paʻa ka lani	Until the earth is a brace holding firm the sky
O lewa ke au, ia Kumulipo ka po	When space lifts through time in the night of Kumulipo
Po no.	It is yet night.
Hanau kama a ka Powehiwehi	Born the child of Powehiwehi
Hoʻoleilei ka lana a ka Pouliuli	To grace the stature of Pouliuli with a wreath
O Mahiuma, o Maʻapuia	Of Mahiuma, of Maʻapuia
O noho i ka ʻaina o Pohomiluamea	Dwelling in the land of Pohomiluamea,
Kukala mai ka Haipuaalamea	Proclaiming the fragrant stem of Mea,
O naha wilu ke au o Uliuli	The split elegance of the branch of Uliuli,
O hoʻohewahewa a kumalamala	Unrecognized and splintered;
O pohouli a pohoʻeleʻele	In the night that darkens and blackens
O na wai ehiku e lana wale	Through seven currents he floats;
Hanau kama a hilu a holo	Born child of the gentle wrasse he swims,
O ka hilu ia pewa lala kau	The hilu whose tail fin marks
O kau(l)ana a Pouliuli	The renown of Pouliuli;
O kuemiemi a Powehiwehi	Powehiwehi shrinks away in respect (from the presence of a chief),
O Pouliuli ke kane	Pouliuli the male
O Powehiwehi ka wahine	Powehiwehi the female
Hanau ka iʻa, hanau ka Naiʻa i ke kai la holo	Born the fish, born the porpoise swimming there in the sea
Hanau ka mano, hanau ka Moano i ke kai la holo	Born the shark, born the goatfish swimming there in the sea
Hanau ka Mau, hanau ka Maumau i ke kai la holo	Born the mau fish, born the maumau swimming there in the sea
Hanau ka Nana, hanau ka Mana i ke kai la holo	Born the spawn of yellowfin tuna
	Born the small threadfin swimming there in the sea
. . .	. . .
. . .	. . .
O kane ia Waiʻololi, o ka wahine ia Waiʻolola	Male for the narrow waters, female for the broad waters

Kumulipo (*continued*)

Kawa Akahi	The First Stage
Hanau ka Palaoa noho i kai	Born the sperm whale living in the sea
Kia'i ia e ka Aoa noho i uka	Kept by the sandalwood living on land
He po uhe'e i ka wawa	It is a night gliding through the passage
He nuku, he kai ka 'ai a ka i'a	Of an opening; sea water is the food of fish
O ke Akua ke komo, 'a'oe komo kanaka	It is the god who enters, not as a human does he enter
O ke ka'ina a palaoa e ka'i nei	In the lead the whales proceed,
E kuwili o ha'aha'a i ka moana	Mingling and submerging beneath the sea;
O ka 'opule ka'i loloa	The 'opule advance in the distance;
Manoa wale ke kai ia lakou	The deep ocean is filled with them;
O kumimi o ka lohelohe a pa'a	Like kumini crabs clustered on the reef
O ka'a monimoni i ke ala	They swallow on the way
O ke ala o Kolomio o miomio i hele ai	Along the path of Kolomio, swiftly darting;
Loa'a Pimoe i ke polikua	Pimoe is found at the bosom of the horizon
O Hikawainui, o Hikawaina	Of Hikawainui, the strong current, Of Hikawaina, the calm current,
O pulehulehu hako'ako'a	Where spire myriad corals
Ka mene 'a'ahu wa'awa'a	From the hollows of blunted reef;
O holi ka poki'i ke au ia Uliuli	The youngest is carried by the current into darkness.
Po'ele wale ka moana powehiwehi	Black as night the opaque sea,
He kai ko'ako'a no ka uli o Paliuli	Coral sea in the dark cliffs of Paliuli,
O he'e wale ka 'aina ia lakou	Land that slid away from them,
O kaha uliuli wale i ka po—la	Dark shore passing into night—
Po-no.	It is yet night.

NOTES

1. *Kumulipo*: HAWAIIAN HYMN OF CREATION (Rubelitte Kawena Johnson ed. 1981). The first division of the *Kumulipo* corresponds with the text of the *Wharewananga*, belonging to the eastern coast of New Zealand, as well as to creation chants found in Tahiti, the Marquesas, and the Tuamotus. MARTHA BECKWITH, HAWAIIAN MYTHOLOGY 311–12 (1970). The full chant is more than 2,000 lines; a portion is reprinted at the end of this chapter.
2. *Kumulipo, supra* note 1, at i.
3. *Kumulipo, supra* note 1; BECKWITH, *supra* note 1, at 310–11.
4. Newly elected president Grover Cleveland, in recognition of the international uproar

over the invasion, refused to support annexation in 1893. In 1897, however, when President William McKinley took office, a treaty of annexation between the United States and Hawaii was hurriedly negotiated and signed. Because he was unable to garner the requisite two-thirds vote for ratification in the U.S. Senate, the president proceeded with annexation through a joint resolution of Congress, by only a majority of votes in the House and Senate.

5. *See* Poka Laenui, Cause for Hawaiian Sovereignty (unpublished manuscript); QUEEN LILIUOKALANI, HAWAII'S STORY (1964); LAWRENCE FUCH, HAWAII PONO (1961).

8 The Unfinished Agenda for the Pacific to Protect the Ocean Environment

Mere Pulea

∙∙

THIS CHAPTER ADDRESSES THE UNFINISHED AGENDA for protection of the Pacific Ocean environment. Since this is such a broad topic, it is possible only to isolate and comment on some essential features. The past two decades have seen the initiation of a number of strategies in the Pacific, notably the legal responses developed to protect the environment within the region. One environmental strategy developed in the region to protect the ocean is the Convention for the Protection of the Natural Resources and Environment of the South Pacific Region and related protocols (usually called the SPREP Convention because it was produced by the South Pacific Regional Environment Programme), which is currently the major instrument of the regional governments' environmental protection strategy. Developed through an integrated approach, the SPREP Convention addresses a range of environmental concerns and includes a number of ocean protection provisions. These proposed measures will be examined to assess whether they can adequately ensure protection of the ocean environment.

BACKGROUND TO OCEAN PROTECTION POLICIES

The Pacific Ocean, "the largest feature on the earth's surface,"[1] will always require special attention, as it is integral to the development of both Pacific countries and the world. The degree and level of intervention necessary to protect this environment depends on environmental awareness, political commitment, administrative mechanisms, and the socioeconomic conditions of each country. The ocean environment cannot be separated as a specific component of the environmental portrait because in the small, scattered communities of the

103

Pacific, environmental issues on land are crucially linked with those of the sea. The health of the wider ocean environment cannot be isolated from that of other ecological features, such as coastal waters and inland rivers, because they are interdependent systems.

A number of writers have highlighted some of the problems affecting the Pacific Ocean environment. Arthur L. Dahl[2] and John Brodie and J. Morrison[3] have observed that problems such as dumping of industrial and domestic wastes, discharge of raw and partially treated sewage, increased coastal populations, and development of tourism have all had detrimental effects on coastal waters, lagoons, reefs, and fragile ecosystems. Camillus Narokobi observes that the high seas areas will continue to be a source of potential trouble for fisheries conservation and management within exclusive economic zones (EEZs).[4]

Until recently, the focus of environmental concern by individual Pacific countries had been those areas within their national jurisdiction. The departure from national focus slowly evolved in the early 1950s, with concern being raised through the South Pacific Conference about France's testing of nuclear devices on Moruroa. The South Pacific Conference and, later, the South Pacific Forum provided steady venues for environmental concerns to be raised in a unified way, and environmental issues were gradually added to the regional agenda for positive action.

Regional environmental policies in the Pacific have been developed in the wake of one crisis after another. A number of writers[5] have produced a wealth of material on the environmental effects of a variety of activities in the region, including nuclear testing by the United States in the Marshall Islands and by France on Moruroa; the proposed dumping of nuclear waste by Japan in the North Pacific; incineration of chemical weapons by the United States on Johnston Atoll; overexploitation of marine resources in the Pacific Ocean by distant-water fishing nations; and the problems of large-scale drift net fishing by Japan and Taiwan. Initially, concern focused mainly on the harmful effects of nuclear testing on the region's inhabitants and marine resources. However, as awareness of the consequences of other environmental abuses began to develop, these matters were progressively interlinked until it became clear that appropriate policies and measures needed to be developed within a comprehensive regional environmental framework.

In the 1970s, a number of significant steps were taken. A regional symposium on reefs and lagoons was organized by the South Pacific Commission (SPC) in 1971. In 1974, the region implemented a Special Project on the Conservation of Nature and appointed a regional ecological officer. Consultation between the SPC and the United Nations Environment Program (UNEP) initiated the development of a comprehensive program for environmental management and a proposal for a regional Conference on the Human Environment. In 1976, the

South Pacific Forum decided that the South Pacific Bureau for Economic Co-operation (SPEC) (now the Forum Secretariat) should consult with the SPC to prepare proposals for a coordinated regional approach to environmental management. In the same year, the South Pacific Conference directed that a comprehensive environmental program reflecting the environmental interests of all countries and territories in the region be jointly prepared by SPEC and the SPC. Each country prepared a statement of its environmental problems and made suggestions about what should be included within the proposed program. These country reports provided for the first time a comprehensive statement of the state of the environment in individual Pacific countries, where "some [sixty] areas of environmental assessment, management and law"[6] were identified for action. The Conference on the Human Environment in the South Pacific in 1982 endorsed the SPREP Action Plan and formally expressed the positive directions to be taken by regional governments toward the protection of "their" environment.

THE SPREP ACTION PLAN

The SPREP Action Plan is considered the environmental bible of the region.[7] The principal objective of the Action Plan is to

help the countries of the South Pacific to maintain and improve their shared environment and to enhance their capacity to provide a present and future resource base to support the needs and maintain the quality of life of the people.[8]

The Action Plan provides for four other specific objectives, namely, assessment of the state of the regional environment and the impact of human activities on land, fresh water, lagoons, reefs, and oceans; development of management methods to maintain or enhance environmental quality while utilizing resources on a sustainable basis; improvement of national legislation and regional agreements to provide for responsible and effective management of the environment; and strengthening of national and regional capabilities, institutional arrangements, and financial support, which will enable the Action Plan to be put into effect.

The plan identifies for action a number of areas directly related to the marine environment. The initial areas identified as requiring environmental assessment on a regional basis include the impacts of (1) sediments, tailings, nutrients, and metallic and organic pollutants on river, lagoon, coral, and reef ecosystems; (2) land use and industrial and urban development on mangrove ecosystems; (3) offshore seabed exploration and the processing of marine products on the marine and adjoining ocean environment; (4) marine oil spills on sensitive coastal environments of the region; (5) tourism development on the land, lagoon, and reef ecosystems; and (6) the storage and dumping of hazardous wastes, particularly nuclear wastes, anywhere in the Pacific region as a whole. But the success of regional assessments depends on the capacities of the individual countries to undertake effective local assessment.[9]

Eight years after the endorsement of the Action Plan, efforts are continuing through the South Pacific Regional Environment Programme (SPREP) to accomplish the range of tasks identified for action. But according to S. Chape, "the general state of the environment has continued to decline in most, if not all, Pacific island countries. It is possible that transformation of regional and national policy commitments into constructive action has been a slow process."[10]

REGIONAL LEGAL ARRANGEMENTS: STRATEGY FOR ENVIRONMENTAL PROTECTION

The South Pacific region has entered the 1990s with a number of environmental achievements, notably the coming into force of the Convention on Conservation of Nature in the South Pacific (the Apia Convention), the South Pacific Nuclear Free Zone Treaty (the SPNFZ Treaty), and the SPREP Convention and its two protocols—namely, the Protocol for the Prevention of Pollution of the South Pacific Region by Dumping (the Dumping Protocol) and the Protocol Concerning Co-operation in Combating Pollution Emergencies in the South Pacific Region (the Pollution Emergencies Protocol). Yet despite these regional arrangements and international initiatives, marine and coastal problems have not been greatly alleviated in the past two decades, and "their increasing incidence and scope as well as the general perceptions of the main threats and corresponding solutions have changed markedly on the basis of scientific studies, instigated measures, knowledge and experience accumulated during that period."[11]

A legal response to these problems that provides rules of conduct, rights, and responsibilities is essential in light of the region's increasing modern development and environmental awareness, since traditional customary law provides little guidance. Although a number of statutes with environmental provisions already exist in the Pacific, these provisions are effective only if they are strictly enforced to meet the substantive environmental goals.

There is a tendency to believe that a law is in effect when a statute has been enacted or when a regional convention has entered into force. But formal enactment usually is not followed by universal compliance. There are a number of reasons for noncompliance in the Pacific, particularly if the provisions call for complex procedures, arrangements, and funding. These same problems could also beset existing regional legal arrangements, even though these arrangements reflect an investment in environmental protection through legal action. But such investments also require public participation because without public compliance, regional legal arrangements will have limited effect. Aside from a small number of people in government and international agencies, a good portion of the public is unaware of or indifferent to these arrangements. There is a need for greater public understanding of environmental issues, and a number of provisions can be singled out to provide this function. One example is the environmental impact assessment (EIA) required under article 16 of the SPREP Convention. EIAs could be viewed as time-consuming, costly, and obstructive to progress, but if their requirements continue to be enforced, they will demonstrate a public commitment to environmental goals.

The problem does not lie in the development of new law or more law but rather in effective compliance with existing national legislation and regional legal arrangements. The worst violators are government agencies that speak about promoting environmental protection, but not at the cost of development projects, such as mining and tourism.

Effective administration of environmental provisions requires a commitment to environmental objectives and the values they represent. The requirements of regional legal instruments that are broad in scope and have implications for the economy and behavior are not easily complied with. But the advantage of having legal arrangements that are accepted as beneficial to society in the long term should be sufficient to compel official commitment to the realization of environmental objectives.

THE SPREP CONVENTION

The SPREP Convention, developed through proposals, compromise, and consensus, entered into force in July 1990. The convention is a major contribution to the development of environmental initiatives through a partnership arrangement. If the SPREP Action Plan is viewed as the environmental Bible for the Pacific, the SPREP Convention can be regarded as the constitution for the environment because it is acceptable to a reasonable number of countries in the Pacific region. By adopting it, the contracting parties share principal responsibility for implementing the provisions of the convention.

Although the convention is popularly known as the SPREP Convention, its original text made no reference to the role of SPREP, the organization. The convention speaks of the role of the parties. A range of environmental programs have already been implemented under SPREP, including marine-related projects, and a protocol has been drafted to establish SPREP's role within the convention's framework.

The SPREP Convention addresses a broad range of concerns, including a variety of environmental risks to the ocean environment. The convention also weaves a web of requirements and prohibitions that affect the various components of the ocean environment. Use of the ocean is subject to environmental requirements, and many of these take the form of "appropriate measures to be taken to prevent, reduce and control pollution" from a number of sources, including oil spills, seabed exploration and exploitation, coastal engineering, and hazardous and radioactive wastes.

Pollution Control Provisions

Of the number of pollution control provisions in the SPREP Convention, eight are of particular relevance to protection of the ocean environment.

Article 15 of the SPREP Convention requires parties to develop and promote individual and joint contingency plans for responding to oil spills and leaks, but it does not spell out the scope of liability arising from an oil spill or other emergency or the scope of liability for natural resource damage—particularly if the resource is protected or irreplaceable, such as an endangered species. Where the SPREP Convention is silent on damage caused to marine life, guidelines would need to be developed for replacement or restoration of the injured resource.

Article 8 provides that parties must prevent, reduce, and control pollution resulting directly or indirectly from exploration and exploitation of the seabed and its subsoils. Most Pacific islands do not have the investment capabilities needed to launch seabed mining operations on the scales required, but the United States, Japan, and various European interests are likely to establish partnerships with those South Pacific island states that have commercially viable seabed deposits within their EEZs. Because state government must essentially certify the activities referred to under this article, states will have an important role in its implementation. Further, states would need to establish and enforce ocean protection measures, requiring that state jurisdiction be established over these activities.

Article 13 requires parties to prevent, reduce, and control environmental damage, particularly coastal erosion, caused by coastal engineering, mining activities, sand removal, land reclamation, and dredging. Nearly one-third of the countries in the South Pacific region have reported problems with coastal reclamation and erosion. A major problem identified was lack of coordination among government departments and ministries with coastal zone jurisdiction. Article 13, strictly applied, should contribute to a sharp reduction in coastal erosion and its unacceptable adverse effects on fragile ecosystems. Protection of the coastal zones can be effected through a permit program, which could take into account habitat evaluation, and restrictions on discharges into the marine environment.

Articles 10 and 11, dealing with the disposal of wastes and the storage of toxic and hazardous wastes, respectively, and the protocol on dumping, have important implications for the health of the ocean environment. Article 10 requires parties to take all appropriate measures to prevent, reduce, and control pollution caused by dumping from vessels or human-made structures at sea and to prohibit the dumping of radioactive waste into the seabed and subsoil and the continental shelf beyond the Convention Area. Article 11 requires parties to take all appropriate measures to prevent, reduce, and control pollution resulting from storage of toxic and hazardous wastes. Article 11 also prohibits storage of radioactive wastes or other radioactive matter.

The Pacific region needs a comprehensive waste management strategy that includes both a treatment strategy designed to eliminate contaminants and a strategy designed to prevent off-site migration of pollution. The problems associated with waste management have enormous implications for liability and compensation under article 20 of the convention, which requires parties to formulate and adopt rules and procedures on liability and compensation for pollution damage in the Convention Area.

Article 12, which requires that pollution from nuclear testing must be prevented, received much attention throughout the negotiations. Nuclear accidents have occurred and can occur again, and testing in the Pacific region remains a constant threat. The Pacific region is not capable of dealing with nuclear accidents or pollution from nuclear testing; therefore, response strategies that discuss the magnitude of the problem must be examined.

Finally, Article 18 requires parties to cooperate in the field of technical and other assistance, and article 17 requires cooperative efforts for scientific research, monitoring, and exchange of data. The success of the convention's pollution control provisions depends on such research, exchange of information, environmental monitoring, and scientific and technical cooperation.

High Seas Issues

Under article 2 of the SPREP Convention, the geographical coverage includes "those areas of high seas which are enclosed from all sides by the two hundred nautical mile zones." Extending the Convention Area beyond the 200-mile EEZ was an important achievement for those countries that pressed for an expansive definition of the Convention Area. Although this was less than what was proposed by Pacific Island countries, it was nevertheless a dramatic step forward.

Attention now needs to focus on the ability of the parties to the convention to regulate, manage, and monitor those areas of high seas beyond their national jurisdiction.

First, new and creative policies for a regional environmental trusteeship for natural resources must be developed. The concept of trustees for marine life and other natural resources is not new in the Pacific. As trustees for natural resources, Pacific Islanders have managed, protected, conserved, and regulated natural resources through the application of taboos (prohibitions) imposed by chiefs, priests, and resource managers under customary law. National legislation in all Pacific countries and territories illustrates these governments' roles as stewards or guardians of natural resources. The nature of such a regional trusteeship needs to be well defined within the environmental context, since industrial nations would insist on unrestricted usage of the high seas for military, commercial, and other political interests.

Second, research should be carried out to identify the types of marine life that exist in high seas areas within the Convention Area. Any areas that are of special significance to marine life could be given protection under some form of classification. This would in effect restrict any activity that would adversely affect those areas.

Third, a moratorium could be imposed on the high seas to prohibit dumping and other harmful activities and prevent the taking of threatened and endangered species. Rules would need to be developed by the parties to the convention on the penalties for violations.

OTHER CONSIDERATIONS FOR THE OCEAN PROTECTION AGENDA

The protection and management of the ocean environment based on the provisions of the SPREP Convention are complex and require both legal and technical expertise and political commitment. The Pacific communities have demonstrated their capability to reach out in a cooperative way to protect the ocean environment. But improving the quality of the environment is a continuing process. The SPREP Convention only began the process of establishing a range of ocean protection strategies. Article 5(3) allows for the formulation and adoption of protocols to control pollution and promote environmental management. Since the regional commitment to improve the Pacific environment has recently gained high priority, in the next decade the convention's environmental program should play a major role in the total regional government effort to tackle environmental degradation.

Since the endorsement of the SPREP Action Plan in 1982, some Pacific governments have been making representations in a number of environmental forums specifically related to the ocean environment. For example, Nauru and Kiribati have introduced a resolution to prohibit ocean dumping under the London Dumping Convention. In addition, successive South Pacific Forum communiqués have included a pronouncement on rising sea levels resulting from global warming. The problems associated with climate change and the rise in sea levels have serious implications for atoll countries. At a meeting of Pacific governments in Majuro in July 1989 on the greenhouse effect,[12] participants expressed the need to develop a regional program of research, planning, and policy-making to address these implications.

Other aspects of ocean protection that could be added to the SPREP Action Plan are the impacts of the greenhouse effect on marine ecosystems and fisheries, which have serious implications for the Pacific region, and new measures to manage the living resources of the high seas within the Convention Area and to protect marine mammals from large-scale drift net fishing activities.

Development of new protective measures makes little difference, however, without full compliance with existing measures and in the absence of political commitment. It is difficult for politicians to commit themselves to policies and goals that are realizable only in the distant future. For most politicians, the time factor is the next election, and for those agencies promoting environmental goals, it is unspecified and in the future. Politicians do not usually want to restrict their freedom to promise and deliver a range of politically attractive projects, even if the risk to the environment has been identified. To overcome the lack of political commitment, there is no substitute for informed and marshaled public opinion. Public concern over the fate of the Pacific Ocean environment could lead to codification of a set of principles of environmental stewardship, which would go a long way toward meeting substantive environmental goals.

NOTES

1. Arthur L. Dahl, *Biogeophysical Aspects of Isolation in the South Pacific*, 13 AMBIO 302 (1984).
2. Arthur L. Dahl, *Oceania's Most Pressing Environmental Concerns*, 13 Ambio 296 (1984).
3. J. Brodie & J. Morrison, *The Management and Disposal of Hazardous Wastes in the Pacific Islands*, 13 AMBIO 331 (1984).
4. Camillus Narokobi, *The Law of the Sea and the South Pacific*, 13 AMBIO 375 (1984).
5. *See* Jon M. Van Dyke, Kirk Smith, & Suliana Siwatibau, *Nuclear Activities and the*

Pacific Islanders, 9 ENERGY 733 (1984). *See also* J. Branch, *The Waste Bin: Nuclear Waste Dumping and Storage*, 13 AMBIO 327 (1984); B. Danielsson, *Under a Cloud of Secrecy: The French Nuclear Tests in the South Eastern Pacific*, 13 AMBIO 336 (1984); Reid J. Carew, *The South Pacific Regional Environment Programme*, 13 AMBIO 337 (1984).

6. *See* Carew, *supra* note 5.
7. *Id.* at 337.
8. REPORT OF THE CONFERENCE ON THE HUMAN ENVIRONMENT IN THE SOUTH PACIFIC (South Pacific Commission, Noumeá, 1982).
9. *Id.* at 39–40.
10. S. Chape, Role of Environmental Planning and Management in the Pacific Islands National Development: A Fiji Perspective (paper presented at the Regional Workshop on Environmental Management and Sustainable Development, Suva, Fiji, sponsored by UNDP, Apr. 1990).
11. Protection of the Ocean and All Kinds of Seas, Including Semi-enclosed Seas and of Coastal Areas and the Protection and Rational Use and Development of Their Living Resources (paper prepared by the UN Office for Ocean Affairs and the Law of the Sea, on file with the editors of this book).
12. ISLANDS' BUS., July 1989, at 11.

9 The Convention for the Protection of the Natural Resources and Environment of the South Pacific Region: Its Strengths and Weaknesses

A. V. S. Va'ai

THE CONVENTION FOR THE PROTECTION of the Natural Resources and Environment of the South Pacific Region[1] and its two protocols—one for the prevention of pollution by dumping and the other concerning cooperation in combating pollution emergencies—have been open for signature and ratification or accession since November 24, 1986. These documents are part of the global effort to mobilize and accelerate responsible measures and precautions in managing, developing, and protecting natural resources and the environment. The convention and its two protocols came into force on August 30, 1990, when Western Samoa became the tenth party to the convention.[2]

The convention and its protocols, the first regional effort to establish a legal framework to prevent, reduce, and control pollution, with its wide geographic coverage, has the potential to become an effective instrument in the battle against degradation of the marine environment. The inclusion in the Convention Area of pockets of the high seas enclosed by exclusive economic zones of the parties[3] is an interesting and significant step in an area that is much abused by the large and powerful nations of the world. It may be of little consequence, given the incapacity for surveillance and enforcement and lack of financial resources of the Pacific Island countries and territories that constitute the majority of parties to the agreements. Nevertheless, it should be an important contribution to the development of rules of international law governing the high seas.

POLLUTION

Although the convention purports to be a comprehensive umbrella agreement for the protection, management, and development of the marine and coastal environment of the South Pacific region, it actually focuses on the problem of pollution. It imposes a general obligation on parties to "endeavour" to take appropriate measures to prevent, reduce, and control pollution in the Convention Area from all sources. It then makes specific provisions on particular sources of pollution: pollution from vessels, pollution from land-based sources, pollution from seabed activities, airborne pollution, disposal of wastes, storage of toxic and hazardous wastes, testing of nuclear devices, and mining and coastal erosion. Except for articles 10 (Disposal of Wastes) and 11 (Storage of Toxic and Hazardous Wastes), these pollution provisions do not impose mandatory obligations on the parties but leave it to them to determine what are appropriate measures to take.

Article 10(1), dealing with disposal of wastes, is divided into three parts. The first part requires parties to take appropriate measures to prevent, reduce, and control pollution. The second part stipulates that the parties "agree to prohibit the dumping of radioactive wastes or other radioactive matter" in the Convention Area. The third part provides that the parties agree to prohibit disposal into the seabed and subsoil of radioactive wastes or other radioactive matter; this obligation, however, is "without prejudice" to whether disposal into the seabed and subsoil of wastes or other matter is "dumping" within the meaning of the convention.

The other area in which a mandatory obligation is imposed on parties concerns pollution by storage of toxic and hazardous wastes. Article 11 requires parties to take all appropriate measures to prevent, reduce, and control pollution in the Convention Area resulting from storage of toxic and hazardous wastes. In particular, the storage of radioactive wastes or other radioactive matter in the Convention Area is prohibited. Article 2(b) provides that "dumping" does not include "placement of matter for a purpose other than the mere disposal thereof, provided that such placement is not contrary to the aims of this Convention."

THE REAL PROBLEM

Under international law, all independent and sovereign countries are equal in their relations. Generally, and quite often correctly, states will negotiate and establish their relationships based on an assumption of equality. Nevertheless, there can be instances in which the assumption of equality can be legitimately set aside and a special approach adopted to address the problem; otherwise, the regime established not only may compound the problem but also may prove ineffective.

Pollution in the world today is not just another of the many universal problems requiring concerted action to solve. *It is a question of survival of the human race.* With the vanishing ozone layer and the greenhouse effect, the globe is in imminent danger of literally burning or drowning in its own rubbish. In fact, the small Pacific Islands will be the most seriously affected, and some will be threatened by extinction.

It is my considered view that all agents of pollution originate from the large countries, and most of the pollution in the Convention Area will be caused by their nationals. The convention and its protocols make unreal assumptions that the small island countries and territories on the one hand and their Pacific Rim neighbors on the other are equal in capabilities, capacities, and abilities in causing and preventing the problem being addressed. They are, in fact, most unequal, not only in causation but also in prevention. This assumption of equality is reflected in the provisions contained in the agreements, which impose equal responsibilities on the parties to carry out prevention measures in the Convention Area. In fact, the major part of the responsibility and obligation in controlling pollution should be imposed on the larger countries, which not only are responsible for causing the problem but also have the capabilities to undertake meaningful preventive measures. This is a major weakness that permeates the whole convention and its protocols.

A more fundamental defect is the failure of the convention to take into account the question of causation, which is basic to the whole problem of pollution. Most pollution occurring in the Convention Area need not happen and should not happen, but some of the large countries choose to pollute here rather than in their immediate backyards. Why should they be allowed to do it? And if they do so in spite of strong objections by countries in the region, should they not pay for it? Even if they are technically within their rights to pollute because their activity is in one of their own territories or in an area of the high seas, those rights must nevertheless be subject to the rights of the small island states and territories to remain free from exposure to the damaging effects of these wastes. If they do dump their wastes there, they should also accept the liability for doing so. After all, it is a danger that they themselves choose not to face in their home territories. The people of the South Pacific region also should not be expected to have such an intolerable danger imposed on them.

CONVENTION AREA

The "Convention Area" includes the "two hundred nautical mile zones" of thirteen independent countries, including Australia and New Zealand, two self-governing freely associated states, and nine territories belonging to Australia, New Zealand, France, and the United States of America. It also includes areas of high seas enclosed from all sides by 200-nautical-mile zones of the parties. They include three pockets, one encircled by the Cook Islands, Kiribati, and French Polynesia to the east; one bounded by Fiji, Vanuatu, and the Solomon Islands to the west; and another encircled by the exclusive economic zones of Fiji, Tuvalu, Kiribati, Nauru, the Marshall Islands, the Federated States of Micronesia, Papua New Guinea, and the Solomon Islands to the northwest. It does not include internal waters or the archipelagic waters of the parties, but it allows for further additions to the Convention Area under article 3.

Inclusion of the pockets of high seas in the Convention Area has important implications for the doctrine of freedom of the seas. The freedom of states to use the high seas with reasonable regard for the interests of other states is well established under international law. The parties to the agreements have restricted their freedom to use the high seas for dumping and other environmentally damaging activities by including these pockets of high seas in the Convention Area. States that are not parties are, of course, not bound by these agreements, and an important question that arises is the effectiveness of the agreements in restricting the freedom of these states to pollute and contaminate these pockets of high seas. The undertaking of any activity likely to pollute these pockets of high seas by nonparties must be weighed against the collective interests of the Pacific Islanders in protecting their natural resources and environment from pollution.

The seabed and the subsoil of the high seas are now generally accepted to be the "common heritage" of humanity. It is therefore the responsibility of all to protect them from any kind of abuse. Any activity carried out in the high seas, however reasonable, that may or would be likely to damage the resources of the high seas, the seabed, or its subsoil should be prohibited in the interests of humanity. Indeed, the flow of the damaging effects of polluting activities in the high seas to adjoining territories of the parties ought to be sufficient reason for prohibiting such activities. No activity that results or may result in deadly and irreparable pollution can be regarded as reasonable. It can be described only as murderous and totally immoral.

The self-imposed restriction on the use of the pockets of the high seas is of little significance to the small island states and territories that surround them, as they would have little, if anything, to dump or store. Furthermore, they do not have the opportunity or desire to exercise any freedom in the high seas, either in these pockets or anywhere else. The two relatively large countries that are parties, namely, Australia and New Zealand, as well as France and the United States, which have the most territories in the Pacific, on the other hand, have more than enough areas of high seas surrounding them to exercise their freedom; thus, they have no legitimate interest in these faraway pockets of high seas.

PACIFIC RIM COUNTRIES

Although the convention leaves the door open for accession by any state, most of the Pacific Rim countries that have the capabilities of polluting the Convention Area and that are doing so are not parties. To be an effective participant in the convention, a party must accede not only to the convention but also to at least one of its protocols. The apparent objective of the convention is to promote cooperation in preventing pollution; it is not a set of rules and regulations to control pollution. The use of protocols to regulate specific problems allows a country to be seen to be a party to the convention but permits it to avoid responsibility in any area in which its self-interest is of more importance. The two protocols—one to promote cooperation in combating pollution emergencies and the other to control dumping—are both appropriately worded so that no party is obliged to do anything.

The omission from the convention and its protocols of provisions imposing mandatory obligations on any of the parties should attract the large Pacific Rim countries to participate in the convention and be seen in the international community to be acting responsibly in an area of immediate global concern. Indeed, the very real threat of a rapidly degrading environment warrants immediate and positive action by all countries at the international level. Their slowness or, perhaps, reluctance to join this convention and at least one of the protocols is a most discouraging factor and does not augur well for the future of these agreements. The convention and its protocols will begin to realize their potential only if there is more participation by the larger countries in the Pacific Basin that are and will be mainly responsible for the pollution in the Convention Area. Fifty percent of the parties that took part in the negotiations are also territories or freely associated states of the Pacific Rim countries and France and rely heavily on their assistance. This raises problems of full and effective participation by these parties with regard to implementing the agreements.

SMALL ISLAND STATES

The disadvantaged situation of the small Pacific Island states and territories vis-à-vis the large, developed countries of the world is sometimes globally recognized and accepted. In some situations, this situation is accommodated by special provisions in conventions that take this fact into account. Unfortunately, however, more often than not such provisions do not confer on the small island states any real rights but merely give the large countries opportunities to offer assistance, opportunities that may never be acted on. These provisions do not constitute a genuine recognition of the disadvantaged situation of the Pacific Islands but are merely benevolent gestures.

The convention, notwithstanding the glowing references in its preamble to the uniqueness and special circumstances of the South Pacific region, contains little evidence of a heartfelt concern or a real acknowledgment of the unique situation of the small Pacific Island states. Articles 17 and 18 of the convention make provisions for scientific and technical cooperation and technical and other assistance; but this obligation is imposed on all parties, including the Pacific Island countries and territories, which themselves depend largely on assistance from the large countries.

The small island states and territories do not have the capability to police effectively any prevention measures intended for the Convention Area. This area is so vast that any dumping operation can be done at will and quite deliberately without detection by any of the parties. An oil tanker could run aground on any of the many offshore reefs and small islands inside the exclusive economic zones or the pockets of high seas of the Convention Area and could dump its cargo without any of the small Pacific Island states and territories knowing of the damage.

Dumping resulting from the normal operation of vessels, aircraft, and artificial structures is also exempted by the convention. The protocol on dumping prohibits the dumping of certain substances; other substances are placed into categories requiring different types of licenses to permit dumping. The protocol also gives guidelines on factors that may be taken into account by national authorities in deciding whether to issue licenses. But why should anybody obtain a license that would probably involve much red tape and a possible refusal? They have to dump the stuff anyway, even if a license is refused, so why bother with a license? Because they lack effective means of policing the Convention Area, the small Pacific Island states and territories are probably being used right now for dumping and other polluting activities. Australia and New Zealand do conduct surveillance flights in the region, but the size of the Convention Area limits the effectiveness of these flights.

LIABILITY AND COMPENSATION

The only reference in the agreements to liability and compensation for damages is a provision in the convention requiring cooperation in formulating and adopting appropriate rules and procedures.[4] Any damage caused should be compensated for irrespective of fault, emergency, or *force majeure*. This is imperative to reach the goal of eliminating pollution in the Convention Area. The absence of any provisions regarding liability and compensation in the agreements, particularly the protocols, is therefore a glaring omission and oversight.

The problem of pollution of the Convention Area was, is, and will be caused by the large, developed countries, with their many potential sources of pollution. As the cause of the problem, they should bear the major part of the responsibility for it. Moreover, only they have the resources to carry out an effective prevention program; to control and contain pollution is a matter requiring not only concerted effort but also unselfish and determined action.

CONCLUSION

The majority of parties to the SPREP Convention and its protocols are small Pacific Island states and territories, whose territories also make up the bulk of the Convention Area. Scarcity of resources is a major and common characteristic of these small entities. Any hope for future development lies in prudent exploitation of the resources of the seas, the seabed, and its subsoil. Protection and responsible management of the exclusive economic zones and the adjoining high seas are critical for the future development and existence of these small parties.

The convention and its protocols now form the basis to encourage cooperation in taking appropriate measures to prevent and control pollution in the South Pacific. However, the tardiness and reluctance of parties, particularly the large countries, to be legally bound and to accept responsibility for controlling pollution in the Convention Area is indeed grim for the future survival of the Pacific Islands. Like the frustrated efforts of the Pacific Islands to bring about a significant reduction in the emission of greenhouse gases, the actions by the parties in establishing meaningful measures to protect the natural resources and environment of the South Pacific region are also devoid of committed and responsible political will from the large countries.

The international community faces a bleak future. The Pacific Island states and territories will cease to exist in the future if the status quo continues. There needs to be a marked change in attitudes and morality. The large countries must fully accept responsibility for the pollution in the world today and take appropriate action that will immediately tackle the problem and produce noticeable changes. To carry on in the present fashion will surely spell catastrophe for humanity.

NOTES

1. Convention for the Protection of the Natural Resources and Environment of the South Pacific Region, Nov. 25, 1986, 26 I.L.M. 38 (1987) [hereinafter SPREP Convention].
2. As of late 1992, eleven countries, including the United States, had ratified this convention.
3. SPREP Convention, *supra* note 1, art. 2(a)(ii).
4. *Id*. art. 20.

10 Steps Taken by South Pacific Island States to Preserve and Protect Ocean Resources for Future Generations*

••

Florian Gubon

SOME STEPS HAVE BEEN TAKEN by the South Pacific island states[1] to protect and preserve ocean resources for future generations. These steps were taken under the auspices of regional organizations as a result of cooperation and dialogue among island countries. The obvious reasons for cooperating in taking these steps include the lack of individual resources and capability in individual island communities. Reasons and objectives for cooperation can be found reflected in preambles to the various cooperative agreements that have been concluded and in many of the decisions and resolutions adopted at regional meetings.

The establishment and work of regional organizations are highly important to any cooperative effort in the South Pacific. Sharing of scarce resources and information is considered to be the best strategy for Pacific Island development. The regional organizations facilitate cooperation and play an important role in organizing and facilitating regional meetings. Regional cooperative processes were designed to facilitate the South Pacific island approach to development, management, and conservation of ocean resources. This strategy obviously took into account regional cohesiveness and the apparent lack of individual ability to manage properly the large areas of ocean space that now comes under the jurisdiction of the South Pacific island states. Notable among these regional efforts are the establishment of the South Pacific Forum Fisheries Agency (FFA) in 1979[2] and the South Pacific GeoSciences Commission (SOPAC), successor

* The views expressed here are those of the author and not those of the South Pacific Forum Fisheries Agency, which runs the Ocean Resources Management Programme at the University of the South Pacific.

to the Committee for Coordination of Joint Prospecting for Mineral Resources in the South Pacific Offshore Areas (CCOP/SOPAC), which began operating in 1972. Both of these organizations play an important role in ocean resource management and development in the region.

The main impetus for taking cooperative measures in the field of ocean resources management was the potential the Pacific Island communities saw in deriving much-needed economic benefits from the exploitation of fisheries, especially the valuable tuna resources that were known to occur in their exclusive economic zones (EEZs). The potential for increasing financial benefits is clearly available, but challenges exist to manage these resources in accordance with requirements of international law.

THE PACIFIC ISLAND ENVIRONMENT AND OCEAN RESOURCES

The South Pacific region (which includes the islands of Micronesia in the North Pacific) is enormous. The distance of nearly 10,000 kilometers from Pitcairn Island to Palau is equivalent to the distance from Tromsö, Norway, to Cape Town, South Africa, or from Washington, DC, to the Palmer Peninsula of Antarctica. Although most island groups are no more than 500 to 1,000 kilometers from their nearest neighbor, even these distances are considerable. Near neighbors like Fiji and Vanuatu, for instance, are as far apart as Paris and Vienna.

The South Pacific islands as we know them today resulted from a combination of biogeographic processes and fall into four categories, based on composition and origin: (1) continental islands, (2) volcanic islands, (3) low coral islands or atolls, and (4) elevated coral islands.[3] The South Pacific region is fragmented biogeographically. Different groups of islands are isolated from one another and are subject to different conditions, which produce quite distinctive flora and fauna. A regional survey sponsored by the South Pacific Commission (SPC) and the International Union for Conservation of Nature and Natural Resources (IUCN) identified twenty biogeographic provinces in the South Pacific.[4] It was estimated that combining these with the known habitat types, island types, distinctive biomass or communities, and endemic species, the region contained some 2,000 types of ecosystems.[5]

Because South Pacific ecosystems are highly fragile,[6] serious care should be taken to protect the ocean environment and in turn the ocean resources that occur in them. Raising regional awareness of this problem began with the SPC. Some of the regional environmental programs and projects undertaken in the region through the work of the SPC and other regional organizations are aimed at addressing this problem.

Pacific Islanders call the ocean their home, whereas to many other people the ocean may be just a route for communication, navigation, and transport or a storehouse of riches. To Pacific Island people, the ocean means much more. It provides their food and livelihood; it explains their existence, their rituals, their traditions: in essence, their way of life. Legends and myths of island people reflect the intensity and profoundness of this relationship. Because of these links with the ocean environment, Pacific Islanders normally view its destruction as the destruction of their culture and way of life.

In recent times, however, the ocean and its resources have taken on a new and different dimension in Pacific Island thinking. In obtaining political independence, Pacific Islanders began in earnest to exploit natural resources to sustain national development and, in particular, to look to marine resources for these purposes.[7] Of these resources, fisheries provide the greatest potential for sustainable development and self-sufficiency.[8] This prospect emerged following universal acceptance of the EEZ regime under the law of the sea. The economic potential of other ocean resources, such as offshore oil and mineral deposits, is either nonexistent or yet to be fully ascertained.

For Pacific Islanders, the ocean environment also plays an important role in the growth and development of the tourism industry,[9] which is increasingly relied on for revenue-raising opportunities. Its growth and maintenance are widely recognized to be relative to the warm climate and beautiful marine environment the region offers to outsiders for recreation. Consequently, efforts to preserve and protect the ocean environment while allowing it to be used for recreational purposes will most probably become an issue of serious concern in the region.

STEPS TAKEN BY SOUTH PACIFIC ISLAND STATES

The South Pacific island nations have expressed regional concern over environmental degradation of the marine environment and have adopted policies and passed laws that address these concerns.

Emergence of Regional Concern and Adoption of Regional Environmental Policies

The concern of Pacific Island states for protection and preservation of the ocean environment and its resources began in 1970, when a proposal was made by the SPC to recruit an ecologist for its staff.[10] At the annual meeting of the South Pacific Conference in Guam in 1973, the governments approved the post and a special project on the conservation of nature. In 1974, a regional ecological advisor was recruited. Subsequently, a wide variety of environmental activities were included in the SPC work program, until it was replaced by the South Pacific Regional Environment Programme (SPREP) in 1980.

The preparatory phase of SPREP concluded with the Conference on the Human Environment in the South Pacific, held in Rarotonga, in the Cook Islands, in March 1982,[11] at which the South Pacific Declaration on Natural Resources and the Environment and an Action Plan for Managing the Natural Resources and Environment of the South Pacific were adopted. These actions signaled widespread regional support for the aims of SPREP. Since then, SPREP has served as the framework from which the Pacific Island states and territories have addressed environmental protection and preservation issues. The legal measures discussed next are the results of work and discussions conducted under the auspices of SPREP and the other regional organizations.

Legal Measures Taken by South Pacific Island States

The legal measures that have been taken were aimed largely at implementing the regional environmental polices and principles adopted at the 1982 Rarotonga Conference.[12] The continued French nuclear testing activities at Moruroa Atoll, the Japanese proposal to dump low-level wastes on an abyssal plain halfway between Tokyo and the Northern Marianas, and other, similar events and developments influenced island governments, with pressure from citizens and environmental groups, to take action through legal arrangements as a way of strengthening the regional stance and policy to protect and preserve the regional environment.

The 1982 Rarotonga Conference laid down the basic policies and principles that guided Pacific Island states and territories in drafting the legal arrangements. These policies and principles not only dealt with the subject of nuclear pollution and waste disposal but also included principles relevant to resource development and management. The Rarotonga Conference also resolved that integrated planning and management through enforceable legal arrangements are "a necessary basis for effective integration of environmental concern with the whole of the development process."[13]

The 1976 Convention on Conservation of Nature in the South Pacific. Regional cooperation in addressing environmental issues actually began with the decision in June 1976, at a regional meeting in Apia, Western Samoa, to adopt the Convention on Conservation of Nature in the South Pacific.[14] This resulted from initiatives of IUCN following similar work in other parts of the world and the regional symposium on conservation in 1971. The impetus for drafting this convention appears to have been drawn from the Declaration on the Human

Environment adopted by the United Nations Conference on the Human Environment at Stockholm in June 1972 and from concern about the growing dangers that threaten the natural resources and fragile ecosystems of the region.[15]

The Apia Convention requires parties to encourage the creation of protected areas (national parks and reserves) and to safeguard representative samples of national ecosystems, superlative scenery, striking geological formations and regions, and objects of aesthetic, historic, cultural, or scientific value.[16]

Parties agree to establish and to maintain a list of species of indigenous fauna and flora threatened with extinction found within their countries and to give such species as complete protection as possible.[17] Parties also agree to conduct research relating to the conservation of nature and to cooperate in the exchange of information on the results of research and on the management of protected areas and protected species.[18]

1986 Convention for the Protection of the Natural Resources and Environment of the South Pacific Region and Related Protocols. The 1986 Convention for the Protection of the Natural Resources and Environment of the South Pacific Region[19] and its two protocols are primarily based on principles and policies endorsed by the 1982 Rarotonga Conference. The SPREP Convention makes provisions for preventing, reducing, and controlling pollution from a variety of sources.[20] Because of the special concerns expressed in the region, this convention also provides for appropriate measures to be taken to protect and preserve rare and fragile ecosystems as well as habitats of depleted flora and fauna.[21] Under the convention, parties have a general duty to cooperate with one another and other organizations, particularly in sharing and exchanging scientific and technological data and other information, and jointly to develop research programs.[22] The SPREP Convention has been signed by almost all Pacific Island states and has recently entered into force.

1985 South Pacific Nuclear Free Zone Treaty and Related Protocols. The 1985 South Pacific Nuclear Free Zone Treaty[23] serves as the testimony of Pacific Island states, including Australia and New Zealand, to a nuclear-free Pacific. In the SPNFZ Treaty, the parties agreed to prevent the testing, stationing, manufacturing, and dumping of nuclear weapons and devices within their territories and also to discourage use of the region for nuclear testing and as waste disposal sites.

The sentiments expressed in the preamble to the SPNFZ Treaty reflect the feeling of Pacific Island people toward nuclear activities: The nuclear technology studies and experiments are not in any way beneficial to their needs and aspirations.[24] Islanders view nuclear activities carried out in the region as blatant acts of superpowers in total defiance of the feelings, concerns, and aspirations of weaker states. The SPNFZ Treaty has received wide regional support. Only

Vanuatu has not signed the treaty, arguing that it is not strong enough, particularly concerning the use of nuclear-powered vessels in the region.

The three protocols to the SPNFZ Treaty are aimed at securing the agreement of nuclear power states to (1) refrain from using territories that they may be occupying in the region to manufacture, station, and test nuclear explosive devices; (2) refrain from using or threatening to use any nuclear explosive device against parties to the SPNFZ Treaty; and (3) undertake not to test any nuclear explosive device anywhere within the area defined as "the South Pacific Nuclear Free Zone." The United States, France, and Britain have snubbed the protocols, but China and the former USSR have supported them. The stance of the United States, France, and Britain on this issue has not been well received within the region.

The 1979 South Pacific Forum Fisheries Agency Convention and the Conservation of Living Resources. The members of the South Pacific Forum Fisheries Agency (FFA), which includes the Pacific Island states plus Australia and New Zealand, have agreed in the FFA Convention to cooperate in the management and conservation of fisheries resources. The preamble expresses the parties' recognition of "their common interest in the conservation and optimum utilization of the living resources of the South Pacific region and in particular of the highly migratory species," their desire "to promote regional cooperation and coordination in respect of fisheries policies," and their concern "to secure the maximum benefits from the living resources of the region for their peoples and for the region as a whole." Some of the concerns related to conserving living marine resources have been reinforced through efforts such as that of the Nauru Group[25] in its conclusion of the "Nauru Agreement Concerning Cooperation in the Management of Fisheries of Common Interest" and through such schemes as the FFA's Harmonized Terms and Conditions of Access[26] and the Regional Register of Foreign Fishing Vessels[27] and perhaps also through the recent multilateral fisheries treaty that FFA members concluded with the United States government.[28] These arrangements and agreements can also be viewed as efforts directed at managing and conserving fisheries resources and not only as avenues for FFA member states to derive economic benefits. The arrangements also enable the monitoring of foreign fishing in the EEZs of FFA countries. The conservation and management of fish species harvested in the South Pacific have been successful largely because of the effectiveness of these cooperative measures.

1989 Convention for the Prohibition of Fishing with Long Driftnets in the South Pacific. The impetus for concluding the 1989 Convention for the Prohibition of Fishing with Long Driftnets in the South Pacific[29] came from concerns expressed throughout the region that large amounts of albacore tuna were being harvested through drift netting. Japanese, Koreans, and Taiwanese were involved in this activity. Drift netters were mainly targeting juvenile albacore tuna occurring in the subtropical regions of the South Pacific,[30] and concerns were expressed that if fishing levels remained at those of the 1988-1989 season, the stock would be depleted in a few years.[31]

The 1989 South Pacific Forum meeting held in Tarawa, Kiribati, condemned this activity in the region. It passed a declaration calling for the convening of a regional meeting of appropriate experts "to develop a Convention to give effect to its common resolve to create a zone free of driftnet fishing" and calling on the international community to support and cooperate with their effort to establish this zone. The mandate granted at the Tarawa meeting led to a series of meetings among FFA members, culminating in the signing of what has come to be known as the South Pacific Driftnet Convention.

Under this convention, parties agree to prohibit their nationals and vessels documented under their laws from engaging in drift net fishing activities in the Convention Area.[32] Parties agree to discourage use of drift nets in the Convention Area and to take measures consistent with international law to restrict drift net fishing activities by prohibiting the use of drift nets and the transshipment of drift net catches within areas under their jurisdiction.[33] They also agree to prohibit (1) the landing of drift net catches within their territories; (2) the importation of any fish or fish product, whether processed or not, caught by use of a drift net; (3) the processing of drift net catches in facilities under its jurisdiction; (4) port access and port-serving facilities for drift net fishing vessels; and (5) the possession of drift nets on board any fishing vessel within areas under their fisheries jurisdiction.[34]

CONCLUSION

Three general concluding comments are in order regarding the steps taken by Pacific Island communities to protect and preserve ocean resources. First, these steps can be regarded as comprehensive and responsive on a regional scale. The small island countries have shown their seriousness in addressing and dealing with environmental and ocean resource issues and problems. Much of the impetus for taking the measures they have taken is their own realization of the importance of protecting and preserving their ocean environment and resources. Their efforts obviously reflect the value they attach to their ocean environment.

Second, the policies and resolutions that have been adopted and implemented through regional agreements and arrangements are most useful for the politics and purposes of the region. These actions demonstrate their cooperativeness and their ability to address common concerns. They have found and developed regional cohesiveness in attending to common issues and problems. This kind of cooperation has yet to be developed in other regions of the world. Although others may want to copy the South Pacific examples, it should always be borne in mind that circumstances existing for cooperation among South Pacific islands may not necessarily exist in other regions. Finally, it may be stated that Pacific Islanders have adequately demonstrated that they can be regarded as good stewards of the ocean for future generations. What they have done thus far supports this conclusion. The actions they have taken not only promote their own concerns and aspirations but also have made useful contributions to the global effort that is needed today to protect and preserve ocean resources for future generations.

NOTES

1. For the purposes of this chapter, "South Pacific island states" refers to all of the developing independent and nearly independent island countries in the South Pacific (which generally includes the Micronesian countries in the North Pacific). These countries are the Cook Islands, the Federated States of Micronesia, Fiji, Kiribati, the Marshall Islands, Nauru, Niue, Palau, Papua New Guinea, the Solomon Islands, Tonga, Tuvalu, Vanuatu, and Western Samoa. Australia and New Zealand, which are members of many of the regional organizations to which this group of countries also belongs, are well-developed countries. The nonindependent countries and dependent territories, such as New Caledonia, existing within this region are also involved with some regional cooperative activities but are excluded from this definition because they are not state entities under international law and lack authority to conclude international arrangements on their own accord.

2. For an understanding of the formation and role of this regional organization, see George Kent, *Pacific Island Fisheries Politics*, 2 OCEAN Y.B. 346–83 (E. M. Borgese & N. Ginsburg eds. 1980); Jon M. Van Dyke & Susan Heftel, *Tuna Management in the Pacific: An Analysis of the South Pacific Forum Fisheries Agency*, 3 U. HAW. L. REV. 1–65 (1981); Florian Gubon, *History and Role of the Forum Fisheries Agency*, in TUNA ISSUES AND PERSPECTIVES IN THE PACIFIC ISLANDS REGION 245–56 (D. J. Doulman ed. East-West Center, Honolulu 1987); and David Doulman, *In Pursuit of Fisheries Cooperation: The South Pacific Forum Fisheries Agency*, 10 U. HAW. L. REV. 137–50 (1987).

3. This description was given by Dr. Arthur Dahl, who was appointed as the first ecological advisor for the South Pacific Commission in 1974. See Arthur L. Dahl, *Biogeographical Aspects of Isolation in the Pacific*, 13 AMBIO 302 (1984).

4. ARTHUR L. DAHL, REGIONAL ECOSYSTEMS SURVEY OF THE SOUTH PACIFIC AREA (Technical Paper No. 179, South Pacific Commission, Nouméa, 1980).

5. ARTHUR L. DAHL, PROCEEDINGS OF THE INTERNATIONAL SYMPOSIUM ON

MARINE BIOGEOGRAPHY AND EVOLUTION IN THE SOUTHERN HEMISPHERE 541-46 (New Zealand Department of Scientific and Industrial Research Information No. 137, Wellington, 1979).

6. *See* R. J. Morrison, *The Fragility of Some South Pacific Ecosystems*, 4 PLES 1 (1988).

7. *See, e.g.*, M. P. Hamnett, Marine Resources in Pacific Island Economies (consultancy paper prepared for the Regional Development Office for the South Pacific of the U.S. Agency for International Development, Suva, Fiji, June 1989).

8. M. Sheppard & L. Clark, South Pacific Fisheries Development Needs 5 (consultancy paper prepared on behalf of the FAO and the UNDP, 1984).

9. *See* Nii-K Plange, *Tourism and the Environment in the South Pacific: Some Observations and Considerations*, PROC. REGIONAL WORKSHOP ON ENVTL. MGMT. & SUSTAINABLE DEV. 183-94 (United Nations Development Program, Suva, Fiji, Apr. 1990).

10. Arthur L. Dahl, *The South Pacific Regional Environment Programme*, in ENVIRONMENT AND RESOURCES IN THE PACIFIC 3-6 (UNEP Regional Seas Reports and Studies No. 69, 1985).

11. Conference on the Human Environment in the South Pacific [hereinafter Rarotonga Conference].

12. *See* Mere Pulea, *Legal Measures for Implementation of Environmental Policies in the Pacific Region*, in Dahl, *supra* note 10, at 157-62.

13. Arthur L. Dahl & Ian I. Baumagart, *The State of Environment in the South Pacific, in* SPREP, REPORT OF THE CONFERENCE ON THE HUMAN ENVIRONMENT IN THE SOUTH PACIFIC, RAROTONGA, COOK ISLANDS, 8-11 MARCH 1982, *reprinted as* UNEP REGIONAL SEAS REPORTS AND STUDIES, No. 31 (1983).

14. Convention on Conservation of Nature in the South Pacific, June 12, 1976 [hereinafter Apia Convention].

15. *See* Pulea, *supra* note 12, at 161.

16. Apia Convention, *supra* note 14, art. II.

17. *Id*. art. V.

18. *Id*. art. VII.

19. Convention for the Protection of the Natural Resources and Environment of the South Pacific Region, Nov. 25, 1986, 26 I.L.M. 38 (1987) [hereinafter SPREP Convention].

20. *Id*. arts. 6-13.

21. *Id*. art. 14.

22. *Id*. art. 17.

23. South Pacific Nuclear Free Zone Treaty, Aug. 6, 1985, 24 I.L.M. 1440 (1985) [hereinafter SPNFZ Treaty].

24. See comments made regarding the absence of benefit in Jon Van Dyke, Kirk R. Smith, & Suliana Siwatibau, *Nuclear Activities and the Pacific Islanders*, 9 ENERGY 733, 747-48 (1984).

25. For more detail regarding the formation and objectives of this group, see David J. Doulman, *Fisheries Cooperation: The Case of the Nauru Group*, in TUNA ISSUES AND PERSPECTIVES IN THE PACIFIC ISLANDS REGION 257-72 (D. J. Doulman ed. East-

West Center, Honolulu 1987). Members of the Nauru Group are the Federated States of Micronesia, Kiribati, the Marshall Islands, Nauru, Palau, Papua New Guinea, and the Solomon Islands. Tuvalu has an observer status with the group.

26. FFA members have adopted "Harmonized Minimum Terms and Conditions of Access," which are requirements and conditions that can be imposed on foreign fishermen seeking fisheries access in the EEZs of member countries.

27. FFA members also established the Regional Register of Foreign Fishing Vessels. For details on this scheme, *see* D. J. Doulman & P. Terawasi, *The South Pacific Regional Register of Foreign Fishing Vessels*, 14 MARINE POL'Y 324 (1990).

28. It has been argued that the FFA States–U.S. Fisheries Treaty does serve some conservation objectives fulfilled through management and administration of the treaty, especially the provisions on the observer program and the monitoring and analysis of catch reports. See Florian Gubon, The 1987 Pacific Island States/United States Fisheries Treaty and Its Implications for Fisheries Management and Conservation in the South Pacific Islands Region (independent research paper in partial fulfillment of the requirements of the L.L.M. degree in law and marine affairs at the University of Washington School of Law; copy available at the University of Washington Law Library, Seattle, 1987).

29. Convention for the Prohibition of Fishing with Long Driftnets in the South Pacific, *done* at Wellington, Nov. 24, 1989, and Nouméa, Oct. 20, 1990, 29 I.L.M. 1449 (1990) [hereinafter South Pacific Driftnet Convention].

30. South Pacific Forum Fisheries Agency, The South Pacific Albacore Driftnets Issue: Developments since November, 1988 (update, Jan. 1990) (second consultation on arrangements for South Pacific Albacore Fisheries Management, FFA, Honiara, Solomon Islands, Feb. 28–Mar. 1, 1990).

31. Judith Swan, *Southern Albacore High Seas Driftnet Fishing*, in 1 PROC. N. PAC. DRIFTNET CONF. 140–47 (British Columbia Ministry of Agriculture and Fisheries, Department of Fisheries and Oceans, 1989).

32. The "Convention Area" is defined as "the area lying within 10 degrees North latitude and 50 degrees South latitude and 130 degrees East longitude and 120 degrees West longitude, and shall also include all waters under the fisheries jurisdiction of any Party to the Convention." See South Pacific Driftnet Convention, *supra* note 29, art. 1.

33. *Id.* art. 3(1).

34. *Id.* art. 3.

11 Governance and Stewardship of the Living Resources: The Work of the South Pacific Forum Fisheries Agency*

Gracie Fong

SOUTH PACIFIC STATES ESPOUSE THE NEED for ocean governance and steward-ship in the way that they interpret their rights and duties to conserve and manage the living resources. They have, however, been made to fight every inch of the way to make meaningful the provisions of the 1982 United Nations Convention on the Law of the Sea (1982 Convention). Their achievements, which have been described as impressive,[1] are based on regional solidarity, informed by dynamic secretariats and the ability to invoke the political process quickly.

SOUTH PACIFIC FISHERIES

In 1977, the South Pacific Forum adopted a declaration concerning the law of the sea, which also called for the establishment of a regional fisheries agency. In 1979, the South Pacific Forum Fisheries Agency[2] (FFA) was established to promote regional cooperation in fisheries with the objective of securing maxi-mum benefits from the living marine resources of the region for their peoples and the region as a whole, particularly for the developing countries.

The FFA Member States

Many of the FFA's member states are small, isolated, and newly independent island states, for whom tuna represents the only economic resource available.[3] Almost all of the FFA member states have expressed their intention to develop an

* This chapter is based largely on information contained in various background papers of the South Pacific Forum Fisheries Agency, some of which are confidential to FFA member states.

industrial tuna-fishing capacity, and a few have gone some way toward achieving this goal.[4] However, until recently the primary focus has been on leasing fishing rights to distant-water fishing nations (DWFNs) under conditions aimed at ensuring that as many benefits as possible accrue to FFA member states. Through fisheries access agreements with DWFNs, FFA member states have secured $20 million to $25 million each year in access fees. The allegation has been made that these access fees are excessive[5] and have led to drift netting on the high seas outside EEZs, where no access fees have traditionally been charged. According to Japanese industry sources, however, these fees have accounted for only a small proportion of total vessel operating costs.[6]

The Resource

Tuna and related species, including skipjack, yellowfin, bigeye, albacore, and billfish, are the important pelagic resources commercially exploited in the South Pacific. Tuna from the South Pacific region contributes 35 percent of the total world harvest. In 1989, production was an estimated 703,908,000 tons, with an unprocessed market value of around $1,066 million. Some 90 percent of the catch was taken by DWFNs: Japan is the major DWFN,[7] followed by the United States,[8] Korea,[9] Taiwan,[10] and the Philippines.[11]

Because living marine resources in the South Pacific are migratory, occurring both in EEZs and in adjacent high seas, it is increasingly vital that any effective management arrangements extend outside EEZs to those areas, including the high seas. FFA member states have developed a number of regional initiatives aimed at the conservation and management of tuna resources on the high seas as a condition of access to their exclusive economic zones.

The FFA Secretariat

The technical needs of FFA member states are served mainly by the FFA Secretariat, although the South Pacific Commission (SPC) also contributes some scientific studies. Under the detailed policy and administrative direction of the Forum Fisheries Committee (FFC), the FFA Secretariat implements a wide-ranging work program that includes harmonization of fisheries regimes and access agreements; surveillance and enforcement; information services; tuna fishery development; economic analysis; training and administrative development; a regional register of foreign fishing vessels; and coordination with other regional programs and organizations.

The South Pacific Commission (SPC)

The SPC includes all South Pacific island states and territories as well as the United States, France, and Britain. Its mandate includes the undertaking of scientific research on tuna and related species in the South Pacific. In addition, the SPC provides secretariat services for the South Pacific Albacore Research Group (SPAR), an informally constituted group of scientists and fisheries officers from various countries (including DWFNs) with an interest in the region's albacore tuna resource.

Until recently, FFA member states sought to manage access to the EEZ resources through bilateral agreements containing minimum terms and conditions of access that had been adopted regionwide. Member countries are now interested in multilateral arrangements, which, although more difficult to conclude, appear to promote more stability in fisheries relations. For example, existing instability in Japanese bilateral fisheries relations with FFA states stems primarily from the inability of the high-cost Japanese tuna industry to compete with lower-cost and equally efficient Asian producers.[12]

SOUTH PACIFIC REGIONAL FISHERIES INITIATIVES

FFA member states are now considering the establishment of comprehensive conservation and management regimes for certain species of tuna within their EEZs throughout the Pacific and are seeking to extend the same standards of stewardship applicable in the EEZs to the high seas. Major regional initiatives include the 1990 Harmonized Minimum Terms and Conditions of Access, including the modified Regional Register rules, the Treaty on Fisheries Between the Governments of Certain Pacific Island States and the Government of the United States of America,[13] a western Pacific purse seine management regime, and developments regarding South Pacific albacore.

Harmonized Minimum Terms and Conditions of Access and the Regional Register of Foreign Fishing Vessels

The Harmonized Minimum Terms and Conditions of Access (MTCs) were first adopted in 1982 to strengthen FFA member states' attempts to exercise their sovereign rights in waters under their jurisdiction and to leave less room for divide-and-rule tactics among FFA member states by DWFNs. During negotiations on the FFA States–U.S. Fisheries Treaty, FFA member states elaborated on the 1982 MTCs. Then, in April and May of 1990, the MTCs were revised and the Regional Register rules were modified formally. The South Pacific Forum, in its August 1990 communiqué, recorded that "Forum members agreed to give high priority to the implementation of the revised MTCs as the basic standard of access to the FFA members-states' EEZs."

The major changes to the MTCs seek to mitigate the problems of collection and verification of data that are essential to proper conservation and management. For the first time, they address the high seas problem, in particular providing for control and monitoring of transshipment, including prohibiting transshipment at sea and requiring the maintenance and submission of catch logs on high seas fishing. These provisions are reinforced by the modified Regional Register rules, particularly the addition of powers to suspend good standing, for example, when a vessel operator violates terms and conditions of access, including reporting.

The FFA States–U.S. Fisheries Treaty

Before they negotiated this treaty with the United States, it appears that FFA member states had not given serious consideration to influencing the activities of DWFNs on the high seas in the regions that affect FFA member states' management within their EEZs. Indeed, FFA member states initially rejected the U.S. proposal to include areas of high seas enclosed by, as well as adjacent to, EEZs within the licensing area of the treaty. However, FFA member states eventually saw that doing so would generate more revenue, make enforcement easier, and broaden reporting responsibilities. The United States stated that it wished to remove the incentive on fishers to report the bulk of their catches as having been taken on the high seas in order to avoid access fees. Consequently, the treaty regulates American purse seine activity within the entire South Pacific region, including the high seas.

The FFA States–U.S. Fisheries Treaty has been in force since mid-1968. Problems that have arisen often have been resolved at the annual consultations. At the third annual consultation, held in March 1991, negotiations commenced on schedule 2 of annex II, which will expire on June 14, 1993. This involves the number of licenses to be made available under the treaty and the access fees. In addition, the FFA member states intend to propose amendments to the annexes to incorporate changes regarding the 1990 MTCs and the Regional Register of Foreign Fishing Vessels. Although former U.S. president George Bush recently announced his interest in the "extension" of the treaty, and the recently reauthorized Magnuson Fishery Conservation and Management Act[14] suggested an additional term of ten years "on terms and conditions at least as favorable to vessels of the United States and the Government of the United States,"[15] it remains to be seen how the Clinton administration will view these issues.

Western Pacific Purse Seine Management

The growth in surface tuna fisheries in the western Pacific is of considerable concern. With the rapid expansion of purse seining in the region since 1980, improved technology, and greater familiarity with fishing conditions in the South Pacific, much larger quantities of both juvenile and adult yellowfin tuna are now being taken. Between 1987 and 1989, yellowfin accounted for an average of 26 percent of the total purse seine catch in the EEZs of FFA member states. It is believed that this stock may now have reached its maximum level of sustainable harvest.

At the ninth annual meeting of the subregional group known as the Parties to the Nauru Agreement[16] (PNA) in April 1990, concern about the expanding purse seine fishery in the western Pacific resulted in agreement to regulate the number of purse seine vessels operating in members' EEZs. The meeting of the Standing Committee on Tuna and Billfish in June 1990 underscored this concern. It found that the total catches of tuna had probably been severely underestimated in previous years, by as much as 25 percent, and that "the prospect of further substantial increases in purse seine effort over the coming years and the demonstrated capacity of purse seiners to take advantage of favourable environmental conditions to achieve substantially larger catches (as occurred in 1987)" may render necessary "management of the fishery for biological reasons." A more comprehensive management regime is under consideration.

To ensure that the region's yellowfin stocks are not harmed by excessive fishing effort, skipjack resources will remain underexploited because it is not always possible to harvest one species selectively with purse seiners, since the majority of schools are polyspecific. Consequently, the management regime for the fishery must be geared to protecting the skipjack stock that is most vulnerable to overexploitation. In limiting effort in the fishery, catches will be controlled, and the possibility of an international oversupply situation will also be averted.

Developments Regarding South Pacific Albacore Tuna

South Pacific albacore tuna has historically been taken by vessels using longlines, with the bulk of the catch being taken in a band across the South Pacific between 10 and 30 degrees south latitude. During the past decade, increasing quantities have been taken by troll lines and drift nets in waters west of New Zealand in the Tasman Sea and to the east of New Zealand, and in the Sub-tropical Convergence Zone, an area as far east as 120 degrees west longitude and between 30 and 40 degrees south latitude.

The rapid development of the surface drift net fishery in 1988 and 1989 led to a series of regional meetings, which concluded that a dramatic increase in the surface fishery through the use of drift nets was likely to be detrimental. Two important developments involving South Pacific albacore tuna followed: (1) a convention to end drift net use in the South Pacific and a United Nations General Assembly resolution requiring cessation of the practice on the high seas; and (2) consultations concerning arrangements for South Pacific albacore fisheries management.

The High Seas Drift Net Fishery. The first regional consultation on drift nets occurred in Fiji in November 1988. The consultation concluded that the increased drift net vessel effort threatened the long-term sustainability of the South Pacific albacore resource, posed a serious environmental threat with respect to their by-catch of marine mammals and other living resources, and was a navigation hazard. A plan of action was established, which included a regional embargo on the acceptance of drift net—caught fish by canneries and cold stores, a ban on transshipment of drift net—caught fish within the region, and a ban on the support of vessels involved in drift net fishing. An investigation of existing legislation and international law on drift nets was commenced. South Pacific countries and the United States were asked to raise the issue with Asian states at their regular consultations.

Over the next seven months, issues and strategies for persuading the drift net fleets to cease their fishing activities were further addressed by the South Pacific countries. A legal consultation agreed that the practice of drift netting was not consistent with international law governing the environment and fisheries conservation and management on the high seas. At the second SPAR workshop in June 1989, scientists reviewed the available data and concluded that in the current surface fishing effort, it was probable that the resource was being exploited beyond its sustainable level. The group agreed to cooperate to improve the data base available for analysis, to consider appropriate methods to assess the status of the stock, and to devise a research plan for improving information on estimates for the stock's population parameters, on the interaction among fisheries targeting the resource, and on oceanographic conditions where albacore range.

The first meeting, involving the Pacific Island states, Japan, Korea, Canada, the United States, and Taiwan, was held in Fiji in late June of 1989. The delegates were unable to arrive at a common position on the future of drift net fishing. Although Korea agreed to withdraw its drift net vessels from the region, the positions of Taiwan and Japan and the South Pacific states remained markedly divergent. The South Pacific Forum, meeting in Kiribati in July 1989, resolved in the Tarawa Declaration to develop a convention banning drift net fishing from the region and seeking the establishment of a management regime for albacore tuna in the South Pacific.

In response to continuing pressure from the South Pacific, Japan announced in September 1989 that it would reduce its fleet for the 1989–1990 season to twenty vessels and offered to provide a research vessel to examine the effects of drift net fishing and a patrol vessel to monitor such activities in the region. South Pacific countries welcomed Japan's initiative but maintained that only a complete cessation of drift net fishing would be acceptable.

In October 1989, the Tarawa Declaration was in turn endorsed by the South Pacific Conference, which includes Britain, France, and the United States; the South Pacific Conference on Nature Conservation; and the Commonwealth Heads of Government Meeting in Malaysia, which in its Langkawi Declaration urged immediate abandonment of the environmentally damaging practice of drift net fishing. In November 1989, the U.S. House of Representatives, in H.R. 214, supported the Tarawa Declaration and advised the United States to cooperate with South Pacific regional organizations in the formulation of an international convention banning drift net fishing in the South Pacific region.

In November 1989, the United States tabled a resolution in the Second Committee of the United Nations General Assembly, cosponsored by Australia, Papua New Guinea, Fiji, the Solomon Islands, Vanuatu, Zaire, Mauritania, and New Zealand, which called for a complete ban on drift net fishing in the South Pacific region. Japan countered with a resolution calling for conclusive scientific evidence before the possibility of a ban was discussed. The compromise resolution unanimously adopted in December 1989 by the UN General Assembly, titled "Large-Scale Pelagic Driftnet Fishing and Its Impact on the Living Marine Resources of the World's Oceans and Seas," recommended the following:

1. That moratoriums be placed on all large-scale pelagic drift net fishing on the high seas by June 30, 1992, unless effective management and conservation measures are taken, based on statistically sound analysis.
2. That immediate action be taken to reduce large-scale pelagic drift net fishing in the South Pacific and that cessation of this method of fishing in this region occur no later than July 1, 1991, as an interim measure, until appropriate conservation and management for South Pacific albacore tuna are entered into by all concerned parties.
3. That the expansion of the large-scale pelagic drift net fleet fishing on the high seas in the North Pacific cease immediately.

Japan announced in June 1990 that it would cease drift net fishing immediately, and Taiwan's Cabinet announced on August 24, 1990, that it would not use drift nets in the region after June 30, 1991.

In late November of 1989, three meetings to pursue further the objectives of the Tarawa Declaration took place in Wellington, New Zealand. At the first two meetings, FFA member states, Pacific territories, France, Britain, and regional organizations developed and endorsed the Convention for the Prohibition of Fishing with Long Driftnets in the South Pacific (South Pacific Driftnet Convention). During the third meeting, Japan, Korea, the United States, Canada, and the Taiwanese Deepsea Tunaboat Owners and Exporters Association (in place of the Republic of China, which could not participate because of political differences over its sovereign status) were invited to adhere to the South Pacific Driftnet Convention through its protocols. The press reported that Taiwan would reduce its South Pacific drift net fleet to twenty-four boats for the 1989–1990 season and that every effort would be made to cease drift netting in the South Pacific in the near future.

The convention and its protocols and associated instruments draw attention to the importance of living marine resources to the South Pacific countries and prohibits nationals and vessels of parties to the convention from engaging in drift net fishing activities within the Convention Area.[17] It also includes other compulsory and optional measures for parties to take in respect to drift net activities, such as prohibiting transshipment and port entry. It contains mechanisms for information exchange and provision and dissemination of data by the FFA and other regional organizations. The South Pacific Driftnet Convention, also known as the Wellington Convention, has entered into force.

Two protocols were prepared to enable DWFNs and other nations with some interest in joining the convention to accept relevant provisions. An associated instrument, a letter from the director of the FFA to the Taiwan Deepsea Tunaboat Owners and Exporters Association, calls on the association to prevent its vessels from using drift nets in the South Pacific Driftnet Convention area.

In August 1990, a delegation from the South Pacific countries met with representatives from Taiwan in Singapore and visited Korea, Japan, Canada, and the United States. Japan, Taiwan, and Korea indicated that for various reasons they would not associate themselves with the South Pacific Driftnet Convention through its protocols. The United States considered signing the convention and protocol II on behalf of American Samoa (and therefore saw no need to sign protocol I). Canada was contemplating signing protocol II but was less positive about protocol I. Both protocols I and II were finalized and opened for signature on October 20, 1990. At that time, thirteen states had signed the South Pacific Driftnet Convention, and the United States (on behalf of American Samoa) and the Solomon Islands had already announced their intention to sign. A number of countries are in the process of enacting legislation prior to ratification.

Management Arrangements for South Pacific Albacore. Early meetings in 1989 aimed at assessing the impact of drift net fishing on the albacore stock concluded that there was a need to develop a conservation and management regime for the resource. So far, there have been three consultations to this end.

The First Consultation took place in Wellington in late November of 1989, but little progress was made. At the Second Consultation, held in Honiara in March 1990, the Pacific states and territories agreed that the arrangements should include data requirements, scientific analysis, information exchange, a defined area, membership, management processes, enforcement, and dispute settlement procedures.

The Third Consultation, in Nouméa in October 1990, received the consensus report of the preceding third SPAR group meeting, which had found that albacore in the North and South Pacific constitute discrete stocks and that the effective fishing area for the South Pacific stock lies between the equator and 50 degrees south latitude and between 90 degrees west and 140 degrees east longitude. This report acknowledged that "[c]oncerns about driftnet fishery expansion resulted in a significant reduction of Japan and Taiwan's South Pacific driftnet fleets during the 1989-90 fishing season." Although reducing the drift net catch lessens the risk of overfishing the albacore stock, the report stated that even a relatively modest expansion of the surface fishery for albacore reduces potential yields in the longline fishery, but "because the longline fishery catches older albacore, on average, than the surface fisheries, in most longline fishing areas it will take two or three years before the impacts of the higher surface fishery catches can be directly measured."

The Pacific delegations tabled a draft arrangement that was described by both the Japanese and American delegations as a valuable contribution to the work of the consultation. The Japanese delegation also tabled a draft convention that is modeled on existing management organizations that the Pacific delegations consider inappropriate for the South Pacific region. Although positions are widely divergent on a number of issues, some progress was made on the following points: All South Pacific states and territories and all fishers of South Pacific albacore have a legitimate interest in participating in the management arrangement; existing regional organizations have a role to play in the arrangement; all participants should provide relevant scientific data in a timely fashion; and the principle of flag state responsibility is an important consideration in the arrangement.

The report of the third SPAR group meeting concluded that "[g]iven the reductions in the driftnet fleets and the total surface catch in the 1989-90 fishing season, and the expectation of further significant reductions in the number of driftnet vessels operating during the 1990-91 season, there appears to be no immediate need for management action to further reduce fishing effort. However, there is still great uncertainty about South Pacific albacore yield potential and fishery interactions." The problem is that our knowledge of the South Pacific albacore stock comes almost entirely from the longline fishery. Although historical data and surplus production models have provided an estimate of maximum sustainable yield for the longline fishery, these estimates alone are insufficient because they include a significant surface fishery that exploits younger age classes at a different latitude with different gear.

CONCLUSION

South Pacific tuna fisheries are facing an unprecedented period of growth and expansion. The global patterns of fishing, processing, and marketing of tuna are undergoing marked changes that suggest considerable uncertainty for the future.[18] In an attempt to mitigate the negative impact of these changes on the region, South Pacific states need to strengthen regionally coordinated fisheries management.

The FFA is the medium for regional cooperation on fisheries. Through harmonization of policy and law and by the creation of innovative tools to suit the realities of mostly small islands with contiguous, large EEZs, FFA member states seek to maximize the benefits flowing from the 1982 Convention for their peoples, the world community, and future generations. FFA member states are conscious that they have been entrusted with responsibility for the preservation and administration of the oceanic resources and environment of their region; stewardship is a tradition to which they still adhere.

NOTES

1. Their achievements are "impressive" even to the point of setting international precedents. *See, e.g.*, Report to Plenary of the Third Technical Subcommittee Meeting of the Forum Fisheries Committee (Honiara, Sept. 21, 1989); William T. Burke & Francis T. Christy, Jr., Options for the Management of Tuna Fisheries in the Indian Ocean (draft, Aug. 6, 1990).

2. The term "South Pacific Forum Fisheries Agency" is used in its constituent convention to refer to the member states whose representatives constitute the Forum Fisheries Committee (FFC) and to the FFA Secretariat. This chapter follows that usage and, after brief descriptions of the FFA member states, the FFA Secretariat, the South Pacific Commission's (SPC's) scientific input, and the fishery, goes on to focus on those recent developments that illustrate the FFA member states' commitment to stewardship of living resources, even on the high seas.

3. The sixteen member countries of the FFA are Australia, the Cook Islands, the Federated States of Micronesia, Fiji, Kiribati, the Marshall Islands, Nauru, New Zealand, Niue, Palau, Papua New Guinea, the Solomon Islands, Tonga, Tuvalu, Vanuatu, and Western Samoa.

4. For example, in the late 1970s, South Pacific countries, including Fiji, the Solomon Islands, Tonga, Vanuatu, and New Caledonia, began developing their own tuna longline capacity. Australia, Fiji, Kiribati, New Zealand, the Solomon Islands, and Tuvalu have pole and line fleets. Fiji, New Zealand, and French Polynesia have troll vessels targeting albacore.

5. *See* Douglas M. Johnston, *The Driftnetting Problem in the Pacific Ocean: Legal Considerations and Diplomatic Options*, 21 OCEAN DEV. & INT'L L. 5, 8 (1990).

6. David J. Doulman, *Japanese Distant-Water Fishing in the South Pacific*, 4 PAC. ECON. BULL. 2, at 22, 25 (somewhere between 0.2 and 6 percent, depending on vessel sizes and gear types).

7. Japan accounted for an estimated 839,195 tons of tuna, or 44 percent of the total catch, in 1989.

8. The United States accounted for 426,497 tons, or 23 percent of the total catch, in 1989.

9. Korea harvested 356,601 tons, or 19 percent, in 1989.

10. Taiwan's 1989 catch was 209,125 tons, or 11 percent.

11. The Philippines accounted for 53,438 tons, or 5 percent of the 1989 total.

12. Doulman, *supra* note 6, at 25, 27.

13. Treaty on Fisheries Between the Governments of Certain Pacific Island States and the Government of the United States of America, Apr. 2, 1987, 26 I.L.M. 1048 (1987).

14. Magnuson Fishery Conservation and Management Act, 16 U.S.C. § 1801 *et seq.* (1988), *amended by* Pub. L. No. 101-627, 104 Stat. 4436 (1990).

15. 16 U.S.C. § 1822(e)(5) (1988).

16. The PNA members are the Federated States of Micronesia, Kiribati, Papua New Guinea, Palau, the Marshall Islands, Nauru, and the Solomon Islands. Tuvalu is in the process of accession.

17. The Convention Area is within 10 degrees north and 50 degrees south latitude and 130 degrees east and 120 degrees west longitude, including all waters under the fisheries jurisdiction of any party.

18. These include the rise and expansion of the purse seine method over other gear; the relocation of purse seine fleets to the western Pacific and conversion of large fleets of drift net vessels to the longline method; the huge investment in fishing technology, which is likely to result in overharvesting and may cause world tuna prices to collapse; major changes in ownership and siting of canneries; and the separation of loining and canning operations.

III

Controlling Ocean Pollution

SINCE TIME IMMEMORIAL, the oceans have been used as a dump site, and the freedom to use the ocean in this fashion has been thought to be an inherent part of the "freedom of the seas." Today, however, the assimilative capacity of the oceans is reaching its saturation point, and many parts of the oceans are severely polluted. In addition to the pollution caused by open ocean dumping and by ocean vessels, the runoff from land-based pollution creates substantial problems, particularly in coastal regions. Today, as Jackson Davis's chapter explains, no part of the oceans can be thought of as "clean," and risks to human health and the health of the marine ecosystem lurk everywhere. Because complex ecosystems are "integrated and dynamically interactive," one component cannot be affected without affecting all others.

The chapters in this part present suggestions to address these threats. Jackson Davis examines the recent reports of the GESAMP scientists and concludes that all parts of the oceans can be considered polluted, especially with dangerous heavy metals. He then argues for a comprehensive global treaty that will unify all the regional and item-specific approaches with one standard that will apply to all. Christopher Stone argues that a "global commons trust fund" would assist in forcing nations of the industrialized world to recognize the burdens their wastes place on others and to internalize the costs of their development. Clifton Curtis examines the existing treaties and explains how they can be improved. Miranda Wecker and Dolores Wesson examine the transportation of hazardous wastes on the seas and explain what new steps need to be taken to prevent accidents. Jon Van Dyke's chapter looks at some of the recent efforts to protect fragile marine environments and argues that protection should be given to all low-lying atolls.

These chapters present a rich trove of ideas designed to encourage increased activity to protect the marine environment from pollution and hazards.

12 The Need for a New Global Ocean Governance System*

W. Jackson Davis

THIS CHAPTER DEVELOPS THE THESIS that a new, integrated system of ocean governance is needed to protect and preserve the marine environment. The thesis is based on three premises. First, the oceans are central to human survival, and therefore, the larger social good is served by preserving them. Second, the oceanic life support system is subject to an unknown (and probably unknowable) but nonetheless potentially significant threat from existing levels of contamination. Third, regional approaches, although essential as a component of a global system, are inadequate to the task of global regulation when used alone.

After developing these premises, this chapter outlines a new ocean governance system based on global coordination of regional conventions in parallel with centralized global standards and enforcement.[1] The central feature of the global system envisioned is regulation of all sources of ocean pollution ("comprehensiveness"), incorporating the scientific reality that all portions of the biosphere are interdependent ("interconnectedness") and can be protected only through anticipatory action (the "precautionary" principle). I will develop the position that existing models of international regulation cannot suffice because they are built on political rather than scientific foundations and that environmental philosophies evolved by indigenous peoples have much to offer in designing new systems of ocean governance.

* Research for this paper was done during a sabbatical leave from the University of California, Santa Cruz, and was supported by a grant from the Environmental Studies Institute. I am grateful to Mr. Peter Taylor Cedrowen and Ms. Boyce Thorne Miller for helpful comments and discussion of an earlier draft of this paper. The analysis and views expressed herein are my sole responsibility and do not necessarily reflect the views of the people or institutions with which I am affiliated.

THE OCEANS AS A PLANETARY LIFE SUPPORT SYSTEM

It is said that about half of the world's human population lives within a few tens of kilometers of ocean waters. And yet laypersons often view the oceans as little more than a source of seafood. Even informed policymakers frequently consider the seas as an aesthetic footnote to the concerns of a largely terrestrial *Homo sapiens*.

Perhaps in part because of this lack of appreciation of the role of the oceans in human survival, our species has treated the seas as a gigantic planetary waste bin with an unlimited "assimilative capacity" to absorb wastes. According to this view, the oceans comprise a remote location to dispose of wastes that would otherwise harm the terrestrial environment and the human populations that reside there. This view minimizes the role of the oceans in human survival and reflects a lack of awareness about the fragility of the marine environment in the face of threats posed by contemporary industrial civilization. Heightened awareness by policymakers might facilitate effective protection of the marine environment. The oceans play a role in three natural cycles on which all life on earth depends: the oxygen cycle, the carbon cycle, and the hydrologic cycle. The oxygen cycle is the source of the fresh air we breathe. Tiny ocean plants called phytoplankton that live in the surface waters of the coastal and open ocean are crucial to this cycle. Each spring, these unicellular life forms participate in the annual phytoplankton bloom, a miraculous global event made visible to the human eye only recently by satellite photographs. The phytoplankton bloom expels a massive yearly pulse of oxygen into the earth's atmosphere, like a gigantic global breath. Our own breathing as individual organisms is tightly coupled to, and dependent on, this planetary respiratory rhythm.

Land plants also release oxygen, of course, but the uptake and release of atmospheric oxygen by land biota are in approximate balance. That is, the oxygen that land plants release by means of photosynthesis is fully utilized by the respiration of land plants and animals. Moreover, global deforestation may be altering the balance so that terrestrial biota are becoming net "consumers" of oxygen.[2] In contrast, oceanic phytoplankton produce more oxygen than is used by marine biota, resulting in a net input of oxygen into the atmosphere. The extra oxygen pumped into the atmosphere each year by the bloom is estimated at 300 million metric tons.[3] In lay terms, the oceans (along with the rain forests) serve as the lungs of the earth. They are as important to planetary life as human lungs are to our individual lives.

The oceans play an equally central role in the related carbon cycle, on which human well-being likewise depends. When the phytoplankton bloom, they release oxygen in proportion to their absorption of carbon dioxide. The carbon that is "fixed" in the structure of the phytoplankton rains to the ocean floor when the plankton die, where it remains sequestered, perhaps for millennia. Whereas the phytoplankton are a net producer of oxygen, they are a net consumer of carbon. Our understanding of the role of the oceans in the carbon cycle is imperfect, but it is estimated that at least 75 percent of the global carbon budget is tied up in the oceans by chemical and biological processes. Without the phytoplankton, carbon dioxide would accumulate more rapidly in the atmosphere, warming the planet through the well-known greenhouse effect. Rapid climate change could, in turn, disrupt significantly the life support system on which we depend. A healthy ocean stands between us and such climatic catastrophe.

Finally, the oceans deliver fresh water to all of humanity through the water, or hydrologic, cycle. Water evaporates from the sea, precipitates on land, and returns to the sea, with an average cycle time estimated at about one month. This freshwater delivery system is in a state of delicate balance and can be altered quickly, as suggested by current droughts in the western United States and sub-Saharan Africa. The intricate interplay among oceans, climate, and water cycle also is imperfectly understood, but scientists do not doubt the linkages, nor the human capacity to influence them.

In sum, the oceans—particularly the phytoplankton—are of immediate and fundamental relevance to the survival of humanity. Phytoplankton, and especially their juvenile forms, are among the most sensitive of ocean life forms to human pollution. The web of ocean life is robust, but it is also delicate; and human activities may be fully capable of disrupting, and even destroying, the web.

THE STATE OF THE OCEANS

But do human activities in fact pose a significant threat at present to the ocean's life support systems? If not, the issue of regulation becomes moot. There is a widespread perception that although coastal waters may be polluted significantly, the open oceans are still relatively clean and undamaged. This "dirty coasts, clean seas" hypothesis reduces the motivation for global regulation of the marine environment by promulgating the notion that there is no problem in need of solution. Moreover, the dirty coasts, clean seas hypothesis is self-perpetuating because it discourages the very monitoring and research on the open ocean that could test its accuracy. The perception that the open oceans are still relatively unimpacted by human activities therefore has important policy and scientific ramifications, and consequently it merits careful scrutiny.

The dirty coasts, clean seas hypothesis originates in large part from the recent analysis of the health of the marine environment by the authoritative Group of

Experts on the Scientific Aspects of Marine Pollution (GESAMP), sponsored by several United Nations agencies and issued by the United Nations Environment Program (UNEP).[4] This report is notable not only owing to its origin but also because it is the most comprehensive scientific analysis of the state of the marine environment and the first since E. D. Goldberg's pioneering effort in 1976.[5] The report notes that coastal regions are deteriorating seriously worldwide: "Destruction of beaches, coral reefs and wetlands, including mangrove forests, as well as increasing erosion of the shore, are evident all over the world." In contrast, the GESAMP report concluded in its summary that "the open sea is still relatively clean."

The data base on which this summary conclusion is based is presented in two technical annexes to the GESAMP report[6] and includes an excellent review of the literature by S. W. Fowler.[7] Exactly the same data are graphed here in figures 12.1 through 12.9, according to the species of contaminant.[8] Polychlorinated biphenyls (PCBs) appear, from the limited data available, to be more concentrated in open ocean waters than in coastal waters (see figure 12.1). Average concentrations range from less than one-tenth of a nanogram (one ten-billionth of a gram) per liter of seawater (Kanechlor in the Antarctic) to more than 75 nanograms per liter (Aroclor 1254 in the Atlantic and Pacific). The higher concentrations in the open ocean may reflect the atmospheric input of PCBs to ocean water.[9]

The distribution of DDT and its metabolites follows a different pattern from that of the PCBs. In the limited number of cases in which comparisons are possible between concentrations in coastal water and those in open ocean surface water, the values are approximately equal (see figure 12.2). In about half of the comparisons, concentrations are greater in the open oceans than in coastal waters. The greatest concentrations by far are found in the Northeast Atlantic oceans, where concentrations of approximately 500 nanograms per liter dwarf concentrations in comparable coastal zones. As a generality, concentrations appear larger in the open ocean in those regions where terrestrial use of DDTs has been curtailed (for example, in the northern hemisphere), while coastal concentrations are greater in regions where terrestrial use persists (for example, in the southern hemisphere). Such a distribution pattern could reflect the input of DDTs from riverine sources. Remote regions, where terrestrial DDT use has never occurred (such as in the Antarctic), also show higher concentrations in open oceans, although not as high as in industrialized regions of the globe, consistent with incomplete ocean mixing. Assuming a relatively long residence time in comparison with ocean mixing rates, DDT concentrations in remote regions may be expected to increase over time as mixing becomes more complete.

A comparable analysis for other chlorinated hydrocarbons illustrates that the concentration in seawater is generally higher by a factor of about two (North

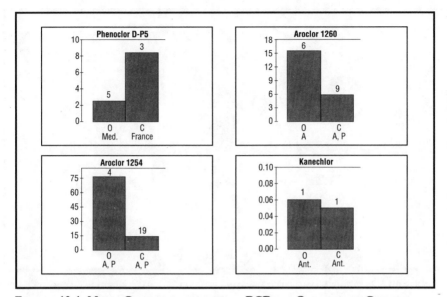

FIGURE 12.1 MEAN CONCENTRATIONS OF PCBS IN OCEAN AND COASTAL SURFACE WATERS *Abbreviations: O, ocean; C, coastal; Med., Mediterranean Sea; A, Atlantic; P, Pacific; Ant., Antarctic. Data are in nanograms per liter and are means of values presented in tables of S. W. Fowler,* Concentrations of Selected Contaminants in Water, Sediments and Living Organisms, *in 1, 2 UNEP:* TECHNICAL ANNEXES TO THE REPORT ON THE STATE OF THE MARINE ENVIRONMENT *(UNEP Regional Seas Reports and Studies No. 114/2, 1990). Numbers above each bar represent the sample size (where individual samples are, in some cases, means of several measurements).*

Pacific and Northwest Atlantic) to ten (Mediterranean Sea) in coastal waters than in open ocean waters (see figure 12.3). Only in remote regions, such as the Antarctic, are concentrations of other chlorinated hydrocarbons higher in open ocean waters. This distribution pattern would suggest continued input from riverine sources, again accompanied by incomplete ocean mixing. Under this scenario, and again assuming a relatively long residence time, concentrations would be expected to increase over time in open ocean waters.

The final class of pollutants examined is the heavy metals, including cadmium, lead, and mercury. Generally, cadmium is moderately to substantially more concentrated in coastal zones than in the open ocean (see figure 12.4). Quantitative analysis shows, however, that over the entire globe the concentration is approximately 2.3 times higher in coastal waters. Analysis of the sources of cadmium (see figure 12.5) illustrates that it originates largely from local coastal "hot spots,"[10] such as the New York Bight in the Atlantic and various North Sea

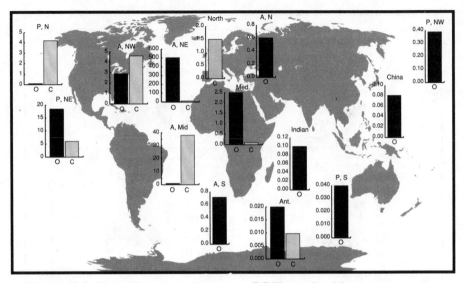

FIGURE 12.2 MEAN CONCENTRATIONS OF DDT AND ITS METABOLITES IN OPEN OCEAN AND COASTAL SURFACE WATERS *Abbreviations: O, concentrations in ocean surface waters; C, concentrations in coastal surface waters. Histograms are positioned in the approximate areas and regions where the corresponding measurements were made. Abbreviations: N, north; NW, northwest; S, south. Additional abbreviations are as in figure 12.1. Data are nanograms per liter and are means of values presented in tables of* S. W. Fowler, Concentrations of Selected Contaminants in Water, Sediments and Living Organisms, *in 1, 2 UNEP: TECHNICAL ANNEXES TO THE REPORT ON THE STATE OF THE MARINE ENVIRONMENT (UNEP Regional Seas Reports and Studies No. 114/2, 1990).*

and Baltic point sources. When these local hot spots are eliminated from the calculation, the ratio of coastal to open ocean concentration drops to 1.6. That is, when point sources are excluded, the contamination of cadmium in coastal waters is less than twice as great as in open ocean waters. Given the uncertainties in the data, this difference is not likely to be significant. The data are consistent with a dynamic source model in which cadmium originates largely from land-based sources and is introduced into coastal regions largely by runoff but also potentially by atmospheric input.

Lead exhibits a pattern similar to that of cadmium. Lead concentrations are generally higher in coastal waters, although occasional comparisons reveal greater concentrations in the open ocean (see figure 12.6). The ratio of coastal to open ocean concentrations for all sources is approximately 2.4; but when elevated point sources (figure 12.7) are eliminated from the calculation, the ratio drops to 1.6.

Finally, mercury concentrations present a pattern similar to those of cadmium

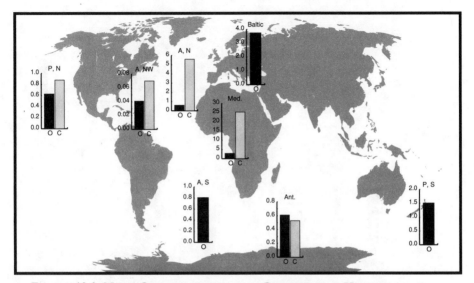

**FIGURE 12.3 MEAN CONCENTRATIONS OF CHLORINATED HYDROCARBONS
OTHER THAN PCBs AND DDT AND ITS METABOLITES (DESCRIBED
SEPARATELY IN FIGURES 12.1 AND 12.2, RESPECTIVELY)** *Concentrations (in
nanograms per liter) are shown for open ocean (O) and coastal (C) surface waters.
Abbreviations: P, Pacific; N, north; A, Atlantic; NW, northwest; Med., Mediterranean
Sea; S, south; Ant., Antarctic. Data are means of values in the tables of S. W. Fowler,*
Concentrations of Selected Contaminants in Water, Sediments and Living
Organisms, *1, 2 UNEP:* TECHNICAL ANNEXES TO THE REPORT ON THE STATE
OF THE MARINE ENVIRONMENT *(UNEP Regional Seas Reports and Studies No.
114/2, 1990).*

and lead. Concentrations are generally, but not always, higher in coastal waters,
though the difference is generally moderate (see figure 12.8). Elimination of
localized hot spots (see figure 12.9) again yields roughly equivalent concentrations in coastal and open ocean waters.

Perhaps the most striking aspect of the data analyzed by GESAMP and
reanalyzed here is their incompleteness. Of the thousands of chemical substances
entering the marine environment, monitoring data on ocean concentrations are
available for only a small fraction, and even those data are incomplete and
inadequate. The inadequacy of the data base is readily conceded in the
GESAMP report, but at the same time the conclusion is drawn that the open
ocean is "relatively clean." It is unclear how such a firm scientific conclusion
with such profound policy implications could be drawn on the basis of incomplete or inadequate data.

Moreover, when the admittedly incomplete data are examined quantitatively,

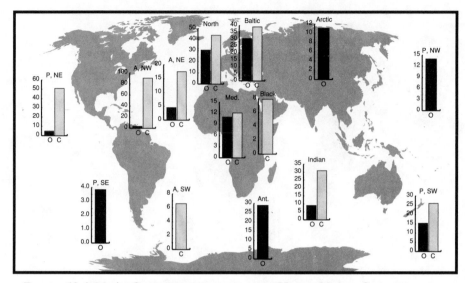

FIGURE 12.4 MEAN CONCENTRATIONS OF THE HEAVY METAL CADMIUM
*Concentrations (in nanograms per liter) are shown for open ocean (O) and coastal (C)
surface waters. Abbreviations: P, Pacific; NE, northeast; A, Atlantic; NW, northwest; SE,
southeast; SW, southwest; Med., Mediterranean Sea; Ant., Antarctic. Data are means of
values in the tables of S. W. Fowler,* Concentrations of Selected Contaminants in
Water, Sediments and Living Organisms, *in 1, 2* UNEP: TECHNICAL ANNEXES
TO THE REPORT ON THE STATE OF THE MARINE ENVIRONMENT *(UNEP
Regional Seas Reports and Studies No. 114/2, 1990).*

as has been done here, the conclusions differ significantly from those drawn by
GESAMP from the same data. Whereas the GESAMP report on the state of the
oceans proclaims a marked difference between contamination of coastal waters
and that of open ocean waters, the data base on which this conclusion is based
suggests that ocean contaminants can be divided into three broad categories.
The first (including PCBs) is more concentrated in the open ocean; the second
(DDT and its metabolites) is equally concentrated in coastal and open ocean
waters; and the third (other chlorinated hydrocarbons and heavy metals) is more
concentrated in coastal waters, although elimination of point sources of high
concentration reduces the difference to probable insignificance. These distinct
distribution patterns would appear to reflect different use histories, different
input pathways, and (potentially) different residence times. Rather than supporting the simplified dirty coasts, clean seas hypothesis, the data suggest a more
complex pattern of contamination that is substance dependent and that differs
depending on use history, input source, and residence time.

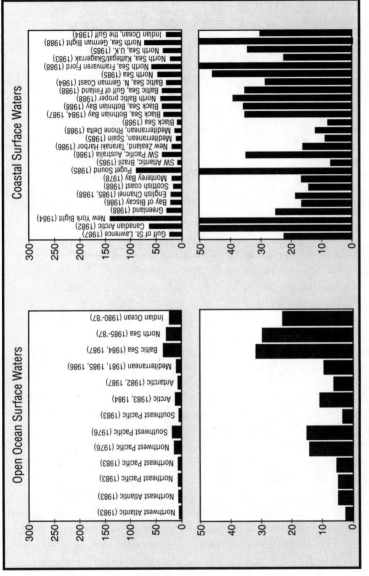

FIGURE 12.5 HISTOGRAMS SHOWING CONCENTRATIONS OF THE HEAVY METAL CADMIUM *Concentrations (in nanograms per liter), shown for open ocean (left panels) and coastal (right panels) surface waters, by location, illustrate locally high concentrations (point sources, or "hot spots"). Lower graphs depict the same data as the upper graphs, but on an expanded vertical scale. Data are means of values in the tables of S. W. Fowler, Concentrations of Selected Contaminants in Water, Sediments and Living Organisms, in 1, 2 UNEP: TECHNICAL ANNEXES TO THE REPORT ON THE STATE OF THE MARINE ENVIRONMENT (UNEP Regional Seas Reports and Studies No. 114/2, 1990).*

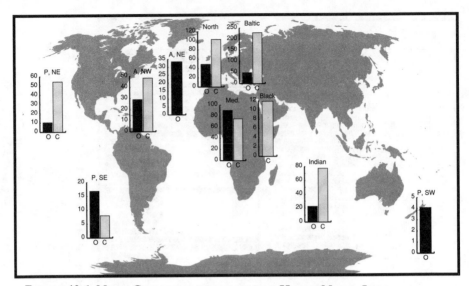

FIGURE 12.6 MEAN CONCENTRATIONS OF THE HEAVY METAL LEAD
Concentrations (in nanograms per liter) are shown for open ocean (O) and coastal (C)
surface waters. Abbreviations: P, Pacific; NE, northeast; A, Atlantic; NW, northwest; SE,
southeast; Med., Mediterranean Sea; SW, southwest. Data are means of values in the
tables of S. W. Fowler, Concentrations of Selected Contaminants in Water,
Sediments and Living Organisms, *in 1, 2 UNEP:* TECHNICAL ANNEXES TO THE
REPORT ON THE STATE OF THE MARINE ENVIRONMENT (*UNEP Regional Seas*
Reports and Studies No. 114/2, 1990).

A highly simplified statement of this complex picture is that open ocean
waters are approximately as contaminated as coastal waters, at least with respect
to the contaminants examined. Therefore, if coastal waters are considered to be
seriously impacted (as concluded in the GESAMP report), it follows that the
open oceans are likewise seriously impacted.

Even if this point is granted, however, it is necessary to ask whether existing
levels of contamination are cause for concern; that is, can they lead to damage of
the marine ecosystem? Unfortunately, data necessary to answer this question are
not available and are unlikely to be available before the time for effective
correction action has passed. There is, however, ample reason for concern, for
the following seven interrelated reasons.

First, most biological productivity (90 percent, by some estimates) occurs in
the open oceans, owing to their greater total surface area, and in the surface and
near-surface waters, where sunlight is available.

Second, the concentration of contaminants in open ocean waters is approximately the same as in coastal waters for some substances examined, but the

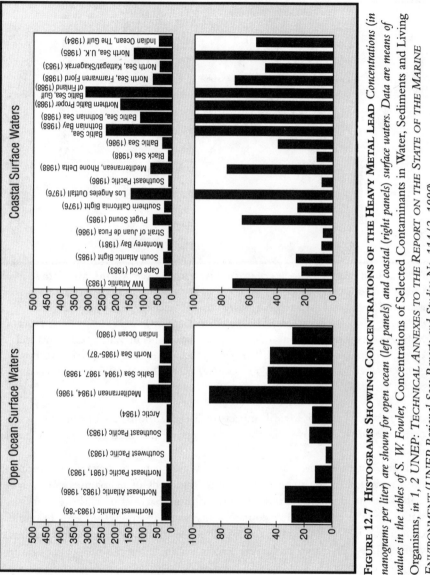

FIGURE 12.7 **HISTOGRAMS SHOWING CONCENTRATIONS OF THE HEAVY METAL LEAD** *Concentrations (in nanograms per liter) are shown for open ocean (left panels) and coastal (right panels) surface waters. Data are means of values in the tables of S. W. Fowler,* Concentrations of Selected Contaminants in Water, Sediments and Living Organisms, *in 1, 2 UNEP:* TECHNICAL ANNEXES TO THE REPORT ON THE STATE OF THE MARINE ENVIRONMENT *(UNEP Regional Seas Reports and Studies No. 114/2, 1990).*

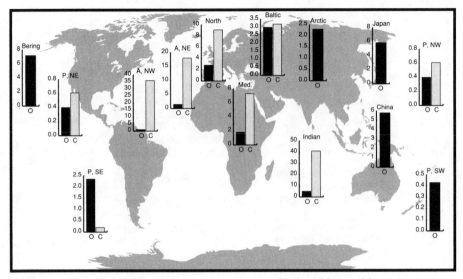

FIGURE 12.8 MEAN CONCENTRATIONS OF THE HEAVY METAL MERCURY
Concentrations (in nanograms per liter) are for open ocean (O) and coastal (C) surface waters. Abbreviations: P, Pacific; NE, northeast; SE, southeast; A, Atlantic; NW, northwest; Med., Mediterranean; SW, southwest. Data are means of values in the tables of S. W. Fowler, Concentrations of Selected Contaminants in Water, Sediments and Living Organisms, *in 1, 2 UNEP: TECHNICAL ANNEXES TO THE REPORT ON THE STATE OF THE MARINE ENVIRONMENT (UNEP Regional Seas Reports and Studies No. 114/2, 1990).*

concentrations and biological impacts of literally thousands of putative ocean contaminants is unknown because the necessary measurements have not been made.

Third, for those few contaminants for which data are available, concentrations recorded in the open ocean (up to hundreds of nanograms per liter of water) are far greater than those known to cause harm at much lower concentrations in the case of other contaminants, such as tributyl tin. For example, concentrations of 500 nanograms per liter are found for some contaminants, while tributyl tin causes biological defects at concentrations of 1–2 nanograms per liter of water.[11]

Fourth, existing measures of biological harm are, at best, crude. For example, a commonly used criterion, LD-50, measures the concentration at which a contaminant is lethal to 50 percent of the population. A more meaningful criterion is whether an existing concentration significantly affects reproductive success; but this criterion can seldom be established with accuracy in laboratory tests. Therefore, by introducing contaminants into the marine environment on

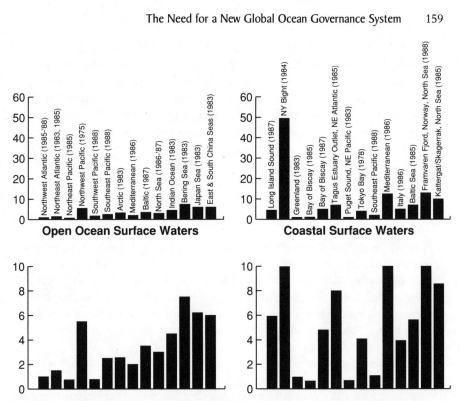

FIGURE 12.9 HISTOGRAMS SHOWING CONCENTRATIONS OF THE HEAVY
METAL MERCURY *Concentrations (in nanograms per liter) are for open ocean (left panels)
and coastal (right panels) surface waters. Data are means of values in the tables of S. W.
Fowler,* Concentrations of Selected Contaminants in Water, Sediments and Living
Organisms, *in 1, 2 UNEP:* TECHNICAL ANNEXES TO THE REPORT ON THE
STATE OF THE MARINE ENVIRONMENT *(UNEP Regional Seas Reports and Studies
No. 114/2, 1990).*

an uncontrolled and unregulated basis, humanity in essence is performing a
large, uncontrolled, and irreversible field experiment.

Fifth, the interactive impacts of ocean contaminants are virtually unknown. It
is possible, for example, that the presence of one contaminant significantly
reduces resistance to another or to a natural pathogen, as has been postulated for
seal deaths in the North Sea. Similarly, it is possible that an unexpected biological
stressor, such as increased ultraviolet radiation from ozone depletion, increases
the impact of other stressors, such as ocean contaminants. Understanding such
interactions is generally beyond the existing capacity of marine sciences.

Sixth, the concentrations measured are generally for gross samples of surface
water. Microregional concentrations are generally unknown. This paucity of
data is especially significant because the critical surface microlayer of the sea—

the millimeter-thin "skin" of the ocean, where numerous life forms undergo developmental stages—is lipophilic and hence concentrates organic contaminants such as chlorinated hydrocarbons. Moreover, nearly one-third of ocean pollution is believed to arrive through this microlayer from the atmosphere.[12] Measurements of the concentration of contaminants in the surface microlayer have been made only recently.[13] These measurements tend to confirm these concerns by showing that concentrations of heavy metals are 10 to 100 times greater in the surface microlayer than in the subsurface waters, and concentrations of pesticides are 1 million times higher. Additional concentration of contaminants would be expected to result from the "concentration factors" of unicellular life forms, which range to several hundred thousand times that of ambient water concentrations.

Seventh, potential damage to the marine environment may not occur in a linear, graded fashion that would enable proportionate corrective action. Experience has shown that the collapse of natural ecosystems often manifests in a discontinuous, binary, or threshold fashion, even though the underlying cause may accumulate steadily for years or even decades. Once such a collapse occurs, it may be irreversible because of the loss of species and the long time lags inherent in natural systems.

Despite its relatively sanguine summary conclusions, even the GESAMP report warns of the need for concerted international action within the next decade if serious and potentially irreversible degradation of the oceans is to be averted. The existing data, combined with the seven concerns enumerated here, suggests that the oceanic life web on which humanity depends for its survival may be more vulnerable than is widely appreciated.

THE REGIONAL APPROACH TO OCEAN GOVERNANCE

Given that the oceans are crucial to human survival and that present threats to their integrity may be significant, how are these threats best met? There are at least two distinct, but not mutually exclusive, regulatory options: the regional approach and the global approach.

Supporters of the regional approach to ocean governance argue that unique local problems demand unique local solutions. They point further to the fact that decentralized regulatory systems are more responsive and efficient and, in any case, are better established in the existing regional ocean protection treaties and conventions.

These are valid arguments. Tropical ocean regions, for example, generally border on developing countries and have vastly different ecologies from the oceans of temperate zones, including the coral reefs and mangrove forests where biological productivity is focused. Temperate ocean regions, in contrast, generally border on industrialized regions of the globe and are characterized by sea grasses and kelp forest coastal ecologies, with localization of biological productivity in both coastal zones and the open oceans.

Supporters of the regional approach also point to an existing network of regional seas treaties and conventions that have arisen independently in response to local needs (e.g., the Helsinki, Oslo, and Paris conventions) or that have grown regionally under the umbrella of the United Nations Environment Program (UNEP) (see figure 12.10). In contrast, there is no global convention now in place to regulate the land-based sources that constitute an estimated 77 percent of ocean contamination, although two global conventions effectively regulate ship discharges (which produce 12 percent of ocean pollution)[14] and pollution from deliberate dumping at sea (which in the past has produced 10 percent of ocean pollution).[15]

Perhaps the greatest asset of the regional approach is its contribution to building institutional capacity and awareness at the regional level, where implementation must take place for both regional and global regulation. Regional programs can serve ultimately as the building blocks of an effective global mechanism of ocean regulation.

The regional approach has existed now for twenty years, however, and some

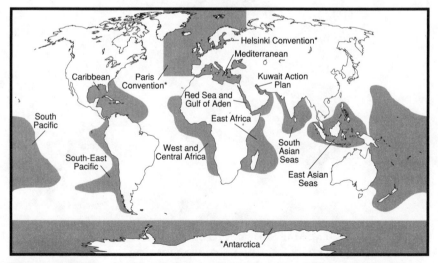

FIGURE 12.10 GLOBAL MAP SHOWING REGIONS OF THE WORLD'S OCEANS REGULATED BY REGIONAL CONVENTIONS OR TREATIES *Asterisks designate conventions formed outside of the UNEP Regional Seas Programme.*

of its limitations are also evident. First, most forms of ocean pollution are global in scope, including contamination from ubiquitous PCBs, other chlorinated hydrocarbons, and heavy metals (see figures 12.1 through 12.9). Local actions cannot effectively address these global sources. Instead, action is required at the origin—and the origin is typically all of the industrialized countries and, increasingly, developing countries.

Second, because most forms of ocean pollution are global in scope, global coordination is required to establish uniform standards and enforcement norms. To approach such standards on the regional level in the absence of global coordination would risk the creation of a redundant and inconsistent regulatory maze. To a large extent, the problems of different regions are interrelated; to this extent, solutions must also be interrelated.

Third, existing regional conventions do not address the primary source of ocean pollution, namely, land-based sources. Only three regional conventions have such protocols at present (see figure 12.11), covering but a small fraction of the oceans and an even smaller fraction of the pollution source. Moreover, only one regional convention, covering the Mediterranean, has an active land-based protocol.

Fourth, a detailed reading of extant regional seas agreements reveals that they

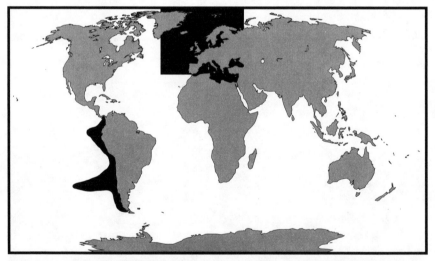

FIGURE 12.11 GLOBAL MAP SHOWING REGIONS OF THE WORLD'S OCEANS (SHADED) REGULATED BY REGIONAL CONVENTIONS OR TREATIES HAVING LAND-BASED SOURCE PROTOCOLS (1990) *The names of the corresponding treaties are shown in figure 12.10. Note that land-based source protocols cover less than 10 percent of the world's oceans.*

are in essence identical. In the history of regional seas agreements, a single model convention has been duplicated, with little modification, and promulgated chronologically from one region of the world to the next, depending on political readiness to implement such agreements. The regional seas network does not consist of separate agreements custom-tailored to address differentiated regional problems but rather is the regional iteration of a single basic model throughout the globe. This fact further weakens the argument that regional agreements are required to solve unique regional problems.

Fifth, and perhaps most telling, in two decades of ocean pollution the regional seas network has not curbed ocean pollution. As noted earlier, only a single UNEP regional seas convention, covering the Mediterranean, possesses an active (funded) protocol that addresses the major source of ocean pollution, namely, land-based sources. Even this most successful and mature of the regional seas programs, therefore, has failed to reverse pollution of the Mediterranean, although the pollution would presumably be much worse without it.

The historical inability of the regional model to reverse ocean pollution is, in part, a failure of financing. When the land-based protocol to the Mediterranean Regional Sea Convention entered into force, it was estimated that full implementation would cost $15 billion. To date, less than 10 percent of this amount has been spent.[16] The funding failure is a weakness that is intrinsic to the regional model. Construction of tertiary sewage treatment systems is politically unrealistic in countries where large fractions of the population are undernourished, undereducated, or underfulfilled, as is the case in many developing countries. Relegation of solutions to regions that cannot afford to implement them is doomed to failure from the outset. Funding must originate from the wealthier, industrial states; and such interregional transfers can be accomplished only by global mechanisms.

THE GLOBAL APPROACH TO OCEAN GOVERNANCE

The foregoing arguments suggest that the regional approach is structurally incapable of reversing ocean pollution. A global mechanism would appear to be required, on the following grounds.

First, most regions of the ocean are contaminated by substances originating outside the region. Reduction of such contamination cannot be achieved by isolated actions within the affected region. The problems must be approached at the source, and this may require an integrated global effort.

Second, although the industrialized world has the know-how and machinery to clean up the seas, no global mechanism presently exists to deliver these to the regions where they are most needed. Such transfers of knowledge and technology need to address the principal source of ocean pollution, namely, land-based sources, and must therefore include the means to achieve clean production. Transfer of knowledge and technology from sources to sinks requires global coordination.

Third, the necessary funding commitments and mechanisms do not exist at present. As illustrated by the case of the Mediterranean Sea, the problems of limiting land-based sources of ocean pollution are not likely to find satisfactory solution without transfer of funds from wealthier countries. These will, in turn, require the perception in the industrialized world that such transfers are necessary and mutually beneficial.

Fourth, a significant source of pressure on coastal zones and, therefore, on the open seas is the burgeoning populations of developing countries. There is presently no mechanism in place that could provide the incentive to limit population growth or to link such incentives with the transfers of knowledge, technology, and funding needed to stem ocean pollution and deterioration of the coastal zone. Such sensitive issues also may best be approached on an integrated, global basis rather than on a piecemeal, regional basis.

The conclusion that a global governance system is essential by no means excludes a role for regional organizations and agreements. On the contrary, regional agreements could be a necessary precursor to an effective global mechanism. I have suggested elsewhere that an effective global mechanism of ocean protection could take the form of an overarching system to coordinate, exchange, fund, and regulate an expanded system of regional agreements.[17]

The precise form of any such global agreement can emerge only when the need is clear and the political will therefore exists. Certainly, the existing and historic 1982 United Nations Convention on the Law of the Sea, which evolved over such a long period through the patient efforts of so many, provides a superb foundation. In its present form, however, the 1982 Convention is intended only to establish general obligations and provide broad guidance to the process of global ocean governance. The means of implementation remains weak. Furthermore, the 1982 Convention lacks critical properties of the integrated global management system that is now required.

What are the necessary and sufficient properties of an effective global ocean governance system? First, it must be anticipatory rather than reactionary. The time lags between cause and effect that are inherent in the global life support system are measured in decades, centuries, and even millennia. Humanity cannot wait for definitive scientific proof of harm to the oceanic life support system before acting, because the proof itself will foreclose options by impeding or preventing corrective action. The long time lags in the global life support system demand anticipatory (precautionary) rather than reactionary approaches.

Second, an effective global ocean governance system must incorporate and reflect the interconnectedness of the global life support system. The land, the air, and the seas all are integral components of a single system, analogous to the organs of the human body. Just as the lungs could not function without the kidneys and heart, so too the oceans cannot function in isolation from the atmosphere and terrestrial environment. For both the body and the globe, the health of the parts depends on the health of the whole, and conversely.

International governance of the commons has never before incorporated interconnectedness. Instead, it has been based on the treaty model, by which a collective or mutual problem is recognized and then addressed by means of a specialized solution. The solution typically takes the form of a treaty or convention that is agreed to by several nation-states. This approach is by design reactionary, compartmentalized, and limited in scope. The utility of the model is that its very narrowness, specificity, and applicability to a manifest problem lessens political objections and thus increases the likelihood of consensus.

Effective global ocean governance, however, cannot now be achieved by utilizing this model. Consider the linkages that render the model outmoded. First, two-thirds of harvested fish species hatch in coastal zones. Therefore, comprehensive regulation of fisheries cannot be undertaken independently of coastal zone management. Second, coastal zones throughout the world are being degraded by sedimentation and erosion from mismanagement of land resources, notably deforestation. Therefore, fisheries, land use, and forest policies are inextricably interwoven. Third, an estimated one-third of ocean pollutants are transmitted via the atmosphere.[18] Therefore, effective ocean regulation cannot be divorced from regulation of atmospheric inputs. Fourth, atmospheric inputs originate largely from transportation and manufacturing sectors of industrialized economies. Therefore, ocean policies are linked with production and consumption systems. Fifth, an estimated 44 percent of ocean pollution arises from runoff from land sources, mainly artificial fertilizers, pesticides, and herbicides. Therefore, ocean regulation cannot be implemented independently of agricultural policies.

Ocean, climate, agricultural, and energy policies—all are intertwined in a single fabric, the warp and woof of which defines human life-style. The goal cannot be to regulate the threads, as in the past, but rather must be to approach the fabric as an integral whole. These realities are recognized increasingly in international treaty organizations, with application of the precautionary principle and calls for holistic approaches. The treaty model is of limited use in achieving these ends. What may be required now is a broader, system model of environmental stewardship that addresses the interconnectedness of the global life support system with anticipatory rather than reactionary policies.

The obstacles to such a system are numerous and daunting. First, the anticipatory or precautionary approach demands action before the need is scientifically proven. And yet pragmatic politicians are understandably loath to allocate increasingly scarce resources before the need is established beyond a reasonable doubt. A way must be found to make precautionary action politically feasible.

Second, an integrated approach that incorporates interconnectedness demands action simultaneously on several fronts, from oceans to climate to production and consumption systems. Presently, national and global institutions are not sufficient for this task, nor does the political will for such a drastic overhaul in the international system currently exist. Moreover, the intellectual, political, and economic tools for dealing with complex systems are not widely available.

Third, effective global governance requires that nation-states surrender a measure of their traditional sovereignty for the broader social good. Although this has been accomplished widely on local and national levels, the world is only now taking the first tentative steps toward supranational organizations to which nation-states relinquish significant power and authority.

Fourth, effective governance of the international commons will require massive allocation of resources during a period of increasing scarcity. Moreover, these allocations will inevitably transfer wealth (and therefore power) from the rich to the poor. Never before in history has such a transfer taken place peacefully.

Fifth, developing nations, understandably impatient with the cycle of poverty in which they are mired, are increasingly unwilling to entertain conditionality in such transfers. And yet conditionality is indispensable in at least two arenas: population growth and developmental pathway. An effective global environmental regime requires that population growth be controlled. Developing nations must also willingly foreclose the same path to prosperity pioneered by the current industrial states, for this path has depended on abundant resources and has led to unsustainable patterns of production and consumption.

Sixth, even assuming a willing global political leadership, the global population must accept the need for major change in order to accept the apparent sacrifices implied in new life-styles. Effective global solutions require enthusiastic and informed participation by individuals, which is inconceivable when people are undernourished, undereducated, and underfulfilled.

Seventh, in some ways the most daunting obstacle, virtually every established world religion portrays the relation between humanity and nature as one of dominion. Contemporary humans are accustomed to thinking of themselves as the masters of nature. Effective global governance requires a different philosophy, one based on partnership and union with nature.

Despite these and other obstacles that can be envisioned, little is served by abandoning hope. It is possible to imagine an integrated global system for managing the earth's life support systems, on which all people depend. The need for interconnectedness requires that any such system not be limited to the oceans but instead that it encompass global regulation of land, sea, and air. A revitalized United Nations could provide the administrative framework. The European Community provides an evolving model for the negotiated surrender of a degree of national sovereignty in the interest of the common good. National defense programs provide a historical precedent for anticipatory action, and the reduction of these programs with the end of the cold war could free funds on the massive scale required for global environmental defense. An enlightened global environmental awareness, born perhaps of looming catastrophes such as ozone depletion and global climate change, could provide the educational force and political will.

At present, such ideas seem infeasibly visionary. We can only hope that they will become practical and realistic before the opportunity to implement them is forever missed.

SCIENCE, POLITICS, AND INDIGENOUS BELIEF SYSTEMS

Traditional models of international environmental governance often are highly regarded for their strong scientific foundation. For example, the only existing global convention for regulating ocean dumping, the London Dumping Convention, was founded on scientific principles, and its annexes can be amended only on "scientific and technical" grounds. Politicians and even scientists themselves frequently appear to believe that science can provide purely objective answers to questions on environmental management options.

When viewed in the broad perspective of contemporary ecological sciences, existing and proposed treaties and conventions are less scientific than political instruments. They ignore interconnectedness in the natural world and are instead compartmentalized: One reacts to ocean pollution; another, to air pollution; yet another addresses depletion of fisheries; a fourth responds to climate; a fifth, to biodiversity; a sixth, to forests; and so forth. Such instruments are isomorphic with human disciplinary divisions and political organization rather than with the natural world. Similarly, existing treaties and conventions have until recently systematically ignored the time lags in the large, global systems they seek to regulate, by adopting reactionary rather than anticipatory approaches.

What modern science tells us about complex ecosystems, however, is that they are integrated and dynamically interactive. Ecology teaches that one component of a natural system cannot be perturbed without affecting others. Physics instructs that the inertia built into complex natural systems demands regulation by anticipation rather than reaction. Moreover, despite the confidence of some scientists, it is impossible to view scientific issues independently of their political context. Science is nothing more nor less than a methodology; its application is inevitably shaped by the social, political, economic, and philosophical context. The very questions that science seeks to answer are determined and then prioritized by nonscientific considerations. Science is far from a purely objective system capable of foreseeing all of the complex interactions within natural systems induced by a particular management option and then asserting with confidence their benevolence.

The current system of international environmental governance, based on centuries of Western legal precedent and evolved political institutions, has not caught up to the new scientific realities. To this degree, current governance of the international commons is emphatically nonscientific.

In contrast, the ancient beliefs of indigenous peoples throughout the world, such as Native Americans and Maoris—articulated in this book by Moana Jackson and others—view the land, sea, and air as part of a single creation. Such beliefs are founded on the precept that human beings do not hold dominion over nature but rather are an integral part of natural cycles. Moreover, these same belief systems incorporate foresight into human relations with nature. Certain Native American peoples, for example, evaluate human actions in the present in terms of their effects on the seventh generation in the future. Such belief systems are isomorphic with the natural world rather than with evolved political institutions.

It is a potent and poignant irony that "primitive" spiritual belief systems are far more consistent with modern scientific thinking than is the existing international system of treaties and conventions. We have evidently come full circle. Perhaps we must now consider how best to return to the beginning.

NOTES

1. W. Jackson Davis, *Global Aspects of Marine Pollution Policy: The Need for a New International Convention*, 14 MARINE POL'Y 191–97 (1990).
2. Oxygen is literally neither produced nor consumed in the conventional meaning of those words but instead is cycled through various planetary reservoirs, including the atmosphere. The terms "production" and "consumption" are utilized here to designate, respectively, net inputs into and withdrawals from the atmospheric reservoir.
3. M. O. Andreae, *The Oceans as a Source of Biogenic Gases*, 29 OCEANUS 27–35 (1986).
4. IMO/FAO/UNESCO/WMO/WHO/IAEA/UN/UNEP JOINT GROUP OF EXPERTS ON THE SCIENTIFIC ASPECTS OF MARINE POLLUTION (GESAMP), THE STATE OF THE MARINE ENVIRONMENT (Rep. Stud. GESAMP No. 39, 111, 1990) [hereinafter GESAMP REPORT].

5. E. D. GOLDBERG, THE HEALTH OF THE OCEANS (UNESCO, Paris, 1976).

6. 1, 2 UNEP: Technical Annexes to the Report on the State of the Marine Environment (UNEP Regional Seas Reports and Studies No. 114/2, 1990) [hereinafter UNEP: Technical Annexes].

7. S. W. Fowler, *Concentrations of Selected Contaminants in Water, Sediments and Living Organisms*, in UNEP: Technical Annexes, *supra* note 6, at 143–208.

8. This analysis is confined to data on the concentration of marine contaminants in coastal and open ocean surface waters, as presented in Fowler's review. *See* Fowler, *supra* note 7. Data published in the literature and reflected in Fowler's review are generally based not on systematic, multiple sampling but rather on random or pseudorandom sampling, especially in regard to the open oceans. Moreover, data in Fowler's tables are presented frequently as means of several measurements from a single locale, with no reference to population variance. It is therefore impossible to subject these data to proper statistical analysis. Instead, the recourse adopted here is to compare means computed for coastal waters with those computed for open ocean waters. This procedure does not enable statistically valid conclusions and instead provides only a broad impression of the state of the oceans. It is emphasized, however, that the approach used here is identical to that used by GESAMP. Therefore, although the approach does not satisfy normal standards of scientific confidence, it does permit critical evaluation of GESAMP's conclusions. The intent of the present analysis was limited to determining whether the admittedly inadequate data base supports the summary conclusions reached by GESAMP, namely, that coastal waters are more contaminated than the open oceans.

9. IMO/FAO/UNESCO/WMO/WHO/IAEA/UN/UNEP JOINT GROUP OF EXPERTS ON THE SCIENTIFIC ASPECTS OF MARINE POLLUTION (GESAMP), THE ATMOSPHERIC INPUT OF TRACE SPECIES TO THE WORLD OCEAN (GESAMP Rep. Stud. No. 38, 1989) [hereinafter GESAMP ATMOSPHERIC REPORT].

10. Point sources, or "hot spots," are defined, for the present analysis, as regions in which the local concentration exceeds twice the mean value computed for all sources.

11. B. S. Smith, *Male Characteristics in Female Snails Caused by Anti-fouling Bottom Paints*, 1 J. APPLIED TOXICOLOGY 22–25 (1981); E. D. Goldberg, *Selected Contaminants: Tributyl Tin and Chlorinated Hydrocarbon Biocides*, in UNEP: Technical Annexes, *supra* note 6, at 209–30.

12. GESAMP REPORT, *supra* note 4; GESAMP ATMOSPHERIC REPORT, *supra* note 9.

13. J. T. Hardy, *Where the Sea Meets the Sky*, NAT. HIST., May 1991, at 59–65.

14. International Convention for the Prevention of Pollution from Ships, I.M.C.O. Doc. MP/CONF/WP.21/Add.4 (1973), *reprinted in* 12 I.L.M. 1319 (1973) *as modified by* Protocol of 1978 Relating to the International Convention for the Prevention of Pollution from Ships, 1973, *opened for signature* June 1, 1978, I.M.C.O. Doc. TSPP/CONF/11 (1973), *reprinted in* 17 I.L.M. 546 (1978) [hereinafter MARPOL 73/78]. MARPOL 73/78 is administered by the International Maritime Organization of the United Nations.

15. Convention on the Prevention of Marine Pollution by Dumping of Wastes and

Other Matter, Dec. 29, 1972, 26 U.S.T. 2403, T.I.A.S. No. 8165, 1064 U.N.T.S. 120.

16. Personal communication, Dr. Stefan Keskes, former director, UNEP Regional Seas Program.
17. Davis, *supra* note 1.
18. GESAMP Report, *supra* note 4; GESAMP Atmospheric Report, *supra* note 9.

13 Mending the Seas through a Global Commons Trust Fund

Christopher D. Stone

THE OCEANS ARE BEING ASSAULTED with pollutants and overfished at a worrisome pace, but the plight of the seas is just one facet of a larger set of interlocking problems. The whole global environment is in peril. Forests are being stripped, stressed, and burned. Lands once arable are yielding to desert. The atmosphere is under siege. Wetlands are vanishing. We are perturbing the climate and decimating species.

All this has worked its way into the public consciousness and echoes in the press every day. There is no reason to recite all the grim prophesies here. What we need now are, if not ultimate solutions, at least some promising strategies to sort through as a start. I have two related proposals. First is the institutionalization of a system of Ocean Guardians; second, and more fundamentally, is a Global Commons Trust Fund that would help underwrite the expenses of the cadre of international organizations that are leading the worldwide environmental effort but are now financially strapped.

THE PREDICAMENT OF THE COMMONS

The "global commons" refers to those sectors of the planet above and beyond the generally recognized sovereign territory of any nation. They include the atmosphere, the high seas and their seabeds, and the zones of space most ideally suited for the orbit of satellites and space stations. Currently, because these areas are in effect "unowned," anyone can use and abuse them with relative impunity.

Of course, much of the planetary environment is imperiled, not merely the commons areas. What makes the plight of the commons special has less to do with any distinct ecological features of the commons than with the special vulnerability of commons areas in international law.

In commons areas, absent some special motivation and effort, there are not apt to be any monitors exercising the scrutiny that nations apply to their own territories. Even if monitors are deployed, and they identify substantial and worrisome changes in the environment, under conventional principles of international law it is not clear that there is any "injured party" empowered to make diplomatic protest or to sue when the commons is harmed.

Indeed, it is not clear what changes in the quality of the commons might constitute the sort of legal damage that would legitimate a lawsuit. Classic principles of international law conceive damage if not in terms of economic loss, at least in terms of affronts to sovereignty. If someone were to sail out on the high seas and simply shoot sea turtles for sport, no internationally recognized person would suffer any legally cognizable damage; even if the turtles had commercial value, so that their killing were a contingent economic loss, it would not be a recognizable economic loss to any nation, nor an affront to any nation's sovereignty, because resources on the high seas are "unowned."[1]

This unownedness is one key factor contributing to the degradation of the global commons. Customary international law (international law unsupplemented by special treaty) has taken the position, in effect, that just as the commons is unowned for purposes of wealth exploitation—anyone can seize a favorable satellite position or scoop up deep seabed minerals without answering to the world community—so, too, does no one have to answer for deterioration. The ocean, like the atmosphere, can simply be appropriated as a free sink for sewage disposal by any nation that finds it convenient to do so. The pressures worsen as the more advanced nations run out of low-cost waste disposal sites on land. As a consequence, the cheapest means of disposal for a coastal state— cheapest, at least, from its own point of view and in the short run—is to pour and dump and vent its wastes into the commons.

THE POTENTIAL VALUE OF GUARDIANS

As one approach to ameliorating the problem, I have advocated a system in which some public or nongovernmental organization is designated to act as guardian for each hazarded portion of the commons.

Consider, for example, the fact that coastal nations annually dump into their coastal waters millions of tons of sewage sludge, industrial waste, and dredged material, all of which is fated to affect the deep sea. Who is to monitor and speak for the global damage?

Under a guardianship regime, an Ocean Guardian would be designated, perhaps GESAMP, the joint Group of Experts on the Scientific Aspects of Marine Pollution, with a legal staff. The guardian would be authorized (1) to monitor ocean conditions; (2) to appear before the legislatures and administrative agencies of states considering ocean-impacting actions to counsel moderation on behalf of their "client"; (3) to appear as a special intervenor-counsel for the unrepresented "victim" in a variety of bilateral and multilateral disputes; and, perhaps most important, (4) to initiate legal and diplomatic action on behalf of the commons in appropriate situations. In other words, the guardian could be empowered to sue to enjoin commons-damaging activities at least in those cases in which, if the damage were to a sovereign state, the law would afford the state some prospect of relief. The guardian would speak for the otherwise voiceless portion of nature, just as, in familiar legal systems, guardians are commonly designated to speak for persons who cannot represent themselves: infants, the insane, and so on.

The notion is hardly farfetched. Indeed, many guardianship functions are already recognized in U.S. environmental law on a more modest scale. For example, under the Superfund legislation,[2] the National Oceanic and Atmospheric Administration (NOAA) is designated as trustee for fish, marine mammals, and their supporting ecosystems within the U.S. fisheries zone, with express authority to request the attorney general to initiate litigation to recover restoration damages from any party injuring its ward.[3] The Brundtland Commission's legal experts' report, in the course of proposing the idea of an environmental international ombudsman, suggested that ombudsmen might exercise some trustee functions.[4]

A system of commons guardians would be a step forward, but it would still be no panacea for biosphere degradation. The commons areas under guardianships, such as living ocean resources, would be elevated to a legal and diplomatic standing more on a par with conventional nation-states. Although the system would provide the foundation for some judicial review on behalf of the commons, there are several distinct limitations to bear in mind. First, under prevailing principles of international law, the ability of nations to protect their internal environments from ordinary transboundary pollution, although theoretically available, is subject to some significant qualifications. Problems of proof are exacerbating; injunctions are not available (as they are in domestic law); the readiness of the defendant state to submit to jurisdiction is unsteady; litigation is notoriously time-consuming; and damages, where they are ultimately awarded, are (in contrast with U.S. procedures, for example) parsimoniously calculated and difficult to collect.

In other words, as long as the principles of customary international law are left unreformed, putting the commons in the same position as a nation-state (making parts of the commons, in effect, international persons) would leave them both inadequately protected—national territories and commons areas alike. Hence, protection of commons areas depends not only on the establishment of guardianship mechanisms but also on significant changes in the substantive law that the guardian would be empowered to invoke. The oceans not only need their own independent voice; they also need more thoughtful—and more environmentally oriented—law.

Even if the law shifts in these favorable ways, the need for funding will remain critical. Obviously, there would be the need to underwrite the costs of the guardians. And more important, litigation, even when abetted by a guardianship system, and even under the most receptive of legal regimes, is not apt to provide a satisfactory solution to protection of the commons areas. Litigation is almost inevitably reactive—after the fact—and depends on elements of proof that are time-consuming and complex, particularly where, as is typical in global-scale pollution, there are major questions of causal linkage, joint causality, damage measurement, and reciprocal (mis)conduct.

I do not wish to discourage the continuous fostering, through litigation, of a body of international environmental law aimed at challenging transfrontier pollution in court and through formal diplomatic channels. It is not at all unrealistic to suppose that international damage suits can be forged into an instrument to provide compensation for, and prospectively to control, abrupt incidents such as sudden, clearly wrongful chemical spills that can be traced unambiguously to a definite point of origin and that are associated with agreed-on risks. But unfortunately, most biosphere degradation occurs too insidiously, too gradually, too "innocently," and too ubiquitously to be stanched by the principles that contemporary international law makes available to plaintiffs. Indeed, even improvements in that body of law by specially tailored bilateral and multilateral treaties ("improvements," from the plaintiffs' point of view), although welcome and foreseeable, are not likely to transform transboundary litigation into an ideal line of defense for the biosphere's major maladies.[5]

SERVICES A GLOBAL COMMONS TRUST FUND WOULD UNDERWRITE

What we return to, then, is the need for funds: funds to drive litigation (and the threat of litigation) but, more important, funds to provide for the cooperative, managerial functions that the maintenance of the global commons requires. More specifically, to what uses might the resources of a global fund be put?

The fund would finance a host of measures most observers regard as critical: improvement in global monitoring and modeling; fostering (and transferring) of adaptive technologies, such as fast-growing trees; enforcement of safer methods of disposing of hazardous wastes in the commons; gathering and storage of genetic material; institution of a global environmental patrol force, with the capability to respond quickly to environmental disasters such as major oil spills and nuclear accidents; drafting and lobbying for new treaty agreements; and promotion of improved enforcement of treaties already in force, such as the Convention on International Trade in Endangered Species of Wild Fauna and Flora. The fund could also defer the costs of compliance with international regulations designed to remedy the ills of the commons. In the context of ozone negotiations, for example, many developing nations, most vocally India, have pointed out that they are not responsible for the buildup of ozone-depleting agents and that if they are to comply with regulations now being proposed to alleviate further deposits, necessitating more expensive, environmentally benign technology, they deserve help in the form of freely transferred technology and shouldered costs.[6]

On the high seas, the fund, in addition to combating pollution, could underwrite marine research and support forceful monitoring of ocean dumping and incineration—to the extent that those activities would be permitted at all. It also could underwrite the policing of fishing agreements, which presently rely all too heavily on the fishing fleet's "self-monitoring."

The infrastructure to do those jobs already exists. It includes, along with the United Nations Environment Program (UNEP), such organizations as the World Meteorological Organization, the World Wildlife Fund, and the International Union for Conservation of Nature and Natural Resources, which serves as an umbrella for hundreds of government and nongovernmental organizations operating in these fields. These organizations are capable of doing first-rate work. What they need now is money.

SOURCES OF REVENUES FOR THE FUND

Where would the funding come from? One source of funds—I think of it as the seedbed—has already been hinted at: the global commons areas themselves. As I observed earlier, at present all these areas can be used and abused with relative impunity—free of charge.

If we were to rectify this practice and charge even a fraction of the fair worth[7] for the various uses to which nation-states put the global commons, we would advance two goals at once. The charges would dampen the intensity of abuse and would at the same time underwrite the costs of providing public order, such as marine resource management and repair of environmental damage, for example, restoring wetlands and restocking impaired areas with new life.

Let me just play with some figures to convey the magnitude of what we are talking about.[8] Start with the oceans. Two hundred billion pounds of fish are harvested annually. An ocean use tax of only one-half of 1 percent of the value would raise $250 million. The same token rate on offshore oil and gas would yield $375 million.

There is another, dirtier use of the oceans: as a sewer. It is often much cheaper (and in some circumstances, relatively less harmful) to dispose of toxic refuse in the sea, particularly because there are no neighbors to complain. Officially reported ocean dumpings, almost certainly understated, amount to more than 200 million metric tons of sewage sludge, industrial waste, and dredged material yearly. Much of this takes place within traditionally territorial waters rather than on the high seas commons as such. But the repercussions are inevitably ocean-wide. A tax of only $1 per ton would raise another $200 million.

Nations use the atmosphere as they use the oceans—as a cost-free sewer for pollutants that will wind up as someone else's problem. By burning fossil fuels and living forests, humans pump 7 billion tons of carbon into the atmosphere annually. A carbon tax of only ten cents per ton—*a dime per ton*—would raise $700 million, more than ten times the current budget of UNEP. The same dime-a-ton tax on sulfur dioxide and nitrous oxides would produce another $30 million.

The total thus far: more than $1.5 billion. And that is before adding a tax on chlorofluorocarbons, halons, the incineration of toxins at sea, or the trillions of gallons of liquid wastes that are discharged into the oceans as runoff. Nor does it include many potential levies for several nonpolluting uses of common heritage assets, akin to fishing and oil. Consider royalties for the minerals that will someday be taken from the seabed and fees for the uses of space: Why should a needy global community give away to the first grabber, rather than sell or lease at auction, limited resources such as positions for geosynchronous and earth-orbiting satellites and frequencies on the radio spectrum? The current practice is a multibillion-dollar giveaway.

Many persons will object to the pollution charge component, arguing that it is outrageous to allow anyone to pollute on the condition only that they pay for it. Yet some pollution is inevitable, and it is more of an outrage that polluters get away with it, as they presently do, scot-free.

Indeed, if there is a real objection to this proposal, it is that the initial rates I have suggested are too paltry. I worry that viewed as a way to combat pollution, the rates suggested here do not confront the polluting nations with the full costs of the harm they are causing and therefore would fail to elicit a high enough investment in conservation and pollution control. Alternatively, viewed as a strategy for maximizing revenues for the environmental infrastructure, they fall short of extracting the full value of what users would pay (for their ocean dumping, for example) were they required to bid for the rights at auction. And even the bottom lines appear too modest. The United States's share, which, not surprisingly, is the largest, would amount to only several hundred million dollars per year, not even as much as the cost of one Stealth bomber.[9]

Another way to bolster the fund would be to designate it as the receptacle for legal assessments under various commons-protecting treaties, such as those pertaining to ocean pollution. Consider, for example, a situation in which a company or nation is responsible for pollution of a commons area; there has been, say, an accidental oil spill on the high seas. No individual nation can claim to have been damaged, but applicable ocean pollution treaties could easily be qualified to provide that "damages" (on some basis) be paid into the trust fund and made available for general research and repair. There are certainly illustrations of this technique domestically. After the spill of the highly dangerous pesticide Kepone into the James River in the United States, Allied Chemical established such a fund for the James; and more recently, after the terrible incident at the Sandoz factory in Basel, Switzerland, in 1986, Sandoz established a $10 million fund to further the ecological recovery of the Rhine. Part of the fund is being used to establish an interactive data base on the Rhine ecosystem.[10] The establishment of such funds should not be ad hoc; the procedure should be extended and institutionalized.

A billion dollars annually is not an unrealistic estimate of what the fund could quickly rise to, even if every country does not blithely submit. Not all will. There will be resistance among the developing nations. But they do not face the highest levies, and the fund therefore does not depend on them.

Some countries will object to any tax on activities within their territories or, in the case of the coastal states, within their self-proclaimed exclusive economic zones (EEZs). But the charges are not for what nations do on the "inside"; it is for the effects of their activities on the "outside" world. Moreover, many noncoastal, landlocked nations—along with many scholars—continue to regard as semilegitimate, at best, the coastal states' proclaimed EEZs of 200 miles and more from their coasts. Allowing the coastal states to supply exclusive management across these zones makes a certain amount of sense; some management of ocean resources is better than none. But allowing the coastal states to snatch all the wealth without any accounting to the rest of the world, just because it happens to be closer to them, is indefensible. That is why I would include taxes on resources such as fish and oil taken beyond the traditional territorial boundaries of 3 or 12 miles from the actual coastline.

Most important, if a Global Commons Trust Fund (or funds, per commons area) could be established as a recognized world body, it could become the vehicle—a magnet—for nations willing to contribute funds to environmental defense voluntarily and independently of their level of use of the global commons. Although such voluntary contributions might be less dependable, in magnitude they are potentially every bit as large as what would be produced by the use fund I am stressing here.[11]

A BRIEF HISTORY OF COMPARABLE PROPOSALS AND THEIR FATES

Certainly, considering the apparent common sense of the proposal, particularly in the light of growing public alarm over global deterioration, one would suppose it to be an idea that could win quick and widespread backing.

On the other hand, ideas about global funds have floated around for a long time and have never garnered sufficient support.[12] Why should we expect a warmer reception for this one?

The idea I am presenting may not gain momentum either; but I would like to point out some distinctions between it and its various predecessors. Indeed, many of the previous proposals have been so ill defined, even casual, that it is hard to be certain exactly what was contemplated.

One sort of proposal looks to the wealth of nations, generally, as a source of revenues. Under the Planet Protection Fund that the late Indian prime minister Rajiv Gandhi advanced briefly in 1989, each nation would contribute one one-thousandth of its GNP, the proceeds ($18 billion per year, by his calculations) to be used to help Third World countries adopt and develop environmentally friendly technologies, which would be given to them free of charge.[13]

A second sort of proposal links taxes not to each nation's wealth, per se, but to its environmentally degrading activities. For example, proposals have been made in the ozone protocol negotiations at Montreal and since then for a carbon use tax, as a way of both dampening fossil fuel use and raising funds. Of course, carbon use is notoriously a proxy for national wealth and industrialization, so the effective difference between a tax on wealth and a tax on carbon use is uncertain.[14] Equally unclear is the use to which the funds would be applied under these proposals. There is some suggestion that the funds would be distributed for the benefit of less developed nations in meeting the challenges of industrialization in an environmentally benign manner. But the level of discourse has been distinctly vague about exactly how the funds would be earmarked for specifically *environmental* action by the recipient states. Considering the problems of exerting international influence over internal expenditures, it is certainly possible to suspect that most of those backing a world climate fund envision it principally as a conduit for transferring wealth and technology from the industrialized states to the less industrialized world.[15] In other words, there is some hint that these proposals enjoy whatever life they live because they are perceived as strategies for wealth redistribution (a critical, but separable, global issue), piggybacking on the environmental consciousness movement.[16]

A third sort of fund is more frankly wealth-redistributionist. These funds look to the commons as a source of wealth for the benefit of, and to be divisible among, nations, the payouts to be employed at the recipient nation's discretion. The idea is not scandalously revolutionary or even new. As early as 1970, President Richard Nixon, in proposing an extension of coastal states' administration over their adjacent seabeds (from the 200-meter isobath to the edge of the continental slope) had proposed a wealth redistribution fund as a part of the package. From considerations of fairness, and undoubtedly in the hope that it would quiet some objections by the landlocked states, Nixon proposed that some percentage of the wealth be set aside for the benefit of developing countries.[17] And indeed, the 1982 United Nations Convention on the Law of the Sea's mining provision (not in force) would endorse, in a safely vague way, some equitable sharing of the wealth of the commons, "taking into particular consideration the interests and needs of developing States and peoples who have not attained full independence."[18]

Note that the Global Commons Trust Fund proposed here is a fourth sort of fund, one that can be distinguished from all of the others.[19] Unlike the others, it would look to the commons both as the principal source and the principal beneficiary of funds. The most closely related predecessor is Elisabeth Borgese's proposal, decades ago, for an ocean development tax. This was to include

> a 1 percent tax . . . on fish caught, oil extracted, minerals produced, goods and persons shipped, water desalinated, recreation enjoyed, waste dumped, pipelines laid, and installations built. . . . This functional, not territorial, tax would be levied by Governments and paid over to the competent ocean institutions (e.g., FAO, UNEP, IOC, IMO, International Seabed Authority) for the purpose of building and improving ocean services (e.g., navigational aids, scientific infrastructure, environmental monitoring, search and rescue, disaster relief, etc.).[20]

Dr. Borgese's recommendation appears to have been the inspiration for a few tangible diplomatic proposals in the negotiations for the 1982 Convention, but it is interesting—perhaps instructive?—to observe that if so, her "supporters" appear to have annexed to her version, at least as they went along, elements of the other types of funds. Originally, if I understand her position correctly, Dr. Borgese was motivated principally by the idea of advancing world government and secondarily, perhaps contingently, by the idea of wealth redistribution. She saw that as a source for supporting world government institutions, voluntary funds were not as dependable as funds "directly and tangibly linked to useful services."[21] If the world were prepared to accept Arvid Pardo's proposal that the high seas and seabeds be considered the common heritage of humanity, why not tax the common heritage to underwrite those expenses that were for humanity's common (that is, indivisible) benefit, such as the maintenance and upkeep of the commons?

In this interpretation, the originally intended beneficiaries of the fund Borgese proposed for the oceans, in parallel with mine for the entire global commons, were government institutions, with priority given to those agencies whose functions were most closely related to maintenance of the commons region generating the revenues.[22] But at least some of those who supported a tax on exploitation of the commons saw an opportunity to raise funds for the benefit not of the commons but of the less developed countries.

That, at least, seems to have been the thrust of the ocean development tax in the form introduced by J. Allen Beesley of Canada in the Sea-Bed Committee in March 1971. His proposal would have taxed the value of minerals extracted from the seabed area beyond the outer limit of any nation's claimed territorial waters. The revenues were to be paid to a conspicuously functional ocean management agency—the projected Sea-Bed Committee. But the uses of the funds were not to be restricted to support of the law of the sea institutional machinery or any other world agencies. A tax of 1 percent, Beesley suggested, "could produce many millions of dollars for the benefit of the international community and the developing countries in particular, as much as fifteen million dollars a month, according to some sources."[23] In other words, the emphasis shifted from (or perhaps was broadened beyond) support of international institutions, from which all benefited indivisibly,[24] to reallocation of wealth to the most needy countries.

One guesses that this new slant—the hint of direct benefits to the neediest states—was influenced by the hope of bringing some of the New International Economic Order (NIEO) membership (whose vigor was then on the rise) into a congenial coalition. Whatever the reasoning, it did not work: The Beesley proposal died a quick and quiet death in the Sea-Bed Committee.[25]

Notwithstanding the short shrift given to the Canadian proposal, during the Third United Nations Conference on the Law of the Sea the Nepalese delegation proposed taxing the removal of nonliving resources from the seabed, with the revenues to go into a Common Heritage Fund.[26] The essential function of their fund seems to have been to compensate the landlocked nations for the fact that the extension of the EEZ by coastal states—or, viewed from the other side of the coin, the gobbling up of the open ocean—had magnified wealth disparities between the already predominantly wealthy coastal states and the predominantly poor landlocked nations.

Against this background, the Nepalese proposal garnered considerable cosponsorship and support. Although it failed to make its way into the final draft, the vague convention provisions cited here are certainly expansive and congenial enough to dangle comparable hopes before the Third World—although the 1982 Convention, which has not yet entered into force, is not good international law on these points.

THE LESSONS OF THIS BRIEF HISTORY: WHAT ARE THE PRACTICAL PROSPECTS?

As we have seen, none of the common heritage and global environment-type proposals, with the prominent exception of Borgese's, have emphasized the funding of institutions directly responsible for maintenance and repair of the global commons. The typical pitch has been wealth-redistributional, with the promise of doling out funds to states either with no strings attached or subject to unspecified (and, frankly, hard to specify) conditions that they be applied for the improvement, in the first instance, of the recipient's own internal environment.[27]

Thus, if we want to impose an optimistic spin on this brief history, there is nothing in the record to prove that a proposal to tax the uses of the commons for the benefit of the commons would not gain the headway denied its predecessors. Nothing quite like it has been put forward, really, for twenty years, and the time (in terms of contemporary public attitude) has never been more favorable. The symbolic and practical logic of realizing and applying the wealth of the commons for the management and repair of the commons strikes me as highly marketable politically.

Yet there may be some unpromising indications in this history that cannot be overlooked. After all, we have to ask why it is that managerial versions such as Borgese's have not gotten as far as those with a redistributionist flavor.

Political theorizing can supply some good guesses. International arrangements, such as global funds, have to be hammered out by, and prove acceptable to, representatives of states. And many, if not most, national leaders are likely to prefer securing marginal resources for their own unfettered disposal rather than leaving them in the hands of more remotely accountable (and potentially self-serving) transnational functionaries.

Indeed, more has been involved here than a tug-of-war between competing self-interests. It would appear perfectly legitimate for the leader of some nation to question the fund I am proposing in the following way. Suppose that Nepal starts with the conception that the commons is in some sense everyone's (its and ours) and that the extraction of part of it by one nation constitutes a compensable appropriation. Why, at that moment of division, should each nation not get its fair share of what has been taken out? If each use (particularly a consumptive use) of the commons is a division of commons wealth, why is it any more just that the extractor, assuming we make it pay, pay *into* the commons rather than *out* pro rata, to each of the cotenants? Nepal will say that if it is pressed for cash for domestic use, and those pressing domestic demands dominate the value to *it* of increasing fish stocks or restricting ozone depletion, what right have the rest of us to impose our agenda on Nepal? If at the margin nations A, B, and C want to leave their moneys in the commons account for the improvement of the commons, they should be free to do so; if the highest use for any of them involves a withdrawal, they should be free to withdraw.

The question of legitimacy does not disappear even if someone can convince the objecting nation that the global commons is not subject to division like some cotenancy under Anglo-American law—that somehow the wealth of the commons is inherently and inalienably the heritage of humankind. For even then there would remain the question: On what theory should the funds be reserved for repair of the environment rather than for other world government problems—UN-authorized peacekeeping actions, famine relief, and so on?

I consider these very difficult questions that merit further discussion. Part of my answer, which I cannot flesh out in this space and which I am not sure can be fleshed out to everyone's satisfaction in any circumstance, involves standard public goods analysis. Briefly, nations can look after their own internal interests and even most of their mutual interests *relatively* well by independent action— negotiating bilateral transboundary pollution accords, for example. But successful solutions to the problems of the commons require bringing together the largest number of players. Without enforcement of a cooperative solution to the free-rider problem,[28] the commons[29] is apt to suffer most severely.

But this, and questions about the governance structure of such a fund[30] and the strategies it might pursue,[31] are certainly among the many issues that remain to be explored further.

NOTES

1. The text is referring to general principles of customary international law that nations are obliged to respect independently of treaty. I am not suggesting that nations might not—for they certainly should be encouraged to—provide by treaty for the prohibition of such acts. But a treaty would not (directly) restrict the conduct of nations that declined to sign; nonsignatories would not be bound.
2. Comprehensive Environmental Response, Compensation and Liability Act of 1980 (CERCLA), 42 U.S.C. §§ 9601-75 (1982).
3. *See* 40 C.F.R. §§ 300.600, 300.615(a)(1) (1990).

4. *See* EXPERTS' GROUP ON ENVIRONMENTAL LAW OF THE WORLD COMMISSION ON ENVIRONMENT AND DEVELOPMENT, ENVIRONMENTAL PROTECTION AND SUSTAINABLE DEVELOPMENT: LEGAL PRINCIPLES AND RECOMMENDATIONS (1986).

5. For further discussion, see Durwood Zaelke & James Cameron, *Global Warming and Climate Change: An Overview of the International Legal Process*, 5 AM. U. J. INT'L L. & POL'Y 249 (1990) (discussing possible strategies for using the International Court of Justice to address another problem of the commons—global warming); Christopher D. Stone, *Afterword: The Global Warming Crisis, If There Is One, and the Law*, 5 AM. U. J. INT'L L. & POL'Y 497 (1990) (discussing the limitations of state responsibility and international litigation to address global warming).

6. In response to these concerns, parties to the protocol established a $160 million to $240 million Interim Multilateral Fund to finance the incremental costs to developing countries of complying with the protocol. Contributions to the three-year fund are based on the United Nations' scale of assessments (with no country paying more than 25 percent). *See* Report of the Second Meeting of the Parties to the Montreal Protocol on Substances That Deplete the Ozone Layer, Decision II/8, UNEP/OzL.Pro./2.3; *see also* D. GOLDBERG, TECHNOLOGICAL COOPERATION AND THE MONTREAL PROTOCOL MULTILATERAL FUND: A BRIEF DESCRIPTION (CIEL-US Working Papers on Technology and Global Change, No. 1, 1991). A number of developing countries, including India, have since indicated that they intend to become parties to the protocol. *See* Draft Report of the Third Meeting of the Parties to the Montreal Protocol on Substances That Deplete the Ozone Layer (Nairobi, June 19-21, 1991).

7. The fair worth (essentially the opportunity costs to the user) is not the same as the "damage." Both types of user charge are available, and the choice between them raises questions of considerable theoretical interest.

8. The figures in the text are current as of approximately 1988-1989.

9. One problem in designing an effective fund is that there is no way to ensure that moneys placed in the fund will augment, rather than supplement, contributions nations are presently making voluntarily. In the worst-case scenario, given the complications of administering international organizations, it is possible that the creation of a fund would simply shift resources from domestic environmental programs to international ones and that because of relative inefficiencies at the international level, the funds would be applied less effectively than if they had been left for handling through the then-drained domestic programs.

10. *Sandoz to Contribute to WWF Project to Restore Rhine River's Flora, Fauna*, 11 INT'L ENVTL. REP. (BNA) No. 2 (Feb. 10, 1988), at 108.

11. For example, even without any world formula, present U.S. funding for oceanic research and the like is $421 million (1989, estimated); *see* UNITED STATES DEPARTMENT OF COMMERCE, STATISTICAL ABSTRACT OF THE UNITED STATES 588 (1990).

12. There is a considerable history of proposals for general-purpose international funds, aside from the World Bank. In the 1950s, an international funding agency,

SUNFED (Special United Nations Fund For Economic Development), was proposed to underwrite nonliquidating projects in less developed countries that did not meet World Bank criteria. SUNFED never got off the ground. Proposals ensued for various other developmental agencies, typically to be funded by levies proportioned to the industrialized nations' gross national product, but they also made little headway. The history of these is reviewed in KRASNER, STRUCTURAL CONFLICT 164 (1985). In 1966, the United Nations General Assembly called for the establishment of a United Nations Capital Development Fund (UNCDF), contributions to be drawn from a fixed proportion of nations' wealth—a proposal that also stalled. A similar fate has met proposals for the Fund for Science and Technology (1979) and a UN Trust Fund for Assistance to Colonial Countries and Peoples, among others. There is a meagerly financed Commodities Fund. Krasner suggests that the Third World regards it as more feasible to obtain political control of these UN-type funds than to control the development banks, under which the principal donors have the strongest hand in framing the rules.

More recently, a pilot fund administered jointly by the World Bank, UNEP, and UNDP (the United Nations Development Program) has been established. The $1.4 billion fund, known as the Global Environmental Trust Fund, is to help developing countries implement programs in four areas: protection of the ozone layer, reduction of greenhouse gases, protection of biodiversity, and protection of international waters. Contributions to the fund are, for the most part, voluntary, although the fund's ozone layer component is based on the UN scale of assessments. See WORLD BANK, ESTABLISHMENT OF THE GLOBAL ENVIRONMENTAL FACILITY (Feb. 1991); see also D. GOLDBERG, TECHNOLOGICAL COOPERATION AND THE GLOBAL ENVIRONMENTAL FACILITY: A BRIEF DESCRIPTION (CIEL-US Working Papers on Technology and Global Change, No. 2, 1991).

13. See Gandhi Calls for $18 Billion Fund to Fight Pollution of Atmosphere, L.A. Times, Sept. 6, 1989, pt. 1, at 8. The Gandhi proposal was considered in October 1990 at a Commonwealth meeting in Malaysia, but it ran into British opposition. A resolution passed at the meeting called for strengthening of existing institutions, such as the World Bank. See Britain Stands Pat Against Sanctions, Chicago Tribune, Oct. 22, 1989, at 27. In April 1989, Gro Brundtland stated that Norway was prepared to contribute $90 million to an international climate fund under the auspices of the United Nations, conditioned on receipt of matching support from other industrial nations. There was little detail—"environmental organizations dismissed Brundtland's programme as vague and" self-promoting—and nothing ever came of it. See Norway's Brundtland Proposes International Climate Fund, Reuter Libr. Rep., Apr. 28, 1989 (dateline: Oslo).

14. There is the difference that depending on the elasticity of carbon use to the proposed tax at any margin, the carbon tax could constrain the use of carbon. Although some doubt the capacity of any politically achievable tax to put a nontrivial damper on carbon use, see UNITED STATES ENVIRONMENTAL PROTECTION AGENCY, CAN WE DELAY A GREENHOUSE WARMING? (1983). The Saudis are reportedly taking some of this seriously, opposing global efforts to restrict carbon dioxide emissions for

fear of the impact on oil demand. *See* Eliyahu Kanovsky, *The Coming Oil Glut*, Wall St. J., Nov. 30, 1990, at A-18.

15. *See* Special Report, *Scientists from 48 Countries View Tax on Fossil Fuel Consumption as a Way of Helping Pay for Action Plan to Safeguard Global Atmosphere*, 11 INT'L ENVTL. REP. (BNA) No. 7, July 13, 1988, at 414-18. William A. Nitze, in advocating a climate convention, asks that it include a fund "to meet all or part of the hard currency costs of preparing and updating the developing countries' national plans" to create energy-efficient systems, and so on. William A. Nitze, *A Proposed Structure for an International Convention on Climate Change*, 249 SCIENCE, Aug. 10, 1990, at 607, 608. The source of the funds is not identified. William Ruckelshaus proposes a "climate protection tax" but apparently only as a means of constraining carbon use; application of tax revenues is not mentioned. William Ruckleshaus, *Toward a Sustainable World*, SCI. AM., Sept. 1989, at 166, 172.

16. Of course, different levels of global scale wealth redistribution can be defended on many moral and pragmatic grounds, independently of environmental concerns. That is something I am not here to argue. While there is some linkage (a nation that is relieved of debt pressures may be able to reduce its short-term stresses on the environment), there is no reason to believe that a policy that optimized (if one can put it that way) a just wealth distribution would be congruent with a policy that optimized global environmental concerns. Decertification and deforestation might diminish with wealth. But some globe-scanning environmental problems correlate positively with wealth and industrialization, as we all too sadly know.

17. *See* Announcement by President Nixon on United States Oceans Policy, May 23, 1970, 9 I.L.M. 807, 808 (1970).

18. United Nations Convention on the Law of the Sea, Dec. 20, 1982, U.N. Doc. A/CONF.62/122, 21 I.L.M. 1261 (1982), art. 160(2)(f)(i); *see id.* art. 140.

19. It is distinct from the others except insofar as it would look to atmospheric effluent charges as a source of funds, whether the emissions originate from the taxpaying nation's territory or from its activities conducted in the commons areas. This yields some overlap with the second sort of fund reviewed in the text.

20. ELISABETH MANN BORGESE, THE FUTURE OF THE OCEANS: A REPORT TO THE CLUB OF ROME 63 (1986).

21. *Id.*

22. In Borgese's version, obligations incurred could be discharged by performing ocean management services rather than exclusively in cash.

23. *See* BORGESE, *supra* note 20, at 64-65 (quoting BEESLEY, in PACEM IN MARIBUS).

24. Although to benefit indivisibly is not necessarily to benefit proportionately to a nation's needs or population.

25. For a fuller discussion of the 1973 United Nations Sea-Bed Committee, see LAW OF THE SEA INSTITUTE, REPORTS OF THE UNITED NATIONS 38-55 (1983).

26. *See* BORGESE, *supra* note 20, at 65.

27. The difficulties of enforcing restrictions on the recipient's use of funds are considerable. Suppose nation A is contemplating the authorization of $10 million in its budget for environmental protection (somehow defined) but has not yet voted on

the budget. Prior to enactment, it receives $5 million from the fund, earmarked for environmental protection. It then appropriates $10 million for environmental protection (its original internal proposal) and adds $5 million to its original defense appropriation. Would the funding agency be able to eliminate this sort of maneuver?

28. In game theory terms, this would be called a "prisoner's dilemma" or, in some imaginable commons situations, even "chicken."

29. Read, if you insist on a reductionist (methodologically individualist) account, "the indivisible interest of each nation's citizens in the commons."

30. In the ozone negotiations in London in June 1990, the United States displayed a strong preference that all required fund management be handled by the World (or other developmental) Bank. Under what circumstances is it possible and preferable to avoid the creation of new, fund-devouring world agencies? Should the largest contributors to such funds retain the greatest control over discretionary allocation, or should allocation be on a one nation, one vote basis? These are critical issues destined to affect the amount of resources the wealthy nations are willing to pump into any such funds.

31. Should such a fund be authorized to auction marketable rights to pollute? In setting charges for, say, ocean dumping, should the fund price according to its best estimate of the *damage* the dumping does or as a monopolist, maximizing fund revenues? As charges for any polluting activities are raised, what would be the effect on the rate of unlawful (unpermitted, unpaid-for) discharges?

14 International Ocean Protection Agreements: What Is Needed?*

Clifton E. Curtis

IN 1987, THE WORLD COMMISSION on Environment and Development (Brundtland Commission), chaired by Gro Brundtland of Norway, released *Our Common Future*,[1] a global examination of critical environmental and development issues. As part of its brief, the commission's findings and recommendations concerning the oceans stressed the urgent need for decision makers to come to grips with the inadequate existing measures to protect marine and coastal ecosystems. Drawing on extensive written submissions and advice from people all over the world, the Brundtland Commission found the following:

> Looking to the next century, the Commission is convinced that sustainable development, if not survival itself, depends on significant advances in the management of the oceans. Considerable changes will be required in our institutions and policies and more resources will have to be committed to ocean management.[2]

Building on that basic finding, the Brundtland Commission identified three imperatives for ocean management:

> The underlying unity of the oceans requires effective global management regimes. The shared resource characteristics of many regional seas make forms of regional management mandatory. The major land-based threats to the oceans require effective national actions based on international co-operation.[3]

* This chapter is a refinement of statements presented by Greenpeace International (GPI) in early 1991 in related discussions within the London Dumping Convention and the Nairobi United Nations Conference on Environment and Development Preparatory Committee session. Although the author takes sole responsibility for the views contained herein, he is indebted to GPI staff members, particularly Remi Parmentier and Kevin Stairs, for their contributions to this evolving document.

Although recognizing that "significant gains have been made in past decades, nationally and internationally, and many essential components have been put in place," the Brundtland Commission concluded that "they do not add up to a system that reflects the imperatives mentioned above."[4]

Now, several years later, this book provides a timely opportunity once again to "sound the alarm" while also charting a more specific ocean agenda as we look toward the next century. To save our oceans and protect their resources, it is necessary to reach agreement on a mix of environment *and* development strategies, including effective "cross-sectoral" technology transfer and financing measures to assist less developed countries.

These concerns need to be pushed in every forum possible because, with few exceptions, the oceans continue to get the short end of the stick. In this regard, two basic beliefs buttress everything explored in this chapter. First, the oceans, although vast and covering more than 70 percent of our planet, are a vulnerable and complex environment. Second, there is substantial and growing evidence that ocean and coastal resources are being seriously mismanaged and abused. This chapter examines the international mechanisms, present and future, that can be employed to protect the marine environment.

EXISTING SUPPORT STRUCTURES

When considering what is needed to improve ocean governance and stewardship in the twenty-first century, one needs to be careful not to reinvent the proverbial wheel. There are existing global and regional mechanisms in place that can and should be built on, where possible. In this regard, several existing support structures come to mind.

First and foremost is the 1982 United Nations Convention on the Law of the Sea.[5] As noted by the Brundtland Commission:

> The ... Convention represents a major step towards an integrated management regime for the oceans. ... Indeed, the most significant initial action that nations can take in the interests of the oceans' threatened life support system is to ratify the Law of the Sea Convention.[6]

From an environmental perspective, the 1982 Convention's articles, and part XII in particular, establish standards and obligations for prevention, reduction, and control of pollution from all sources, as well as obligations for protection and conservation of living marine resources. At the same time, numerous articles interspersed throughout the convention would facilitate technical assistance, technology transfer, and international cooperation, including funding, to assist less developed countries in such efforts.

That is not to say that the 1982 Convention is perfect. One of its principal deficiencies is its legitimization of increased national sovereignty over ocean resources, a somewhat odd result given that a principal impetus for negotiating the convention was the concern of Ambassador Pardo and others over ill-considered encroachments on "the common heritage" of humankind. When viewed as a whole, though, the benefits of the 1982 Convention—for all nations—far outweigh any real or theoretical disadvantages. These benefits are especially significant with respect to part XI's deep seabed mining provisions, whose deficiencies have been blown far out of proportion by the United States and a few other key industrialized maritime states. Those concerns need to be resolved as soon as possible so that the convention can be brought into force. Once that task is accomplished, substantial energy can be devoted to using the 1982 Convention as a key framework for pursuing the kind of system of ocean governance and stewardship that is needed.

Second, there are numerous existing issue-focused treaties and institutions at the global level. The London Dumping Convention (LDC)[7] is one such treaty, as is the International Convention for the Prevention of Pollution from Ships.[8] Both treaties have evolved over many years, with their contracting parties attempting to shape them to address existing and new issues requiring attention. As with the 1982 Convention, neither the LDC or MARPOL 73/78 is perfect, but both of those regimes have worked fairly well, showing some flexibility in adjusting to new situations.

An example of an institutional structure at the global level that has an important role to play in strengthening ocean governance and stewardship is the International Maritime Organization (IMO), which serves as secretariat for both the LDC and MARPOL, as well as many others. Other institutions with equally important roles include the United Nations Environment Program's (UNEP's) Oceans and Coastal Areas Program Activity Center (OCA/PAC), the United Nations' Law of the Sea Office, the Food and Agriculture Organization (FAO), the United Nations Educational, Scientific, and Cultural Organization (UNESCO) (including the Intergovernmental Oceanographic Commission), and the World Meteorological Organization.[9]

Third, UNEP's Regional Seas Program, directed by OCA/PAC, offers a regional approach that can help improve ocean governance. Covering ten coastal regions around the globe and more than 120 states, these UNEP-sponsored efforts have stimulated action and cooperation that take into account the special environmental problems, as well as cultural, socioeconomic, and political factors, that exist in the different regions. To be more effective, however, the

Regional Seas Program must correct several weaknesses. For example, some coastal regions do not have a formal regional seas convention or other regionally focused marine and coastal program. And those regional conventions that do exist vary significantly in the extent, detail, and effectiveness of protocols or action plans designed to address particular problems. Nonetheless, it is clear that in most cases regional agreements, like those of the Regional Seas Program, are much better suited than global treaties to taking concrete, hands-on actions, complementing global and national efforts.

WHAT IS NEEDED?

The prior sections detail several international agreements and institutions that could be strengthened to improve governance and stewardship of the oceans. Drawing in part on those findings, recommendations, and actions, the following section highlights some guiding principles and regulatory mechanisms to improve ocean protection.

Key Principles

In considering measures to enhance protection and wise use of the oceans, several underlying principles deserve inclusion as key components of any new or existing treaty, regime, or other mechanism. This list is by no means exhaustive. Some of those principles are as follows.

Ecocentric Perspective. Far too often, decisions concerning the use of the oceans are driven by a human-centered, anthropogenic perspective. This must change. Humans are only one species among millions on planet Earth. We are the dominant species in terms of our ability to direct and influence many events; but this dominance brings with it responsibilities, and our human-centered approach sometimes impedes our ability to meet these responsibilities. The environmental impact assessment process concerning proposed marine activities is a classic example of our misguided perspective. With few exceptions, the principal focus of such assessments is to determine impacts on humans alone, with no regard for impacts on marine ecosystems. As humans sharing the oceans and the planet with other species, we need to be much more careful to think and act in a manner that reflects equal regard for others' rights.

Technology, Training, and Information Exchange. Much more needs to be done to ensure effective exchange of technology, training, and information. We will not succeed in addressing environmental concerns unless and until industrialized states and the private sector take seriously the importance of making these resources available, especially to less developed nations.

Precautionary Approach. Laws and other regulatory instruments, policies, and programs at the global, regional, and national levels need to adopt a precautionary approach. Simply stated, such an approach must be designed to eliminate and prevent emissions of hazardous wastes (broadly defined) where there is reason to believe that damage or harmful effects are likely to be caused, even where there is inadequate or inconclusive evidence to prove a causal link between emissions and effects.

Clean Production. The adoption of a "clean production" principle could lead to the development of ecologically compatible manufacturing processes that use a minimal amount of raw materials, water, and energy and emit no toxic waste or toxic by-products into the environment. In pursuing this objective, clean production initiatives should cover the entire gamut of pertinent activities— from raw material selection, extraction, and processing, through product manufacture and assemblage, to industrial and household use, and, ultimately, to management of the product at the end of its useful life.

Enforcement. Significantly stronger and more effective enforcement mechanisms are essential to ensure that states adhere to the letter and intent of global and regional agreements. Such mechanisms need to include intervention, inspection, response, dispute settlement, imposition of sanctions, and liability and compensation. At present, few of these mechanisms are in use, and for those that are, far too much deference is afforded to national authorities to do what, if anything, they believe is appropriate.

State Liability. Dumping or discharge (both operational and accidental) of dangerous substances and wastes at sea implicates the principle of state liability, in addition to applicable civil liability. Assigning state liability to the consequences of harmful activities is consistent with the overriding concern for controlling potentially dangerous activities. As a result, direct state liability is the most effective way to achieve the necessary control, in contrast to subsequent or indirect liability behind a private actor.

Development Assistance. As a corollary to transfer of technology, training, and information, a key component of effective international regimes is the inclusion of development assistance for environmentally benign production and other resource-related projects.

Effective NGO Participation in Decision Making. Very few international forums concerned with ocean matters provide effective opportunities for nongovernmental organizations (NGOs) to participate in their decision-making
processes. NGOs contribute ideas, information, and perspectives that are invaluable not only in their own right but also as voices for the public at large and for
other species on planet Earth. Where the procedures are overly restrictive, as in
the case of the Antarctic Treaty system, I feel strongly that the product of
government decisions is substantially diminished. In contrast, the LDC meetings, while not perfect, allow significant participation by NGOs and provide a
constructive model for other treaty organizations.

Preventive Strategies

A variety of preventive strategies are needed to advance protection of the marine
environment. Some of those are addressed in the context of the principles listed
earlier. In particular, it is becoming increasingly obvious that the most important
solution relative to hazardous wastes is to concentrate efforts and resources on
clean production and waste prevention. Although each of the following strategies
has its attributes, those that address hazardous wastes should be viewed as interim
measures pending the achievement of effective waste prevention strategies.

Uniform Emission Standards. The use of uniform emission standards (UES)
plays an important role in ongoing efforts to protect the marine environment
from the harmful effects of waste disposal. Although the term is somewhat
ambiguous, since all countries do not apply UES for all industries and since all
countries use a number of pollution management tools, it does accurately
describe the practice of setting maximum limit values for discharges. Numerous
waste disposal regimes, such as the London Dumping Convention, have UES
standards that include blacklisted substances, which cannot be discharged, and
"grey-listed" substances, which can be discharged only under special care procedures.

Environmental Quality Standards. Strategies based on marine quality
standards—often referred to as environmental quality standards (EQS)—relate
directly to the quality of water, biota, or sediments that must be maintained for a
desired level of quality and intended use. As employed in the United Kingdom,
for example, the standards are set on the basis of the declared use of the receiving
waters; the system is not rigid, and objectives may change according to circumstances. Some countries set EQS as closely as possible to natural background
levels, while some apply a critical path, critical organism approach. Where EQS

are based on sufficient scientific evidence, they can be an effective measure to protect the marine environment. If the state of knowledge is insufficient, however, EQS are no solution.

Hierarchical Waste Management Approach. There is growing agreement that successful management of hazardous waste depends on implementation of a hierarchy of techniques. Although the hierarchical approach is expressed in terms that imply that a waste already exists, a significant factor of waste management in a society with changing product demands and evolving technologies is the prior step of evaluating product choices and plant design. In this regard, the hierarchical approach is closely linked, at its upper tier, with the clean production principle. Simply stated, five descending options constitute this approach: (1) elimination or reduction at the source; (2) reuse or recycling; (3) destruction of hazardous constituents by physical, chemical, or biological means; (4) treatment to reduce the hazard; and (5) disposal into land, air, or water. Actions at the top are designed to reduce the volume or the hazard level of the waste; actions at the bottom are seen as the least desirable because they generally have more adverse environmental impacts.

Environmental Impact Assessments (EIAs). Prediction of the impact of new or modified projects on the environment is required by an increasing number of states. The environmental impact assessment (EIA) concept has evolved and been more widely used over the past decade. Nonetheless, its use and value remain very uneven, and EIAs too frequently serve as little more than a "window dressing" justification for a predetermined decision. If used properly, EIAs are important tools to guide decision making, and their "preventive cost"— where ill-conceived projects are brought to light—can represent significant savings over the future costs of environmental reparation, assuming that repair is possible.

Industrial Sectors Approach. Recently, the Paris Commission for the Prevention of Marine Pollution from Land-Based Sources in the Northeast Atlantic initiated an "industrial sectors approach," which consists of taking entire industrial sectors one by one, examining their problems, and defining solutions. This strategy, which also has been employed by some governments on a national or subnational level, has the merit of looking at the global and practical problems of industrial sectors concerned rather than looking only at substances in a somewhat abstract, or impractical, way. It can also constitute a new way to develop international cooperation for resolving problems identified.

Other Strategies. Additional strategies meriting consideration include those pertaining to activity management, use designation, regional planning, coastal zone management, watershed or drainage basin planning, beach access easements, an "ecosystem of the whole" approach to fisheries management that takes account of dependent species and habitat, and specially protected areas.

Issue–Specific Needs

Special attention needs to be given to those problems that at present are not adequately addressed, in particular the problem of land-based sources of marine pollution. Both UNEP's Governing Council and the LDC's consultative meetings have recognized land-based pollution as the major threat to the health of ocean and coastal environments.[10] Although land-based pollution may be best addressed at the national level, far too few states are giving the problem any significant attention.

A global treaty addressing land-based sources of pollution would provide an important vehicle for establishing common principles and obligations that would be applicable to all states, as well as for sharing information, analyses, and expertise. Others have suggested that perhaps an updated statement of principles or a "code of conduct" might be sufficient. Perhaps one of those alternative approaches will be sufficient, but absent a convention-type structure—with some binding obligations—the risk remains real that the unsatisfactory situation now existing under the voluntary 1985 Montreal Guidelines agreed to under UNEP's direction will not be remedied.

Whatever approach is developed at the global level for land-based sources, it can and should be designed in a manner that complements efforts at both the regional and national levels.

There are other issue areas, as well, in which international coverage is extremely deficient. Offshore activities, especially oil and gas development, are one. Seabed burial of nuclear or other toxic wastes, accessed from shore, is another. And fishing activity that risks harm to marine resources or marine ecosystems is another. This latter category includes the physical impact of fishing gear on the seabed, mariculture's biological impacts on marine ecosystems, and the death and maiming of nontarget species caused by the widespread use and discarding of drift nets at sea.

For each of these activities, special attention needs to be devoted to formulating effective global and regional measures. Where possible, existing legal instruments can be strengthened or expanded. For some issues, such as land-based sources, however, it seems likely that a new instrument will be required at the global level.

Improved Coordination

As stated in the first operative paragraph of a recent LDC resolution, there is a need for

> a global mechanism to co-ordinate the protection of the marine environment from pollution from all sources. Taking into account existing international agreements, co-ordination may embrace such matters as: accidental and operational pollution from vessels, dumping, pollution from land-based sources, atmospheric pollution, offshore industry and disposal of wastes into the sea-bed as well as liability and mechanisms for transfer of technology and financial resources, and arrangements concerning liability and compensation. . . .[11]

Wiser people than me can help us figure out the best mechanisms for accomplishing such coordination, but it is obvious that existing structures are not yet doing what is needed. If the 1982 Convention on the Law of the Sea were in force, one could envision a secretariat committed to ensuring that related environmental and development issues addressed in part XII and other pertinent articles would be better integrated and coordinated. In such a hypothetical situation, the secretariats and key parties to the various global and regional regimes dealing with marine pollution issues could be brought together on a regular basis to discuss and coordinate their respective tasks.

I am not suggesting that there is no coordination. UNEP, the United Nations Law of the Sea Office in New York, and individuals in issue-specific forums and at international conferences all contribute to coordination on marine issues. But one plus one, and so forth, does not yet add up to achieve the coordination that scientists, regulators, policymakers, and others concerned about governance and stewardship of the oceans need and deserve.

Another kind of coordination is also needed, one that would ensure that high-level government officials are aware of, and are invested in, the decisions of global and regional marine pollution forums. Two examples offer models for achieving that objective more effectively.

First, the Paris Commission's convening of a ministerial-level meeting in 1992 offers officials dealing with national and regional efforts to reduce and prevent land-based sources of marine pollution the opportunity to participate directly in the important work of that body. Second, the sequential grouping and interaction among NGO, scientific, and ministerial-level meetings in relation to the North Sea has contributed to a more effective exchange of information, data, and views among the NGO community, the scientific community, and ministers of those states bordering the North Sea.

There is yet another kind of coordination—the kind that comes from consolidating related ocean and coastal programs within one institution, or creating a new one if none of the existing ones is up to the task. This issue is a tough one, in part because of the array of protective "turf" considerations that come into play whenever there is even the mildest suggestion of reorganizing or moving elsewhere an institution's programs. Moreover, as noted earlier, major institutional restructuring takes a tremendous toll on all involved and is best avoided unless absolutely essential.

Nonetheless, the adequacy of existing institutional structures and interagency relationships deserves close examination as part of the effort to improve coordination. For issues for which new global agreements are under consideration (such as land-based sources) there exists an opportunity—assuming global initiatives are called for—to look closely at matters like effective coordination as a key element in selecting an institutional home. And the same considerations apply in relation to the creation of a new global coordinating mechanism, though its "home" might well be some umbrella structure that belongs to no one entity.

CONCLUSION

The work that lies ahead consists of pursuing ideas that will make a real difference for ocean governance and stewardship. It is a challenging task, but it is timely for at least two reasons. First, it has to be done, somehow and somewhere, soon. Second, some special opportunities are in the offing in the numerous ongoing global and regional marine pollution forums to lobby for important new initiatives that need to be undertaken.

NOTES

1. WORLD COMMISSION ON ENVIRONMENT AND DEVELOPMENT, OUR COMMON FUTURE (1987) [hereinafter BRUNDTLAND REPORT].
2. *Id.* at 264.
3. *Id.* at 264.
4. *Id.* at 264–65.
5. United Nations Convention on the Law of the Sea, Dec. 10, 1982, U.N. Doc. A/CONF.62/122, 21 I.L.M. 1261 (1982) [hereinafter 1982 Convention].
6. *Brundtland Report, supra* note 1, at 273–74.
7. Convention on the Prevention of Marine Pollution by Dumping of Wastes and Other Matter, Dec. 29, 1972, 26 U.S.T. 2403, T.I.A.S. No. 8165 [hereinafter LDC]. For a fuller discussion of the history of the London Dumping Convention, *see* Clifton E. Curtis, *Legality of Seabed Disposal of High-Level Radioactive Wastes Under the London Dumping Convention,* 14 OCEAN DEV. & INT'L L. 383, 393 (1985); Kaplan, *Into the Abyss: International Regulation of Subseabed Nuclear Waste Disposal,* 139 U. PA.

L. REV. 769 (1991); and Davis & Reitze, *Reconsidering Ocean Incineration as Part of a U.S. Hazardous Waste Management Program: Separating the Rhetoric from the Reality*, 17 B.C. ENVT'L AFF. L. REV. 687 (1990).

8. International Convention for the Prevention of Pollution from Ships, I.M.C.O. Doc. MP/CONF/WP.21/Add 4 (1973), 12 I.L.M. 1319 (1973), *as modified by* Protocol of 1978 Relating to the International Convention for the Prevention of Pollution from Ships, 1973, *opened for signature* June 1, 1978, I.M.C.O. Doc. TSPP/CONF/11 (1973), 17 I.L.M. 546 (1978).

9. There is always a risk in mentioning some treaties or institutions while not mentioning others. And though there are numerous others, my point is made more to suggest that there are existing global and regional treaties and institutions that should, where possible, serve as the point of departure for future efforts.

10. *See* UNEP Governing Council, Decision SS.11/6, Need for Effective Global Protection of Ocean Coastal Ecosystems (Aug. 3, 1990); and Thirteenth Consultative Meeting of Contracting Parties to the Convention of the Prevention of Marine Pollution by Dumping of Wastes and Other Matter, 1972, Resolution LDC.39 (13), The Protection of the Oceans and All Kinds of Seas, Including Enclosed and Semi-Enclosed Seas, and Coastal Areas (London, Oct. 29–Nov. 2, 1990) [hereinafter LDC Resolution].

11. LDC Resolution, *supra* note 10.

15 Seaborne Movements of Hazardous Materials*

Miranda Wecker and Dolores M. Wesson

WHAT IF THE *EXXON VALDEZ* HAD CARRIED HAZARDOUS MATERIALS?

On the night of March 24, 1989, a series of rather ordinary human mistakes occurred on the bridge of the supertanker *Exxon Valdez* that shattered a long-standing climate of complacency regarding the safety of commercial shipping operations. Before the *Valdez* accident, oil spill experts would have considered a spill costing in excess of $500 million to be nearly inconceivable. Since then, the financial stakes associated with the business of shipping hazardous commodities, including oil, have been radically reassessed. As of July 1991, the Exxon Corporation had already spent $2.5 billion in cleanup efforts and more than $300 million in damage claims. In addition, Exxon has been named as a defendant in approximately 200 lawsuits.[1]

What if the *Valdez* had carried substances that were more toxic, more persistent, or more explosive than crude oil? What if such a spill were to take place off the coast of a developing country without the technical expertise and equipment to respond, without information about the contents of the stricken ship, or without the connections to obtain immediate assistance for salvage and cleanup?

ASSESSING THE RISKS

There has been no spill of hazardous materials on the scale of the *Amoco Cadiz* or the *Exxon Valdez*. This is due in part to the more stringent regulatory approach taken toward shipments of hazardous materials and in part to the fact

* The excellent editorial assistance of Sarah Oyer and Stefan Delfs, summer associates at the Center for International Environmental Law (CIEL), is gratefully acknowledged.

that many of the most hazardous substances can be shipped only in smaller quantities that are securely packaged. Overall, the amount of hazardous substances transported at sea is estimated to be substantially less than the quantity of oil. Yet no international organization has the capacity to calculate accurately the volume of hazardous materials transported. Collection and compilation of these data within the United States alone is difficult because differing criteria used by various government agencies prevent reliable data identification and comparison. Experts estimate that more than half of all bulk cargoes are potentially dangerous to human health and the environment. Approximately 10 to 15 percent of containerized cargoes are also considered hazardous.[2]

It is widely recognized that the potential threat to the marine environment posed by spills of noxious, toxic, and persistent materials can be several orders of magnitude greater than that presented by oil spills. Some chemicals are dangerous to marine ecosystems and human life in trace amounts. Some can persist in the marine environment virtually forever, irrevocably contaminating marine resources and endangering human health by despoiling the marine food web. Others can make the water unfit for swimming or other recreational uses, causing significant economic losses for coastal communities.

History has demonstrated that costly accidents give rise to strategies to avoid their recurrence. If a catastrophic spill of hazardous materials were to occur in a densely populated or ecologically valuable area, it is likely that coastal governments around the world would consider enacting laws severely restricting the navigation rights of all vessels carrying hazardous materials. Enactment of such laws by several countries would have the effect of destabilizing a key aspect of international ocean law: freedom of navigation.

Although no such catastrophic spill of hazardous materials has taken place as yet, the potential conflict between the navigation rights of vessels carrying hazardous materials and a coastal state's interests in protecting its environment is of great concern to nations. It can be expected that conflicts between these interests will lead to strenuous efforts to safeguard them both by establishing increasingly more complex regulatory systems.

This chapter presents a review of the existing international regulatory regimes governing the seaborne movement of hazardous materials. It discusses a number of key developments under way in several international forums, including the formulation of a global system for liability and compensation, further refinement of maritime safety codes, and the establishment of regional arrangements to control the movement of hazardous wastes. In conclusion, it emphasizes the importance of reaching a widely accepted agreement balancing the environmental protection interests of coastal states and the navigation rights of all nations.

THE EXISTING LEGAL FRAMEWORK FOR
THE 1982 UNITED NATIONS CONVENTION
ON THE LAW OF THE SEA

Although not yet in force, the 1982 United Nations Convention on the Law of the Sea[3] stands as the most recent codification of widely accepted principles of international law governing the rights and responsibilities of nations as they use the ocean commons. It addresses a comprehensive range of issues, including the navigation rights and duties of vessels in different portions of ocean space and the rights and responsibilities of coastal states in protecting areas under their jurisdiction. A number of articles speak to the rights and duties of vessels carrying hazardous materials as well as to the coastal state's authority in responding to the added threats posed by such ships.[4] The generally accepted international rules referred to in the 1982 Convention are embodied in various international legal regimes developed under the auspices of the International Maritime Organization (IMO) and the United Nations Environment Program (UNEP).

In keeping with the principle of granting a coastal state more extensive authority within its territorial sea than within its exclusive economic zone (EEZ), the treaty affirms the coastal state's right to enact laws to prevent, reduce, and control marine pollution from vessels within its territorial waters, including vessels in innocent passage. Such rules, however, may not hamper innocent passage. To control vessel source pollution in the EEZ, the coastal state may enact rules implementing generally accepted international rules and standards established by the relevant international organization. Further, with the relevant organization's approval, the state may enact additional rules in clearly defined sea areas of special concern. In sum, the 1982 Convention was intended to give coastal states adequate authority to control and monitor the movement of hazardous substances without creating unnecessary impediments to freedom of navigation.

PREVENTION OF POLLUTION .

The SOLAS Convention

The first International Convention on Safety of Life at Sea[5] was adopted in 1914, after the tragic sinking of the passenger liner *Titanic* in 1912. In 1974, SOLAS was substantially revamped to bring it up to date with modern advances in the transportation industry. Although SOLAS was aimed primarily at preventing accidents that injure or take the lives of passengers and crew, it also serves to safeguard the marine environment. The 1974 SOLAS Convention[6] covers a wide range of aspects relating to safety at sea, such as survey and documentation; fire safety; construction, machinery, and electrical installations; lifesaving appliances; communications; and rules of ship operation.

Chapter VII of the SOLAS Convention specifically addresses safety aspects of the transportation of dangerous goods. Nations are required to enact rules on safe packaging and stowage of dangerous goods and to prohibit their carriage by vessels flying the nation's flag except when carried in accordance with the provisions of the convention. The SOLAS Convention covers bulk chemical tankers and carriers of liquified gas by mandating compliance with highly technical guidelines and standards established by the IMO.[7]

The International Maritime Dangerous Goods Code[8] provides a uniform international code of guidance for implementation of regulations under SOLAS dealing with safety standards for dangerous goods. It addresses nine classes of dangerous goods: explosives, gases, flammable liquids, flammable solids, oxidizing substances and organic peroxides, poisonous and infectious substances, radioactive materials,[9] corrosives, and miscellaneous dangerous substances, including marine pollutants. Although no more than hortatory, the IMDG Code has been made mandatory through incorporation into the domestic law of many nations and through insurance companies' practice of scrutinizing vessels' compliance with international standards.

MARPOL 73/78

The primary agreement governing routine pollutant discharges associated with normal ship operations is the 1973 International Convention for the Prevention of Pollution from Ships, as modified by the 1978 protocol relating thereto.[10] In its five annexes,[11] rules concerning different categories of pollutants are set forth:

1. Annex I, Regulations for the Prevention of Pollution by Oil;
2. Annex II, Regulations for the Control of Pollution by Noxious Liquid Substances in Bulk;
3. Annex III, Regulations for the Prevention of Pollution by Harmful Substances Carried by Sea in Packaged Forms, or in Freight Containers, Portable Tanks or Road and Rail Wagons;
4. Annex IV, Regulations for the Prevention of Pollution by Sewage from Ships; and
5. Annex V, Regulations for the Prevention of Pollution by Garbage from Ships.

Under annex II, which entered into force on April 6, 1987, noxious liquids are placed within one of four categories according to their potential for harm. This categorization is based on a method devised by the joint Group of Experts on the Scientific Aspects of Marine Pollution (GESAMP).[12] Taking into consideration the category of substance, parties are required to control noxious liquids carried in bulk by (1) restricting the routine discharge of chemical residues from cargo tanks, line washings, ballast water, and other residues; (2) setting requirements for allowed discharges; and (3) mandating the use of reception facilities in ports for certain categories of substances. Discharges are more limited in the Baltic Sea and the Black Sea, the two special areas recognized under the annex. Although annex II focuses primarily on operational discharge procedures, it also addresses accidental pollution by incorporating by reference the ship construction and equipment guidelines provided in the applicable codes.

Also pertaining to hazardous materials, annex III of MARPOL 73/78 sets forth rules regarding the marking, labeling, packaging, stowage, quantity limitations, exceptions and notifications, and documentation of cargoes containing potential pollutants. Identification of marine pollutants under annex III is based on the same criteria or "hazard profiles" developed by GESAMP for categorization of bulk hazardous materials under annex II. The United States ratified annex III on July 1, 1991, providing the remaining percentage of the world merchant fleet necessary to trigger its entry into force.

International Convention on Standards for Training, Certification, and Watchkeeping

According to U.S. government estimates, human error is a contributing factor in 80 percent of ship accidents. The 1978 International Convention on Standards for Training, Certification, and Watchkeeping[13] provides international training and qualification standards for mariners so that the number of accidents resulting from human factors can be reduced. It provides guidelines regarding qualifications for different crew positions and requirements for masters' and watch officers' training, qualifications, and ratings for service in the chemical, oil, and liquified gas trade. Mariners are required to carry official certificates indicating their qualifications for inspection by port authorities. Provisions for minimal navigational and engineering watchkeeping are also included in the STCW. The convention provides that noncompliance with its terms can result in detainment of the ship. Controversy over the convention's applicability to vessels servicing offshore platforms prevented initial United States ratification. That obstacle was removed recently, and on July 1, 1991, the United States declared its intent to ratify the convention.

RESPONSE TO MARITIME ACCIDENTS

The Intervention Convention

The 1969 International Convention Relating to Intervention on the High Seas in Cases of Oil Pollution Casualties[14] establishes the right of coastal states to take actions on the high seas to prevent or mitigate oil pollution damages from ships that pose a serious threat to their coastlines and resources. A 1973 protocol to the Intervention Convention[15] extends the scope of the agreement to actions taken in response to casualties involving substances other than oil. The substances covered by the Intervention Convention either are listed specifically in the annex or have characteristics similar to those listed. Measures taken by the coastal state must be proportionate to the damage or threat. Actions exceeding those contemplated by the Intervention Convention open the coastal state to liability for compensation for any damages associated with excessive measures.

The Salvage Convention

The 1989 International Convention on Salvage[16] governs the rights and duties of those providing assistance to vessels in danger (i.e., salvors). Previous practice had focused primarily on preservation of human life and property, ignoring environmental considerations. The Salvage Convention remedies some of the inadequacies of previous salvage practices by providing salvors with strong financial incentives to protect the marine environment, in part by making the owner of the vessel or cargo responsible for reimbursing salvors for efforts to minimize environmental damage. The Salvage Convention attempts to provide greater economic stability within the salvage industry by recognizing the right to a reward for all productive salvage operations. The convention encourages efforts to protect the marine environment by calling for special compensation to salvors who fail to earn a reward for saving vessels or cargoes but who succeed in preventing or minimizing damage to the sea. Salvors who fail through negligence to safeguard the marine environment may be deprived of the special compensation, in whole or in part.

The Salvage Convention recognizes the general authority of coastal states to direct salvage operations but takes note of the international interest in providing vessels in distress with access to safe harbors. Thus, in deciding whether to allow vessels in distress admittance to their ports, coastal states must consider the interests of other parties and the need for cooperation among salvors to ensure safety of life and property and prevention of damage to the marine environment.

Oil Pollution Preparedness and Response Convention

The International Convention on Oil Pollution Preparedness and Response (OPPR)[17] was adopted at an IMO conference in November 1990. The OPPR Convention calls for international assistance and cooperation only in relation to spills of oil, but an associated resolution[18] calls for its expansion to include hazardous and noxious substances. The convention covers information exchange regarding nations' incident response capabilities, preparation of oil spill response plans, reporting of significant incidents or incidents affecting coastal states, and research and development into ways of combating pollution in the marine environment. The OPPR Convention further requires vessels and offshore installations to have oil spill response plans and to report any incidents involving discharges into the marine environment. Parties to the convention are required to prepare national contingency plans, and oil spill equipment and personnel are to be ready on a national, bilateral, or multilateral level for emergency purposes.

Draft Convention on Liability and Compensation in Connection with the Carriage of Noxious and Hazardous Substances by Sea (HNS Convention)

Article 235 of the 1982 Convention calls on nations to cooperate in the development of international laws governing liability and compensation for damages caused by pollution of the marine environment. In 1984, a draft convention on the subject was considered by a diplomatic conference, but it did not gain adequate acceptance for adoption. Work on the HNS Convention resumed within the framework of the IMO's Legal Committee in late 1987. The most recent draft of the HNS Convention addresses the following elements of a liability and compensation regime for incidents involving hazardous materials: a new, two-tiered strict and joint liability system for the shipowner and the shipper; limitations on liability; requirements for compulsory insurance; and definitions of substances covered by the convention.

To advance discussion of this complex subject, the Legal Committee decided that several delegations should act as "lead countries" in developing alternative approaches, without committing their nations to any specific approach. Toward this end, the Netherlands, Norway, and the United Kingdom submitted drafts in 1989. In 1990, the United States prepared a submission that outlined the current status of deliberations, including the general terms of the emerging consensus and the major issues still to be resolved.

The U.S. summary of progress to date[19] made a number of points about the status of negotiations on hazardous and noxious substances (HNS). It noted that the debate over the fundamental design of a liability regime for HNS produced a number of variations on a one- or two-tiered system. There is consensus that an acceptable regime must impose strict liability, up to a specified limit, on the owner of a vessel involved in an HNS incident (the first tier). The U.S. summary further observed that a two-tiered system could be modeled on the International Oil Pollution Compensation Fund (IOPCF) but should also incorporate elements tailored for the carriage of HNS. Differences between the carriage of oil and the carriage of HNS cargoes could cause difficulties in establishing and operating the second tier of the compensation fund for HNS. The HNS Convention presents far more complex challenges than IOPCF, in that (1) HNS cargoes are made up of assorted materials, often transported in small quantities, rather than a homogeneous and usually large bulk cargo; (2) the chemical industry is strongly diversified rather than concentrated, as is the oil industry; and (3) a substantial portion of HNS materials are currently transported on small, harder-to-regulate "tramp" ships rather than on the larger liner trade. The hazards presented by unladen or partially laden vessels carrying flammable residues give rise to further complexities.

According to the summary, other unresolved questions include (1) how to determine whether a shipment poses a sufficient threat to trigger the HNS requirements and (2) whether ship and cargo owners will be required to obtain insurance (or contribute to a fund) for the risks associated with the specific amount of a substance shipped on a particular voyage or for a specific period of time for all voyages involving that substance.

Consensus is emerging that the regime's scope of application should be very broad with respect to hazardous substances, the nature and cause of harm, and geographic coverage. There should also be extensive cross-referencing to existing IMO codes and other relevant conventions, such as MARPOL 73/78 and the Convention on the Control of Transboundary Movements of Hazardous Wastes and their Disposal.[20] Despite the progress made, the recent U.S. rejection of the two-tiered liability and compensation regime set forth in the 1984 protocols to the 1969 International Convention on Civil Liability for Oil Pollution. Damage (CLC) and the 1971 International Convention on the Establishment of an International Fund for Compensation for Oil Pollution Damage[21] remains fresh in the minds of IMO delegates. Among the reasons why these protocols were rejected by the Senate in 1990 was the view that the levels of liability and compensation remained too low. Some participants have suggested that the outlook for agreement on the second tier of the HNS regime is not hopeful.

The Basel Convention

The Basel Convention regulates the transboundary trade and disposal of toxic wastes. Transboundary movements of hazardous wastes are allowed only with prior informed consent from importing and transit countries. Exports to and imports from nonparties to the convention are banned in the absence of bilateral agreements at least as stringent as the Basel Convention. Exports are to take place only if the exporting state does not have the capacity to dispose of the wastes in an environmentally sound manner or if the wastes are to be recycled or recovered by the importing state. The convention designates forty classes of wastes that must be controlled but does not list individual substances. Other goals articulated in the convention are waste minimization, recycling, and source country responsibility. By curtailing their access to cheap or unregulated disposal sites in developing nations, the convention will encourage waste producers to reduce transboundary movement and disposal of wastes.

Article 12 of the Basel Convention calls for the adoption of a protocol on liability and compensation for damages resulting from the transboundary movement and disposal of wastes. In addition, the convention provides that the obligation to manage wastes in an environmentally sound manner may not be transferred from the source nation to the importing or transit nations under a liability and compensation regime. Implementation of the Basel Convention will require the harmonization of requirements under several IMO conventions regulating the transport of hazardous and noxious materials, including annexes II and III of MARPOL 73/78 and the 1972 Convention on the Prevention of Marine Pollution by Dumping of Wastes and Other Matter.[22] Contracting parties to the LDC have begun to consider whether transboundary movements for the purpose of dumping at sea fall within the purview of the Basel Convention.

Some environmental activists have criticized the Basel Convention because it does not specify standards for environmentally sound disposal of wastes in the receiving country. Although the treaty requires exporters to ensure the adequacy of disposal sites, there is no mechanism to provide for the enforcement of this requirement. Critics also point to the lack of clear definitions of wastes, which can result in mislabeling of wastes as nonhazardous. They question whether it is advisable to place the burden on the importing nation to discover misrepresentation in labeling. They also point to the potential for fraudulent use of the convention's recycling provisions to permit toxic waste dumping.

Several developing countries have called for the development of regional instruments that would provide more stringent regulations of hazardous waste movements than those authorized in the Basel Convention. In December 1989, as part of the Lomé IV Convention[23] on assistance and trade matters, the

European Community (EC) and sixty-eight African, Caribbean, and Pacific (ACP) nations banned the transboundary movement of wastes between EC and ACP signatories.[24] The parties to the 1983 Convention for the Protection and Development of the Marine Environment of the Wider Caribbean Region[25] have called for a regional agreement regarding hazardous waste movement.

EMERGING LEGAL AND POLITICAL ISSUES

Legal Framework of Rights and Duties

Recently, the traditional balance of rights and responsibilities set forth in the 1982 Convention has been challenged. In developing the Basel Convention, diplomats called for recognition of coastal state pollution control authority exceeding that authorized in the 1982 Convention.

Although language was included preserving the 1982 Convention balance, some nations continue to indicate that they do not consider themselves to be limited by traditional notions of international law. In 1988, for example, Haiti announced that it would prohibit entry into its ports, territorial seas, and exclusive economic zone to vessels transporting wastes, refuse, residues, or any other materials likely to endanger human health or pollute the marine environment. The U.S. government has objected to Haiti's assertion of such authority as violative of international law. Thus far, Haiti has not taken any action to enforce this prohibition.

Actions viewed as infringing on the navigational rights of either commercial or military vessels have long been considered a grave threat to national security by defense strategists within the major maritime nations, including the United States. The "assertion of rights" program conducted by the U.S. Navy illustrates the importance the major maritime nations accord to these rights. Under this program, U.S. warships exercise their navigational rights in order to contest foreign jurisdictional claims viewed as infringements on international laws guaranteeing freedom of navigation. It was an exercise under this program that led to an intentional bumping incident between Soviet and U.S. warships in the Black Sea in 1988.

The U.S. reaction to efforts in the South Pacific and Indian Ocean regions to limit the presence of nuclear weapons has also illustrated U.S. sensitivities regarding navigational freedoms. The 1985 South Pacific Nuclear Free Zone Treaty[26] confirmed the right of nations to bar access to their ports to vessels carrying nuclear weapons or powered by nuclear reactors, but it explicitly guaranteed the navigational freedoms of all vessels—including nuclear-armed and nuclear-powered ones. The United States nevertheless refused to sign the treaty on the grounds that it opposed actions that might lead to limits on the navigational rights of nuclear-powered and nuclear-armed warships.

Definitional Problems

Among the problems facing the international community as it seeks agreements to regulate the handling and movement of hazardous materials are the recurring difficulties associated with the definition of hazardous materials. These difficulties surfaced during the protracted negotiations undertaken in the wider Caribbean region regarding extending the Cartagena Convention's emergency response protocol to cover hazardous materials. Since 1987, the signatory Caribbean nations have been unable to agree on how hazardous substances are to be defined under the amended protocol. The United States contends that because the Caribbean protocol addresses only emergencies, hazardous substances should be defined very broadly. Thus, government officials faced with an emergency would not be forced to refer to the voluminous lists of individual noxious substances typically annexed to regulatory regimes before requesting aid. Other Caribbean nations insist that the relevant materials must be specified in detail before they will agree to such an extension.

The definitional problems faced in the context of the Caribbean protocol have also appeared in other international negotiations. In 1989, the U.S. government proposed the establishment of a global emergency preparedness and response regime (see the earlier discussion of the OPPR Convention). U.S. officials stated that they supported application of such a global regime to hazardous spill emergencies but anticipated protracted negotiations if hazardous substances were to be covered. To avoid delay in establishing global emergency response mechanisms for oil spills, the United States advocated the immediate ratification of the OPPR Convention and then put forward a resolution calling for future extension of the OPPR to substances other than oil.

Lack of Adequate Response Technologies or Expertise

The *Exxon Valdez* spill indicated in unmistakable terms the inadequacy of efforts to prepare for and respond to oil spills. Immediately after the spill, the weather in Prince William Sound presented ideal conditions for spill containment; nonetheless, the system in place was not able to deliver experts and equipment to the scene quickly enough to avert widespread damage. U.S. government officials have worried aloud that the system for responding to spills of hazardous materials is considerably less developed and less effective than that in place for oil spills. There is little doubt that response capabilities in developing countries are far less sophisticated.

The IMO recognizes and addresses the particular needs of developing countries through technical assistance programs. In general, efforts are being made to discover why many developing countries have not become parties to the various pollution prevention treaties and to identify ways to encourage them to do so. The IMO has also cosponsored regional workshops on emergency response strategies in collaboration with the UNEP regional forums. Unfortunately, the lion's share of attention has been aimed at oil spill response, leaving fewer resources to apply to preparedness for spills of hazardous materials.

Progress in the arduous process of developing regulatory regimes for hazardous materials provides some grounds for optimism. The IMO continues to refine and update the highly technical lists annexed to the LDC, the Intervention Convention, and the MARPOL-related codes in order to keep pace with sophisticated pollution control technologies and the development of new hazardous substances.

Liability and Compensation for Damages

The U.S. Congress's recent rejection of the improved international system for oil spill liability and compensation[27] reflects a growing public awareness of the enormous cleanup costs of very large spills. U.S. policymakers considered the limited financial responsibility established through international negotiations to be unconscionably advantageous to a polluting industry. Holding the polluter liable for significantly higher damages became politically necessary following the revelation that the captain of the *Exxon Valdez* had a history of drinking problems. To that end, the Oil Pollution Act of 1990[28] establishes much higher liability limits than those accepted by the international community. Further, the act provides that these liability limits do not apply in cases in which a spill is caused by gross negligence, willful misconduct, or violation of an applicable federal safety, construction, or operating regulation by the responsible party or its agents, employers, or contractors.[29] The liability limits are also inapplicable if the responsible party fails to report the incident, cooperate in removal activities, or comply with certain remediation orders.[30] In raising the liability limits in the Oil Pollution Act, Congress obviously discounted the self-serving and speculative warnings of shipping specialists regarding unlimited liability. Since passage of this legislation, the consequences of the U.S. rejection of an international system of limited liability and supplemental compensation have begun to become apparent. A sizable number of shipping companies, including Royal Dutch, Elf Aquitan, British Petroleum, Petrofina, A. P. Moller, Teekay, Bouchard Transportation, Maersk, and Texaco, have stated that they will no longer transport oil to the United States or certain U.S. ports because the U.S. liability regime creates too great a financial risk. Since a majority of ships in the international trade must serve U.S. ports, the liability system for oil spills has placed intense stress on worldwide insurance capacity. The industry has begun to discuss the possibility of creating separate pools of ships—those that are adequately insured for the U.S. trade and those that are not. The industry would have to schedule these pools separately, causing increases in the costs of shipping.

New Politics of the Waste Trade

The Basel Convention has been subject to harsh criticism by developing countries that sought an outright ban on all exports of hazardous wastes from developed countries. Greenpeace and other environmental groups have also advocated the imposition of such a ban. In part, they pressed for an international ban on waste movements in an attempt to achieve at the international level what they had been unable to persuade individual governments to do. It is clearly within the authority of nations to close their borders to hazardous waste shipments, but large fees for accepting wastes and internal pressures from domestic industries have often blocked efforts to enact domestic policies against hazardous waste imports. The United States holds the view that a total ban on hazardous waste movements would be harmful because not all countries are able to dispose safely of the hazardous wastes they create, thus requiring disposal options in other countries.

Disappointed that they were unable to achieve their aims at the Basel Convention, advocates of the ban have shifted their attention to UNEP's regional forums. Caribbean and Southeast Pacific nations' discussions of the issue illustrate this trend. During the past four years, a number of nations in the Caribbean have called regularly for a regional agreement on transboundary movement of hazardous materials. Other members of the forum have insisted that the region defer to negotiations under way at the global level. In the year following the adoption of the Basel Convention, the Caribbean forum again considered the development of a regional agreement on hazardous waste movements. Further discussion of such negotiations will take place after the parties have had the opportunity to consider a background paper on appropriate regional arrangements that is to be prepared by the UNEP office charged with administering the Basel Convention.

According to U.S. government officials, shifts in policy may be taking place. Even within the ranks of developing countries that expressed disappointment with the Basel Convention's failure to produce a ban on hazardous waste shipments, they have observed a new recognition of the need to continue shipments of wastes to countries with advanced disposal capabilities. At the same time, the United States may be moving toward a policy of banning hazardous waste shipments from U.S. manufacturers to developing countries.

CONCLUSION: EVOLUTION OR DISINTEGRATION OF THE INTERNATIONAL LEGAL ORDER?

The risks posed by seaborne transportation of hazardous materials have never garnered the level of public attention paid to casualties involving oil. If a large-scale accident involving hazardous materials were to occur, however, the maritime industry would be likely to face a tough dilemma: whether to improve environmental protection by accepting new and potentially costly limits on navigation rights or to oppose such politically popular causes. The 1982 Convention provides the most recent international articulation of the legal order governing uses and protection of the oceans. As mentioned earlier, it recognizes traditional navigational freedoms qualified by responsibilities to protect the marine environment. The balance between the world's interests in the free flow of commerce and in the control of pollution is a complex and delicate one. For vessels that pose an extraordinary danger, the balance between the rights of the vessels and the rights of coastal states is even more precarious.

In the absence of a convention establishing a universally accepted and binding maritime law, the practices of individual nations will define international law in the coming years. These practices either will solidify further the current order of the oceans as reflected in the 1982 Convention or will lead to new international principles. In the latter case, an initial lack of consensus on the rights and responsibilities of coastal states and maritime nations would ensue. Lack of consensus on fundamental international legal issues such as navigational freedoms would increase tensions and result in diplomatic protests at best; at worst, it could lead to gunboat diplomacy. Eventually, to secure the interests of all nations, consensus on the fundamental principles of ocean use must emerge. Currently, widespread consensus on and compliance with the principles of law governing ocean uses and protection of the marine environment minimize the likelihood of a resort to aggressive behavior.

NOTES

1. Telephone conversation with Karsten Rodvik, spokesperson for Exxon Alaska Operations (July 24, 1991).
2. C. HENRY, THE CARRIAGE OF DANGEROUS GOODS BY SEA 92 (1985).
3. United Nations Convention on the Law of the Sea, Dec. 10, 1982, U.N. Doc. A/CONF.62/122, 21 I.L.M. 1261 (1982) [hereinafter 1982 Convention].
4. See id. arts. 22(2), 42(1)(b), 194, 195, 211.

5. INTERNATIONAL CONVENTION ON SAFETY OF LIFE AT SEA, LONDON 1913-14, S. Doc. No. 463, 63d Cong., 2d Sess. (1914).

6. International Convention for the Safety of Life at Sea (SOLAS), Nov. 1, 1974, 32 U.S.T. 47, T.I.A.S. No. 9700, 14 I.L.M. 959 (1975).

7. *See* International Code for the Construction and Equipment of Ships Carrying Dangerous Chemicals in Bulk (IBC Code), adopted on June 17, 1983, by the Maritime Safety Committee (MSC), Resolution M.S.C. 4 (48); Code for the Construction and Equipment of Ships Carrying Dangerous Chemicals in Bulk (BCH Code), Resolution M.S.C. 9 (53); and International Code for the Construction and Equipment of Ships Carrying Liquified Gases in Bulk (IGC Code), Resolution M.S.C. 5 (48). The IBC Code applies to chemical tankers constructed after July 1, 1986. The BCH Code applies to tankers constructed before that date. Requirements in the IBC and BCH codes cover design, construction, equipment, and operation of tankers. The IGC Code is also mandatory under SOLAS.

8. I.M.C.O. Resolution A.81 (IV), Approval of the International Maritime Dangerous Goods Code, Sept. 27, 1965.

9. Provisions for radioactive materials are based on recommendations issued by the International Atomic Energy Agency (IAEA).

10. International Convention for the Prevention of Pollution from Ships, I.M.C.O. Doc. MP/CONF/WP.21/Add.4 (1973), 12 I.L.M. 1319 (1973), *as modified by* Protocol of 1978 Relating to the International Convention for the Prevention of Pollution from Ships, 1973, June 1, 1978, I.M.C.O. Doc. TSPP/CONF/11 (1973), 17 I.L.M. 546 (1978) [hereinafter MARPOL 73/78].

11. All parties to MARPOL 73/78 are required to become parties to annexes I through III. Annexes IV and V are optional.

12. GESAMP is an advisory scientific body of experts sponsored by the United Nations, the International Maritime Organization (IMO), the Food and Agriculture Organization (FAO), the United Nations Educational, Scientific, and Cultural Organization (UNESCO), the World Meteorological Organization (WMO), IAEA, UNEP, and the Intergovernmental Oceanographic Commission (IOC).

13. International Convention on Standards of Training, Certification, and Watchkeeping for Seafarers (STCW), July 7, 1978.

14. International Convention Relating to Intervention on the High Seas in Cases of Oil Pollution Casualties, Nov. 29, 1969, 26 U.S.T. 765, T.I.A.S. No. 8068, 9 I.L.M. 25 (1970).

15. Protocol Relating to Intervention on the High Seas in Cases of Pollution by Substances Other Than Oil, Nov. 2, 1973, T.I.A.S. No. 10561, 13 I.L.M. 605 (1973).

16. International Convention on Salvage, U.N. Doc. LEG/CONF/7/27, Apr. 28, 1989.

17. International Convention on Oil Pollution Preparedness, Response and Co-operation (OPPR), Nov. 30, 1990, 30 I.L.M. 733 (1991).

18. Conference Resolution 10, 30 I.L.M. 760 (1991).

19. I.M.O. Doc. LEG/63/3/3, Aug. 17, 1990.

20. The Basel Convention on the Control of Transboundary Movements of Hazardous Wastes and Their Disposal, UNEP Doc. W.G. 190/4, Mar. 22, 1989, 28 I.L.M. 649 (1989).

21. Protocol to the International Convention on Civil Liability for Oil Pollution Damage, May 25, 1984, 23 I.L.M. 177 (1984); Protocol to Amend the Convention on the Establishment of an International Fund for Compensation for Oil Pollution Damage, May 25, 1984, 23 I.L.M. 195 (1984).

22. Convention on the Prevention of Marine Pollution by Dumping of Wastes and Other Matter, Dec. 29, 1972, 26 U.S.T. 2403, T.I.A.S. No. 8165, 1064 U.N.T.S. 120.

23. Fourth African, Caribbean and Pacific States–European Community Convention of Lomé, Dec. 15, 1989, 29 I.L.M. 783 (1990).

24. *Id.* art. 39.

25. Convention for the Protection and Development of the Marine Environment of the Wider Caribbean Region and Protocol, Mar. 24, 1983, 22 I.L.M. 227 (1983).

26. South Pacific Nuclear Free Zone Treaty, Aug. 6, 1985, 24 I.L.M. 1440 (1985).

27. *See supra* note 21 and accompanying text.

28. The Oil Pollution Act of 1990, Pub. L. No. 101-380, § 1004 (1990).

29. *Id.* § 1004(c)(1).

30. *Id.* § 1004(c)(2).

16 Protected Marine Areas and Low-lying Atolls*

Jon M. Van Dyke

BECAUSE OF THE SECURITY and economic needs of the developed world, the isolated low-lying atolls of the Pacific have become inviting targets for hazardous activities and disposal of toxic products that would not be tolerated elsewhere. The Republic of the Marshall Islands, for example, recently proposed a feasibility study on the storage and disposal of U.S. nuclear wastes on the uninhabited atolls of the Marshalls.[1] Possible sites included Erikub Atoll, 175 miles east of Kwajalein Atoll (where the U.S. government tests missiles), and Bikini and Enewetak atolls.[2] In another example, studies were undertaken in mid-1989 to assess the feasibility of bringing millions of tons of garbage from the West Coast of the United States to the Marshalls for landfill and to build causeways connecting islands.[3]

These missions are hazardous to the marine environment and are incompatible with protection of the unique ecosystems found at these atolls. A low-lying atoll cannot be distinguished from its surrounding marine environment and must be thought of as an inherent part of the ocean ecosystem. Particularly when dealing with long-lived radioactive nuclides or highly toxic chemical weapons, it is unrealistic to imagine that these materials could be separated from the marine ecosystem during the entire period that they present dangers.

These testing and disposal activities are inconsistent with the obligation of all nations to protect the marine environment, a responsibility that has been recog-

* I would like to thank Dale Bennett and Brien Nicholas, both graduates of the William S. Richardson School of Law, University of Hawaii, for their assistance in the preparation of this paper; Greenpeace-USA for supporting some of the early research on this topic; and Jack H. Archer, Graham Baines, Norm Buske, Biliana Cicin-Sain, and Robert Knecht for their constructive comments on earlier drafts. A longer version of this chapter appeared in 16 J. OCEAN & SHORELINE MGMT. 87 (1991).

nized repeatedly in recent treaties. The 1986 Convention for the Protection of the Natural Resources and Environment of the South Pacific Region[4] (frequently called the SPREP Convention because it was negotiated through the South Pacific Regional Environment Programme) requires all parties "to prevent, reduce and control pollution of the Convention Area, from any source"[5] and to establish "protected areas" in order "to protect and preserve rare or fragile ecosystems and depleted, threatened or endangered flora and fauna as well as their habitat."[6] Similarly, the 1982 United Nations Convention on the Law of the Sea requires parties to "protect and preserve the marine environment"[7] and "to protect and preserve rare or fragile ecosystems."[8] Because of their structure and vulnerability, all low-lying atolls should be considered "rare or fragile ecosystems." Substantial efforts are now being made at the global, regional, and national levels to identify suitable "specially protected areas" and to define the types of protection that should be provided for these areas. Criteria identified by these efforts reinforce the conclusion that an isolated low-lying atoll is, almost by definition, a rare or fragile ecosystem and should be designated as a specially protected area deserving of careful planning, management, and protection.

LOW-LYING ATOLLS: UNIQUE, FRAGILE ECOSYSTEMS THAT NEED PROTECTION

The creation of a coral atoll from its birth, beginning as magma escaping on the sea floor, to maturity as a fully developed reef, is an extremely long and complex process that occurs over millions of years.[9] Only when coral growth outpaces erosion does an atoll emerge. Factors such as mean annual temperature, winds, ocean currents, tides, and salinity of the water play important roles in the creation and shape of an individual coral atoll.[10] All low-lying atolls are susceptible to dramatic changes when major weather systems pass over them and are subject to constant erosion through the actions of wind and sea. For example, in August 1972 and again in August 1984, *all* of the hundreds of civilian and military personnel on Johnston Atoll were evacuated because of the severe winds and waves that battered the islands.[11] Similarly, unexpected storms smashed into Moruroa in 1980 and 1981, requiring all residents and employees to seek refuge on raised platforms while the winds and rains washed across the atoll.[12] Coral can grow only in highly saline waters supersaturated with calcium carbonate, which provides the ionized calcium vital to the coral-forming process. Fresh water may totally destroy the coral-forming process. For example, a reef on Stone Island, in Queensland, Australia, was killed to a depth of 10 feet below the mean tide level after a cyclone dumped more than 35 inches of rain during one week.[13]

Biological forces may also greatly affect these reef systems. Several predators of coral reefs, including fish and invertebrates, feed on the soft tissues of the reef builders or bore directly into the coral rock. The most well known is the crown-of-thorns starfish, whose growing numbers have recently destroyed vast areas of coral reefs throughout the Southwest Pacific.[14]

Because of their relative isolation, atolls are almost always home to unique flora and fauna. The United States, ironically, recognized this phenomenon as early as 1926, when it designated Johnston Atoll as a wildlife preserve. Nonetheless, the United States has permitted Johnston to be used for military missions that have seriously damaged the environment. The damage caused by these activities led the U.S. Fish and Wildlife Service to identify Johnston Atoll as an area requiring immediate cleanup.[15]

Low-lying atolls are fragile and unique environments requiring specific environmental and climatic conditions. They are at the mercy of complex ocean weather systems that periodically bring typhoons and tsunamis, carrying waves across the atolls that affect all facilities and disperse land-based materials into the ocean environment. Low-lying atolls are not, therefore, appropriate sites for hazardous activities.

LOW-LYING ATOLLS: AN INHERENT PART OF THE MARINE ENVIRONMENT

Recent treaties have established broad obligations to protect and preserve the "marine environment." Is the concept of the marine environment limited to the ocean waters alone, or does it also inevitably include the land areas that merge with the ocean? Both a textual analysis of the governing treaties and common sense lead to the conclusion that the term "marine environment" must include atolls and coastal areas with immediate contact with the ocean waters. In the 1986 Convention for the Protection of the Natural Resources and Environment of the South Pacific Region,[16] for instance, the words "marine environment" are not defined, but they are used in the definition of pollution in a way that clearly states that areas in addition to the ocean waters are to be included:

> "[P]ollution" means the introduction by man, directly or indirectly, of substances or energy *into the marine environment* (including estuaries) which results or is likely to result in such deleterious effects as harm to living resources and marine life, hazards to human health, hindrance to marine activities, including fishing and other legitimate uses of the sea, [and] impairment of quality for use of sea water and reduction of amenities. . . .[17]

By explicitly including estuaries in the marine environment designed to be protected from pollution, the drafters were stating that coastal areas as well as the open ocean are to be included in the marine environment.

Perhaps even more significant is the repeated use in the text of the terms "the marine and coastal environment." Consider, for instance, paragraphs 5, 8, and 9 of the preamble (stating that the parties are seeking to address threats to the "marine and coastal environment") and article 4(1) ("The Parties shall endeavor to conclude bilateral or multilateral agreements, including regional or sub-regional agreements, for the protection, development and management of the marine and coastal environment of the Convention Area"). All land areas in a low-lying lagoon generally meet the definition of "coastal environment" because such atolls have virtually no elevation above sea level.

Article 11, which specifically addresses the "Storage of Toxic and Hazardous Wastes," also applies to low-lying atolls. Article 11 reads as follows:

> The Parties shall take all appropriate measures to prevent, reduce and control pollution in the Convention Area resulting from the storage of toxic and hazardous wastes. In particular, the parties shall prohibit the storage of radioactive wastes or other radioactive matter in the Convention Area.

The second sentence explicitly prohibits storage of radioactive material "in the Convention Area," but the first sentence is broader and more general. It is designed to limit such storage *anywhere* that might cause "pollution" in the Convention Area, which is defined broadly in article 2(f) to include estuaries.[18] Taken as a whole, these provisions thus indicate that the drafters of this treaty had a broad vision of the marine environment and intended the treaty to apply to activities on land that affected the marine areas.[19]

Common sense also supports the result reached by this textual analysis. A low-lying atoll has virtually no land areas that can be used for farming or gathering or any resource production truly based on the land itself. People living on an atoll are almost totally dependent on the marine environment. Activities on the atoll's limited land areas that affect the surrounding ocean can deprive the residents of the food that they need to live.

As mentioned earlier, ocean waters can and do wash over atolls during storms and tsunamis. Because the fate of residents on atolls is so completely linked to their surrounding sea areas, the atolls themselves must be considered to be part of the marine environment.

GLOBAL AND REGIONAL EFFORTS TO ESTABLISH PROTECTED MARINE AREAS

The SPREP Convention

Article 14 of the 1986 SPREP Convention[20] requires the contracting parties to take "individually or jointly . . . all appropriate measures to protect and preserve rare or fragile ecosystems and depleted, threatened or endangered flora and fauna as well as their habitat in the Convention Area."[21] The mechanism for providing such protection is for the parties to "establish protected areas, such as parks and reserves, and prohibit or regulate any activity likely to have adverse effects on the species, ecosystems or biological processes that such areas are designed to protect."[22] This language is similar to language in the 1982 Convention[23] and other regional seas conventions,[24] and thus the meaning of this provision can be better understood by examining how parties to these other conventions are implementing them and what initiatives are being taken at the national and global levels.

The Caribbean Convention

The Conference of Plenipotentiaries Concerning Specially Protected Areas and Wildlife in the Wider Caribbean Region contains language[25] similar to article 14 of the SPREP Convention, and the Caribbean countries have been active in meeting their obligation to protect threatened ecosystems. The January 1990 Protocol Concerning Specially Protected Areas and Wildlife[26] (Caribbean Protocol), for example, initiated a process for identifying and protecting the important marine areas of the region. This protocol contains a number of ideas of interest to the South Pacific region and can aid in interpretation and implementation.

First, it is clear in the Caribbean Protocol that protected areas can include land as well as ocean areas. The Caribbean Convention had defined the "Wider Caribbean Region" to include ocean waters offshore to a distance of 200 nautical miles from the coast,[27] and article 1(c) of the protocol adds "waters on the landward side of the baseline"[28] and "*such related terrestrial areas* (including watersheds) as may be designated by the Party" creating the protected area.[29]

Second, article 8 of the protocol recognizes that "buffer zones" may need to be established in order to protect the areas of primary concern. Obviously, these areas may include land as well as water.[30]

Third, the Caribbean Protocol articulates broad purposes for establishing specially protected areas, suggesting that all unique and fragile ecosystems should be seriously considered for "protected" status:[31] Such areas shall be established in order to conserve, maintain, and restore, in particular:

1. Representative types of coastal and marine ecosystems of adequate size to ensure their long-term viability and to maintain biological and genetic diversity;
2. Habitats and their associated ecosystems critical to the survival and recovery of endangered, threatened or endemic species of flora or fauna;
3. The productivity of ecosystems and natural resources that provide economic or social benefits and upon which the welfare of local inhabitants is dependent; and
4. Areas of special biological, ecological, educational, scientific, historic, cultural, recreational, archeological, aesthetic, or economic values, including in particular areas whose ecological and biological processes are essential to functioning of the Wider Caribbean ecosystems.[32]

"Special Areas" under the International Convention for the Prevention of Pollution from Ships

Annexes I, II, and V of the International Convention for the Prevention of Pollution from Ships[33] encourage countries to identify "special areas" where the need to control pollution is much greater than elsewhere.[34] In these areas, discharge of oil into the sea is banned from all ships of 400 gross registered tons and above, facilities for the reception of oily wastes in ports must be provided, and solid wastes must also be brought to port. Thus far, the Mediterranean, Baltic, Black, and Red seas and the Persian, Oman, and Aden gulf areas have been designated as special areas for this purpose. The North Sea ministerial conference has recently initiated action to designate the North Sea as an annex V special area, and the United States has taken similar action regarding the Gulf of Mexico. In response to this U.S. initiative, Mexico and Cuba have argued that the special area designation should also include the Caribbean area.[35] Criteria now used to identify a special area include, among others, oceanographic and ecological conditions, the existence of rare or fragile ecosystems, and vessel traffic patterns.[36] It is worth noting that coral reefs are given as the first example of rare or fragile ecosystems.

"Particularly Sensitive Sea Areas" or "Ecologically Sensitive Areas"

The international community has also been developing criteria to identify a second category of marine areas needing protection. Some documents have used the term "particularly sensitive sea areas,"[37] and others are now using the term "ecologically sensitive areas."[38] The International Maritime Organization's (IMO's) Marine Environment Protection Committee has said that to meet this category an area should meet one or more of the following criteria: rare or fragile ecosystem, dependence, representativeness, diversity, and productivity.[39] The IMO's Group of Experts on Effects of Pollution (GEEP) sought to add three additional criteria proposed by the International Union for Conservation of Nature and Natural Resources (IUCN)—naturalness, integrity, and vulnerability—stressing that integrity was particularly important.[40] These criteria all point toward the protection of low-lying atolls, which are functional units (thus meeting the integrity requirement); which are natural (if they have not yet been altered through development); and which are certainly vulnerable.

"Areas to Be Avoided"

The IMO has already identified a number of "areas to be avoided" by large vessels carrying oil and other hazardous material.[41] The criteria used to designate these areas are fragility of the ecosystem, presence of endangered species, scientific interest, other measures for protection already taken, and potential for damage by ships.[42] Proposed areas are identified by the Maritime Safety Committee, and if the areas are formally adopted by the IMO at an assembly meeting, avoidance of them becomes obligatory for large vessels of member nations.

The IMO's identification of 50-mile radii around eight of the northwestern Hawaiian Islands is particularly significant because the decision was made primarily to protect many unique seabirds, the endangered monk seal and green sea turtle, and the insects endemic to these small islets.[43] Low-lying atolls in the South Pacific often provide habitat for similarly unique and endangered fauna.

CLASSIFICATION OF LOW-LYING ATOLLS

These precedents and efforts, as applied to the Pacific, should lead to an active program to designate protected areas and would support the proposal that *all* low-lying atolls should be designated as protected areas.[44] The goals of such programs are to protect unique areas and thus to protect the diversity of the world's resources. The marine and bird life of the Pacific atolls is unique, and the introduction of hazardous activities to these atolls has invariably been harmful to these living resources. If these areas are designated as protected zones, an effective surveillance system should be established to monitor environmental quality in these areas, and a regional program should be established to regulate and restrict dumping, incineration, and other activities involving hazardous materials there.[45]

The regime of "specially protected areas" is not necessarily a regime intended to keep the designated locations as inviolate and pristine areas, as are wilderness regions.[46] Rather, this regime provides a planning mechanism whereby competing resource uses can be weighed against the values recognized in designating the areas as a protected zone. In particular, of course, no hazardous activities and no dumping that would detrimentally affect it should be permitted in the protected area.[47]

Protecting Atolls from Hazardous Activities and Wastes

Many types of environmental burdens have been imposed on or proposed for Pacific Island atolls. Nuclear explosions—the most serious environmental insult humans have yet created—have taken place in, over, or under the atoll lagoons of Bikini, Enewetak, Moruroa, Fangataufa, Johnston, and Christmas atolls, and these atolls have all been contaminated.[48] A major superport was proposed for Palau that, if built, would have permanently altered the pristine reef environment that exists there.[49] Nerve gas is now stored on Johnston, and a chemical incineration plant has been built there to burn these and other wastes that are now stored elsewhere.[50] Agent Orange has also been stored on Johnston, and an area of this atoll remains contaminated with dioxin.[51] The Marshall Islands, and possibly other atoll communities, are considering allowing their uninhabited atolls to be used to store nuclear and other hazardous wastes.[52] Such high-risk activities are clearly inappropriate for these fragile and unique ecosystems.

A final example of the dangers that unplanned development can bring to a tropical reef environment is the proposal of the Japanese government to build a $241 million airport on small Ishigaki Island, 440 kilometers south of Okinawa.[53] Many observers predicted that this project would lead to the destruction within five years of the world's oldest and largest known colony of blue coral because pollutants entering the water would trip the balance of nature in favor of the coral-eating crown-of-thorns starfish, which has already destroyed more than 90 percent of the coral reefs in Okinawa.

After considerable local and international protests against this project, the Japanese government announced in April 1989 that it would establish a marine national park at the location of the blue coral colony and reduce the size of the planned airport and move it to a site 4 kilometers (2.5 miles) north of the reef. The matter may not be finally resolved, however, because the citizens' groups protesting the airport immediately denounced the new plan, saying that the new site is also rich in coral and is close enough to the prime site to have an impact on it.[54] Although this situation remains in the controversial category, it is significant that environmental planners were brought in after the citizen protests and that the Japanese government did alter a development program because of its impact on a fragile ecosystem. Such reliance on environmental planning is still somewhat novel for Japan, but perhaps it will become more common as the long-term worth and fragility of special areas and their species becomes understood.

This example also illustrates once again that coral reef areas have special value and vulnerability and that, at a minimum, atoll environments should not be altered without careful planning and a full understanding of the impact of any changes. These unique ecological systems demand careful protection from all potentially damaging activities.

CONCLUSION

Substantial progress has been made in recent years in articulating the standards that should govern the protection of the marine environment. The 1982 Convention and the regional seas conventions lay down sound principles designed to control pollution and preserve the living resources of the sea. These treaties establish the concept of specially protected zones, but they leave to the present generation the job of developing criteria to designate such areas.

Because all low-lying atolls are so intimately connected with and affected by the surrounding marine environment, it is not logical to treat the land areas of these atolls separately from the marine regions. The limited land areas of these atolls are the ultimate coastal zone because they have no vertical component. They should be seen as an inherent part of the marine environment, and all of the principles that govern the marine environment should also govern these low-lying atolls. This proposition is clearly supported by the text of recent conventions, particularly the 1986 SPREP Convention.

The logic of the SPREP Convention also supports the proposition that all low-lying atolls should be designated as "specially protected areas" pursuant to article 14. By their very nature, atolls have small or no populations and thus will always be underrepresented in human terms when development decisions are made. They thus need the "specially protected" designation to ensure that the particular environmental concerns of these fragile ecosystems are weighed against the interests that seek to use them for some purpose. This approach would not interfere with rational resource development but would promote careful planning and limit dumping and hazardous activities on or near these atolls to protect their unique and important environments.

NOTES

1. Howard Graves, *Islands Want Waste: Contamination Could Turn into an Asset*, Pac. Daily News (Guam), Dec. 25, 1987, at 3; this article also appeared in the Honolulu Star-Bull., Dec. 25, 1987, at A2, col. 2 (*President of Marshalls Wants Nuclear Waste Site*). See also Marshall Islands J., Jan. 1, 1988, at 7, col. 1; and Miller, *Marshalls*

Suggest Developing a Nuclear Dump Site on an Atoll, Honolulu Star-Bull., Feb. 20, 1989, at A-3, col. 5.

2. Erikub Atoll consists of fourteen islets surrounding a lagoon that has a circumference of 20 miles; all of the islets are uninhabited. THE COLUMBIA LIPPINCOTT GA-ZETTEER OF THE WORLD 584 (L. E. Seltzer ed. 1962). For a more complete description, *see* P. Thomas, F. Fosberg, L. Hamilton, D. Herbst, J. Juvik, J. Maragos, J. Naughton, & C. Streck, THE NORTHERN MARSHALL ISLANDS NATURAL DIVERSITY AND PROTECTED AREAS SURVEY 74-80 (SPREP & East-West Center, Honolulu 1989). None of the atolls in the Marshalls rises higher than 20 feet (6.5 meters) at any point. Section 5041 of the U.S. Nuclear Waste Policy Amendments Act of 1987 includes the Republic of the Marshall Islands as a location that can be considered for nuclear wastes. Pub. L. No. 100-203, title V, subtitle A, sec. 5041, 101 Stat. 1330-227, 1330-243 (codified at 42 U.S.C. § 10241 (1988)). Recommendations to protect the unique and pristine atolls in the Marshalls are made in Thomas et al., *supra*.

3. *U.S. Trash for Marshalls*, PAC. ISLANDS MONTHLY, June 1989, at 44; Giff Johnson, *Laying Waste in the Islands*, ISLANDS' BUS., July 1990, at 35.

4. Convention for the Protection of the Natural Resources and Environment of the South Pacific Region, Nov. 25, 1986, art. 5, para. 1, 26 I.L.M. 38, 45 (1987) [hereinafter SPREP Convention]. As of 1992, eleven nations, including the United States, had ratified this treaty.

5. *Id.* art. 5.

6. *Id.* art. 14.

7. United Nations Convention on the Law of the Sea, art. 192, Dec. 10, 1982, U.N. Doc A/CONF.62/122, 21 I.L.M. 1261, 1308 (1982) [hereinafter 1982 Convention]. Although the SPREP Convention, *supra* note 4, does not use the term "marine environment" in its main operative paragraph (art. 5(1)), it does use that term in other parts of the treaty (*see, e.g.*, arts. 2(f) and 5(2)).

8. *Id.* art. 194(5).

9. Scoot & Rotondo, *A Model to Explain the Differences Between Pacific Plate Island-Atoll Types*, 2 CORAL REEFS 140, 142-43 (1983).

10. *Id.*

11. *See* Jon M. Van Dyke, *Protected Marine Areas and Low-Lying Atolls*, 16 J. OCEAN & SHORELINE MGMT. 87, 121 (1991).

12. *Id.* at 113-14.

13. 25 ENCYCLOPAEDIA BRITANNICA 160 (15th ed. 1987).

14. *Id.*

15. Associated Press, *Wildlife Refuges Threatened*, Feb. 4, 1986. This announcement focused on the dioxin and plutonium contamination on Johnston as well as the nerve and mustard gases now on the atoll. Although Johnston Atoll is not within the Convention Area of the SPREP Convention, *supra* note 4, arts. 1 and 2(a), because the United States declined to include it in this treaty region, it is discussed in this chapter because of its proximity to the Convention Area and the similarity between its situation and that of other low-lying atolls.

16. SPREP Convention, *supra* note 4. The South Pacific nations have not yet negotiated a protocol on protected areas, but they have taken steps to develop a comprehensive approach to protecting their sensitive sea areas. See Holthus (SPREP project officer), Marine Protected Areas and Conservation of Marine Resources: The South Pacific Situation—Status and Strategies (Theme Paper No. 8, delivered to the Fourth South Pacific Conference on Nature Conservation and Protected Areas, South Pacific Commission, Port Vila, Vanuatu, Sept. 4-12, 1989), and PROGRESS WITH THE ACTION STRATEGY FOR PROTECTED AREAS AND THE SOUTH PACIFIC REGION (SPREP/4th South Pacific Conf. Nat. Cons./Information Paper No. 3, Aug. 25, 1989).

17. SPREP Convention, *supra* note 4, art. 2(f) (emphasis added).

18. *See supra* text at note 17 (quoting from art. 2(f)).

19. This conclusion is reinforced by an incident that occurred during the November 1985 negotiating session at the Fourth Meeting of Experts in Nouméa New Caledonia. Considerable time was devoted during this session to the phrasing of the title of the treaty, and at one point the question was asked whether this treaty should be the "Convention for the Protection of the Natural Resources and *Marine* Environment of the South Pacific Region" rather than "Convention for the Protection of the Natural Resources and Environment of the South Pacific Region." The delegates were virtually unanimously opposed to the idea of inserting "marine" before "environment," insisting that because the land areas of most Pacific islands are so small and are so intimately linked to the oceans, the people of this region have only one single, unified environment, not separable "land" and "marine" environments. Several delegates were quite eloquent on this issue, indicating a strong desire to draft a treaty that would have broad applicability to both land and ocean areas.

The proper scope of the South Pacific Environmental Convention later came up in a panel discussion at the April 1988 annual meeting of the American Society of International Law. Dr. Kilifoti Eteuati, who represented Western Samoa during the treaty negotiations and also chaired the final negotiating sessions, explicitly stated that storing nuclear wastes on the Marshall Islands' atolls would violate the treaty. *See* 82 PROC. AM. SOC'Y INT'L L. 370 (1988).

20. SPREP Convention, *supra* note 4.

21. *Id. See generally* ACTION STRATEGY FOR PROTECTED AREAS IN THE SOUTH PACIFIC REGION (prepared at Third South Pacific National Parks and Reserves Conference, Apia, Western Samoa, June 24-July 3, 1985) (South Pacific Commission, Nouméa 1985).

22. SPREP Convention, *supra* note 4, art. 14, second sentence.

23. 1982 Convention, *supra* note 7, art. 194(5); *see also* arts. 162(2), 211(6)(a).

24. *See, e.g.*, Convention for the Protection and Development of the Marine Environment of the Wider Caribbean Region, art. 10, Mar. 24, 1983, 22 I.L.M. 227 (1983) [hereinafter Caribbean Convention]; Protocol Concerning Protected Areas and Wild Fauna and Flora in the Eastern Africa Region, art. 8, June 21, 1985 [hereinafter East African Protocol].

25. Caribbean Convention, *supra* note 24, art. 10. The Caribbean region includes

thirty-five countries and territories and the European Economic Community. As of January 1990, sixteen countries had ratified the Caribbean Convention. Final Act of the Conference of Plenipotentiaries Concerning Specially Protected Areas and Wildlife in the Wider Caribbean Region, Kingston, Jamaica, Jan. 15-18, 1990, at 1, para. 3.

26. Protocol Concerning Specially Protected Areas and Wildlife to the Convention for the Protection and Development of the Marine Environment of the Wider Caribbean Region, Jan. 18, 1990, 21 INT'L ENVTL. REP. (BNA), Mar. 1990, at 3261 [hereinafter Caribbean Protocol].

Also recently completed is the Protocol for the Conservation and Management of Protected Marine and Coastal Areas of the South-East Pacific, at Paipa, Colombia, Sept. 21, 1989, 15 L. SEA BULL. 47 (UN Office for Ocean Affairs and the Law of the Sea, May 1990) [hereinafter South-East Pacific Protocol].

27. Caribbean Convention, *supra* note 24, art. 2(1).

28. Caribbean Protocol, *supra* note 26, art. 1(c)(i).

29. *Id.* art. 1(c)(ii) (emphasis added). *See also* article 2 of the Mediterranean Protocol on Specially Protected Areas, at Geneva, Apr. 1, 1982 [hereinafter Mediterranean Protocol]; article 1(a) of the East African Protocol, *supra* note 24; and article I of the South-East Pacific Protocol, *supra* note 26. The Mediterranean Protocol states that the protected area may "include wetlands or *coastal areas* designated by each of the parties" (emphasis added). Article I, paragraph 3, of the South-East Pacific Protocol states that "[t]he coastal zone, where interaction between land, sea and the atmosphere is ecologically apparent, shall be determined by each State Party, in accordance with the relevant scientific and technical criteria."

30. Article 11 of the East African Protocol, *supra* note 24; article 5 of the Mediterranean Protocol, *supra* note 29; and article VI of the South-East Pacific Protocol, *supra* note 26, also recognize "buffer zones."

31. A recent study by the Organization of American States (OAS) reported that ninety-three marine protected areas have been established through national legislation in areas of the Caribbean region other than the United States and its affiliated island entities. Of these, however, only ten are fully operational. A meeting was held in La Parguera, Puerto Rico, September 15-17, 1988, by the University of Puerto Rico Sea Grant Program and the Council on Ocean Law to discuss the social, economic, technical, and legal problems that have prevented these marine areas from receiving the protection they deserve.

Efforts in the Caribbean are continuing; in March 1989, a meeting was held in Dominica as part of the Caribbean Park and Protected Areas Network to identify proposals for protected areas. *Caribbean: Protected Areas Workshop*, The Siren, No. 41, July 1989, at 27 (UNEP). *See also Caribbean Says Yes to Future Strategy*, THE SIREN, No. 38, Oct. 1988, at 1 (UNEP).

32. Caribbean Protocol, *supra* note 26, art. 4(2). The East African Protocol, *supra* note 24, also contains a similar list of broad purposes. The language in the Mediterranean Protocol, *supra* note 29, art. 3, and in the South-East Pacific Protocol, *supra* note 26, art. II, is similar to both the East African Protocol and the Caribbean Protocol but is

more general in phrasing. *See generally* R. Salm & J. Clark, MARINE AND COASTAL PROTECTED AREAS: A GUIDE FOR PLANNERS AND MANAGERS (IUCN 1989).

33. Convention for the Prevention of Pollution from Ships, Nov. 2, 1973, T.I.A.S. No. 10,561, 12 I.L.M. 1319; Protocol to the Convention, with Annexes, Feb. 17, 1978, 17 I.L.M. 546 (1978) [hereinafter MARPOL 73/78].

34. This idea is also recognized in article 211(6) of the 1982 Convention, *supra* note 7.

35. *See Law of the Sea, Report of the Secretary General*, at 31, U.N. Doc. A/43/718 (1988); 32 OCEAN SCI. NEWS 9, Mar. 31, 1990, at 1-2.

36. *See* INTERNATIONAL MARITIME ORGANIZATION, MARINE ENVIRONMENT PROTECTION COMMITTEE, IDENTIFICATION OF PARTICULARLY SENSITIVE AREAS, INCLUDING DEVELOPMENT OF GUIDELINES FOR DESIGNATING SPECIAL AREAS UNDER ANNEXES I, II AND V: REPORT TO THE WORKING GROUP, 26th Sess., MEPC 26/WP.10, Annex I, at 2-3 (Sept. 8, 1987). These proposed criteria are still being reviewed. *See also* MEPC 29/14/1, containing EXTRACT FROM THE REPORT OF THE FIFTH SESSION OF THE GROUP OF EXPERTS ON EFFECTS OF POLLUTANTS (London, Apr. 17-20, 1989) [hereinafter 1989 GEEP REPORT].

37. *See, e.g.*, 1989 GEEP REPORT, *supra* note 36.

38. *See* INTERNATIONAL MARITIME ORGANIZATION, MARINE ENVIRONMENT PROTECTION COMMITTEE, IDENTIFICATION OF PARTICULARLY SENSITIVE AREAS, INCLUDING DEVELOPMENT OF GUIDELINES FOR DESIGNATING SPECIAL AREAS UNDER ANNEXES I, II AND V, MEPC 29/13 (Dec. 1989) (containing draft of MANUAL FOR THE PROTECTION OF ECOLOGICALLY SENSITIVE AREAS AGAINST DAMAGE FROM MARITIME ACTIVITIES, prepared by Gerard Peet of Friends of the Earth International [hereinafter cited as PEET DRAFT MANUAL]).

39. *See* MEPC 26/INF.20, in 1989 GEEP REPORT, *supra* note 36, at 3.

40. 1989 GEEP REPORT, *supra* note 36, at 4.

41. These areas are the Rochebonne Shelf (France), Cape Terpeniya (Sakhalin, USSR), Nantucket Shoals (United States), 50-mile radii around eight of the northwestern Hawaiian Islands (United States), the Great Barrier Reef (Australia), and the Bermuda Islands, and between Smalls Lighthouse and Grassholme Island (United Kingdom). *See* PEET DRAFT MANUAL, *supra* note 38, at 52-54.

42. *Id.* at 55, para. 5.4.3.2.

43. For a description of these islets, *see* Jon M. Van Dyke, Joseph Morgan, & Jonathan Gurish, *The Exclusive Economic Zone of the Northwestern Hawaiian Islands: When Do Uninhabited Islands Generate an EEZ?* 25 SAN DIEGO L. REV. 425, 465-82 (1988); Craig Harrison, *A Marine Sanctuary in the Northwestern Hawaiian Islands: An Idea Whose Time Has Come*, 25 NAT. RESOURCES J. 317 (1985).

44. The International Maritime Organization is now compiling a list of all national and regional efforts to designate protected ocean areas. *See* Letter of Manfred K. Nauke, Head of the Marine Science Section of the IMO Marine Environment Division, to G. E. Hemmen of the Scientific Committee on Antarctic Research, Dec. 8, 1986.

45. *See* John W. Kindt, 4 MARINE POLLUTION & L. SEA 1917 (1986).

46. *See* Diane Schenke, *The Marine Protection, Research, and Sanctuaries Act: The Conflict Between Marine Protection and Oil and Gas Development*, 18 HOUS. L. REV. 987, 1013

(1981) (discussing the U.S. Marine Sanctuaries Program). Congressman Thomas Pelly, talking of the U.S. Marine Sanctuaries Program, said: "A sanctuary is not meant to be a marine wilderness where man will not enter. Its designation will insure simply a balance between uses." Quoted in Kifer, *NOAA's Marine Sanctuary Program*, 2 COASTAL ZONE MGMT. J. 177, 178 (1977).

47. The regime for protected areas is described in article 7 of the Mediterranean Protocol, *supra* note 29; article 10 of the East African Protocol, *supra* note 24; and article 5 of the Caribbean Protocol, *supra* note 26.

48. *See* Van Dyke, *supra* note 11, at 104-27.

49. *Id.* at 127-33.

50. *Id.* at 122.

51. *Id.* at 121.

52. *See supra* notes 1-3 and accompanying text.

53. *See* Schoenberger, *Airport Row Splits People of Idyllic Isle*, L.A. Times, July 21, 1988, at 1, col. 1; Sanger, *Is Paradise Still a Paradise After It's Paved Over?*, N.Y. Times (nat'l ed.), Apr. 18, 1989, at A4, col. 1.

54. Agence France Presse, *Japan to Move Airport Site to Save Coral Reef*, Apr. 26, 1989, 11:50:41; *Environmentalists Save Japan Blue Coral Reef*, Honolulu Star-Bull., Apr. 27, 1989, at A-11, col. 1.

IV

The Living Resources

FOR AT LEAST A CENTURY, nations have recognized the need to use the living resources of the high seas in a manner that will maintain their bounty for future generations. Nonetheless, international fisheries policy has not yet produced "sustainable" results. Fisheries commissions still function as "user clubs" that are unable to halt the serious depletion of global fishery stocks. The United Nations Food and Agriculture Organization reports that in 1990 global fish harvests declined for the first time ever and that most traditional marine fish stocks were unlikely to yield increased harvests even with intensified fishing effort. Further, following the collapse or depletion of northern fisheries and subsequent relocation of fishing pressure, southern oceans are now being overfished. In her chapter, Claudia Carr describes the significant role fisheries aid has played in this process. Even more alarming are the possible long-term effects of overfishing: Contrary to traditional conservation assumptions, fishery resources may not be "renewable" because overfishing a stock could lead to its permanent displacement from the ecosystem. Compounding the lack of strict regulation by international management commissions, overcapitalized fishing fleets contribute significantly to declines in global fisheries.

Protecting the ocean commons requires international cooperation because of the sheer scale of the task and because it is unfair and unrealistic to expect one or two nations to do the work that benefits all. In the 1970s, some participants in the United Nations Law of the Sea Conference sought to curb these destructive patterns by ending the freedom to fish and eradicating the species approach to fisheries management. These goals were not realized: Nations would not relinquish their sovereignty or the access to resources they gained by joining high-value species conventions. The compromise management measure contained in the 1982 United Nations Convention on the Law of the Sea was the creation of the 200-mile exclusive economic zone (EEZ).

Because more than 90 percent of commercial fisheries are found within 200 miles of the coast, this extension of national jurisdiction from 12 to 200 miles "resolved" many high seas fishery problems by making them domestic problems. Aside from this display of "creeping jurisdiction," nations have made few changes to the way they manage high seas stocks: "Ownership" of resources (except anadromous stocks) on the high seas belongs to the captor; voluntary cooperation among states is the sole management tool; and virtually the same high seas management regimes established prior to the Law of the Sea Conference are still in place. Indeed, in the decade since the 1982 Convention was signed, cooperative management regimes for international fisheries have failed to overcome the central problems of monitoring and enforcement. Moreover, the EEZ "solution" has not set an encouraging precedent for future international cooperation over common resources.

Compounding the legal and political obstacles to sustainable management, fisheries science often lacks the data necessary to create reliable predictive models. Yet science, in concert with efforts of both government and nongovernmental groups, can positively influence international fisheries management. The very conservatism of the scientific community can galvanize rapid international action when it proclaims the existence of a threat. Recent action against the use of high seas drift nets illustrates the ability of regional groups and the international community to respond to a threat of serious resource depletion even without "proof" but because of widespread scientific alarm over preliminary by-catch mortality data. James Carr and Matthew Gianni in their chapter discuss the ravaging effects of high seas drift netting and actions taken to end the practice.

The United Nations General Assembly Resolution 44/225 moratorium on drift nets longer than 2.5 kilometers—first in the South Pacific, then in all high seas areas—broke new ground in other ways. As Catherine Floit discusses in her chapter on whether the practice is permitted under international law, the UN Driftnet Resolution reversed traditional burdens of proof, requiring user groups to take a precautionary approach and show that their actions would not harm ocean resources *before* permitting the practice to continue. The resolution recommended that moratoriums be implemented on June 30, 1992, unless nations take "*effective* conservation and management measures . . . based on statistically sound analysis . . . to prevent unacceptable impacts" of the practice (emphasis added). In December 1990, the General Assembly adopted Resolution 45/197, which reinforced international support for Resolution 44/225 and described actions taken during 1989-1990 to end drift netting.

Kazuo Sumi writes on the Japanese view regarding whether drift netting can be effectively managed, arguing that economic pressures to end drift netting amount to coercion and violate international law. Japan, Korea, and Taiwan, the major drift netting nations, had resisted meeting the June 30, 1992 deadline— although they did meet the 1991 deadline in the South Pacific—citing the need for further scientific study on whether "effective" conservation measures for drift nets could be implemented. But in late 1991 and in 1992, events began to develop quickly. On November 26, 1991, after Japan and the United States sponsored opposing drift net resolutions in the Second Committee of the United Nations, they agreed to a compromise resolution calling for nations to halve their drift netting effort by June 30, 1992, and to end drift netting entirely by December 31, 1992.

This compromise regime, UN General Assembly Resolution 46/215, was adopted by consensus on December 20, 1991. The Japanese cabinet voted to take this step in response to international pressure, particularly that from Pacific Island nations and the United States (including pending U.S. legislation to impose trade sanctions on countries that do not make a commitment to end the practice). Taiwan (not a UN member) and South Korea were not sponsors of the resolution. Neither has formally announced that it will comply with the deadlines, although the U.S. Department of State reports that these nations are preparing internal compliance plans. European drift netting in the North Atlantic is also ending: In October 1991, the European Community resolved to cease the practice, although it effectively permitted French drift netting until 1993. Although the UN drift net resolutions call for a stop to the spread of drift netting to other oceans, anecdotal evidence has raised strong suspicions that Taiwanese drift net vessels are still operating in the South Atlantic. Adoption of these resolutions does not mark an end to the issue; attention will merely shift to the need for strong monitoring and enforcement. Some proposals now before Congress suggest using U.S. Navy vessels and aircraft to supplement U.S. Coast Guard monitoring and enforcement.

On the high seas, very few enforcement measures are available. Because international trade drives high seas fisheries, the most direct way to implement conservation measures, aside from boarding and seizure of cargo and vessels, is to limit trade in these products. Melinda Chandler discusses the variety of ways trade measures are used to conserve natural resources. The Convention on International Trade in Endangered Species of Wild Fauna and Flora (CITES), for example, uses trade measures to protect endangered and threatened species. Because this approach has been more effective than those of international fisheries regimes, concerned groups attempted (unsuccessfully) in 1992 to list the Atlantic bluefin tuna as endangered under CITES because management measures taken by the International Commission for the Conservation of Atlantic Tunas had failed to arrest the stock's twenty-year decline. The use or threat of trade sanctions to enforce conservation measures, however, is in jeopardy because it may conflict with the General Agreement on Tariffs and Trade (GATT). A 1991 GATT panel report, for instance, rejected the use of mandatory U.S. embargoes against Mexican (and other nations') tuna to bring about reductions in dolphin mortality in the high seas tuna fishery. Still, the United States has not dropped its embargoes; other embargoed nations are threatening to move against the United States on their own GATT challenges; and GATT itself is assessing when, if ever, trade sanctions may be used for protection of global natural resources.

The time is ripe for action by both scientists and policymakers, who must accept that fisheries science is inexact and that conservation in the short term translates to increased benefits in the long term. The concept of the common heritage of humankind, originally proposed three decades ago, is well suited to the emerging view that what is good for all is good for each. Conditions in the commons affect all nations, and they are the responsibility of all. Perhaps in the near future nations will consider adopting proposals that incorporate these concepts, such as Christopher Stone's suggested ocean guardian system. Some new movement toward resolving issues regarding living marine resources is perceptible: At the March 1992 preparatory meetings for the United Nations Conference on Environment and Development (UNCED), Canada called for nations to strengthen fisheries management regimes. Nongovernmental organizations stressed that this goal would require mandatory dispute settlement procedures or perhaps a global fisheries authority. Precautionary approaches—such as in the UN Driftnet Resolution—should be incorporated more fully in fisheries agreements in order to help restore stocks and, more important, to prevent future "crashes."

William Burke begins this part with a definitive chapter on the conflicts and challenges of international fisheries management under the 1982 Convention. The Agenda 21 UNCED text on the protection of oceans, adopted on April 3, 1992, called for an intergovernmental conference to make recommendations to promote "effective implementation" of the 1982 Convention's provisions on straddling stocks and highly migratory species, and a meeting on the crisis in high seas fisheries took place at the United Nations during the summer of 1993. Although some nations have been active against specific threats, such as the use of destructive fishing methods and depletion or extinction of species, all nations must pursue more aggressive tactics if they wish to preserve the known—and unknown—benefits of responsible use of ocean resources. The authors of these chapters describe the legal, scientific, and policy considerations necessary to craft effective future solutions to the irresponsible use of living ocean resources.

17 Unregulated High Seas Fishing and Ocean Governance

William T. Burke

TYPES OF PROBLEMS ARISING FROM UNREGULATED FISHERIES

Two categories of problems are generated by high seas fishing beyond national jurisdiction. One is the set of conflicting claims to specific fisheries. The other is the need for a decision-making process that provides adequate governance in an area long considered subject only to the decisions of flag states. This chapter provides some observations on both categories.

Conflicting Claims to High Seas Fisheries

Freedom of Fishing. The most fundamental of these claims, which is the root of the difficulties in high seas fisheries, is the assertion that under contemporary international law the living resources of the high seas beyond national jurisdiction are open to free access to all who have the capital and initiative to seek them out. Traditionally, fishing vessels on the high seas are subject to no laws and regulations other than those prescribed or accepted by the flag state.

Straddling Stocks. One counterclaim to this generic claim of freedom to fish is the demand to prohibit, or at least reduce, high seas fishing because it takes stocks that are also found in the exclusive economic zone (EEZ), where they are subject to the sovereign rights of the coastal state.[1] This is the so-called straddling stock problem.

In the high seas area of the central Bering Sea, the dispute is over the catch of pollock by vessels of various flags because the catch allegedly reduces the abundance of pollock in the adjacent EEZs of the former Soviet Union and the United States and is inconsistent with conservation measures established on the same stock within these adjacent zones. In addition to concern over stock abundance within the EEZs, where fishing is controlled by the coastal states, it is now evident that the fishing fleets of the two coastal states want to fish for pollock in the high seas area and believe that the current foreign effort there reduces abundance in that area to the disadvantage of their own fishing fleets. U.S. vessels entered the "donut fishery" during the fall of 1990 because of the early suspension of the fishery within the U.S. EEZ in the Bering Sea.[2]

In the Northwest Atlantic, the catch of cod and other species on the Grand Banks beyond 200 miles has interfered with Canadian conservation and management measures for the same stocks within its fishing zone. Furthermore, in this area a regional fisheries body (the Northwest Atlantic Fisheries Organization) established with the purpose of regulating straddling stocks has had severe problems in gaining acceptance by member states of its high seas conservation measures.[3]

Similar claims are made regarding certain species of fish. Claims by Mexico to tuna in the eastern tropical Pacific linked the exercise of the coastal state's sovereign rights over tuna within 200 miles to an assertion of authority to establish total allowable catches (TACs) and entry fees for harvesting by foreign vessels in the area outside 200 miles.[4] The United States has long contested claims by coastal states to jurisdiction over tuna, and Mexico's claim was contested. Since then, the United States appears to have withdrawn from its historical position that tuna are not subject to coastal state authority within 200 miles; amendments to the Magnuson Fishery Conservation and Management Act (Magnuson Act) recently enacted[5] assert U.S. jurisdiction over tuna within the U.S. EEZ.[6] It does not follow, however, that either of Mexico's claims regarding tuna, inside or outside the 200-mile zone, has been or will be accepted.

Another, similar claim concerns anadromous fish, particularly salmon. Most states from which salmon originate believe that these fish should not be subject to harvesting on the high seas unless the state agrees to it. High seas exploitation of salmon is now uncommon, Japan being the only nation still engaged in this activity, and it is likely that such fishing will be terminated except as expressly agreed by states of origin.[7]

Prohibition of Catch: Marine Mammals. A second type of counterclaim to freedom of fishing is to prohibit the take of certain animals on the high seas because their harvest either is morally wrong, and therefore should be terminated, or endangers their continued existence because population abundance is already at a diminished level in relation to intensity of harvesting. Whales and small cetaceans are instances of controversy about whether any harvest is permissible or ought to be. The arguments about whales, which are protected under a moratorium by the International Whaling Commission (IWC), center on the abundance of particular species, which are arguably large enough to withstand exploitation. There is opposition to any take of whales, irrespective of population abundance.[8]

Limitation of Gear: Drift Nets, Purse Seines, Trawls, and Longlines. A third type of counterclaim centers on limitations on the use of certain gear, thusfar primarily drift nets and purse seines, which take large incidental catches of various species, especially marine mammals. This claim has been advanced in the North Pacific in the salmon, squid, and tuna drift net fisheries prosecuted by Japan, Taiwan, and South Korea. There is also concern about incidental catches elsewhere of other species, including sharks, as well as target catches of drift nets, such as tuna in the South Pacific. The problem in the eastern tropical Pacific, where dolphins are killed incidentally in the purse seine tuna fishery, is well known.[9]

A separate word needs to be said about the concern over large-scale drift net gear and its effects on particular species on the high seas. These effects are still largely unknown because the scope of the fisheries involved is so large and the impacts have not been carefully studied until recently. It seems to be widely assumed that drift nets are a unique hazard in the ocean because the gear is so vast and indiscriminate in harvest. Whether the gear is unique in this sense is not established. Some experts believe that the dangers are exaggerated and that other types of gear are at least as inefficient in by-catch as the drift net.[10] Purse seines are well known for their impact on dolphins. Trawls, especially in the shrimp fishery, probably take a larger ratio of by-catch to target than any other gear.[11] Longlines are also indiscriminate among species of a certain size, not the least being sharks, which are known to be very fragile in the face of intensive harvesting.[12] This gear also is reported to attract and kill albatross.

Whether drift nets are more destructive to nontarget species than other gear seems mostly irrelevant if drift nets take significant by-catches. Such a problem needs to be resolved, whatever the gear. On the other hand, if all large-scale gear is similarly destructive unless regulated adequately, it is difficult to argue for or justify unusually drastic remedies for drift nets alone. One remedy currently recommended by the UN General Assembly would reverse the traditional burden, requiring those who seek to restrict fishing activity to establish the basis for restriction. The notion of reversing the burden of proof to require that the fishing operation be shown not to have "unacceptable impacts" would surely be more difficult to justify if this gear did not differ from other gear in its incidental destructiveness. In this event, the remedy proposed needs to be assessed in terms of all gear having similar effects. This might mean that reversing the burden of proof has a different calculus of costs and benefits.[13]

Conservation: Maintenance of Optimum Yields. A fourth type of counterclaim is the ordinary conservation demand that high seas fishing be reduced in order that there continue to be future harvests at optimum levels. The purpose of this counterclaim is to control high seas catches that bring down stock abundance to a level that is uneconomic or wasteful of resources. This demand can be seen in the Antarctic Ocean concerning overexploitation of particular species in that area[14] and in the expressions of concern now springing up about the squid drift net fishery in the North Pacific.[15] This claim might be advanced in a variety of fisheries, since there is estimated to be more than 400 species of finfish, which are primarily occupants of the high seas.[16]

Claims to Establish a Decision Process

The other major challenge regarding unregulated high seas fisheries is how to create the institutional authority needed to establish measures required for conservation and allocation. Because by definition no single state has authority over the high seas, an agreed mechanism is required to make the needed decisions to attain agreed conservation and allocation goals. Agreement on such goals is indispensable because they shape and determine the needed decision process.

The important claims relate to the way management functions should be fashioned. The differing views concern (1) whether and how to provide for the research, analysis, collection, and dissemination of scientific and other data for management purposes; (2) the scope of authority to provide regulations for conservation and allocation concerning tuna exploitation (including by-catch issues); and (3) the means for achieving effective enforcement through an international body. All of the functional issues are overshadowed by the dominant question of how to maximize benefits and how to allocate them.

In addressing these functional issues, the more detailed contentions are (1) whether an independent or separate component of the new agency should be established to carry out the research function or whether this function should be retained by member nations; (2) whether the agency should have authority to prescribe regulations directly or whether competence to regulate should be restricted to recommendations or restricted by requiring approval or objection by members; and (3) whether the entity should have some or all enforcement functions or whether some or all should be retained by individual nations.

THE INTERESTS AT STAKE

The various categories of claims and counterclaims briefly described involve basic questions of determining the permissible level of catch (and effort) in particular high seas fisheries (including whether any should be allowed either of the target fishery or of incidentally caught species), allocating the benefits and costs of high seas fishing, and establishing an institutional means for making decisions about allocation and conservation. In short, the problems are to determine who gets what and who bears the costs, what standards or criteria should be employed to decide on permissible takings, and how decision functions should be allocated among participant states and international organizations, including regional and multilateral agencies.

These questions may be addressed by assessing what common interests are involved in the high seas fisheries and in straddling stock fisheries.

Common Interests in Conservation and Allocation in a New Regime

Regarding conservation of high seas fisheries other than straddling stocks, the common interests of fishing states and the community of states are in maintaining catches that provide the maximum net benefits from stocks that all agree are appropriate targets of a fishery. But to maintain such catches requires not only a regulatory regime for current participants but also a fishery considered closed to new entrants except under agreed circumstances. The only way to satisfy these conditions in most instances is to provide an institutional entity to regulate the fishery by establishing means for limiting access, through a permit system, an auction, or another approach. Yet the creation of such an entity is extremely difficult to arrange.

For conservation of animals on which participants have differing views about whether they should be harvested at all, the problem is vastly more difficult. For some participants, some species should be considered sacrosanct, beyond any mortality inflicted by humans. States that do not accept this approach have diametrically opposed views. It is difficult to identify any common interests in this context unless one party is willing to defer to the other's view completely.

In contrast, for those who are concerned about protection of certain species but are willing to concede limited mortality, the question is one of agreed limitations on quantity of incidental catch. There is then, at the least, a common concern that the species' abundance not be depleted to the point of endangering it. In some situations, this can extend to the view that discrete populations ought not be endangered even though the species is not threatened. One basis for this is the common interest in maintaining an ecosystem that benefits all. This issue is susceptible to negotiations about the protective measures required. The key appears to be the cost that is acceptable to the restricted party.

Turning to allocation of benefits of purely high seas fishing, it is an obvious but complicated goal that states should seek to maximize or to optimize the benefits. A major difficulty is that benefits are commonly seen differently from one state to another. These differences about benefits can result in different views about management objectives. Such differences must be reconciled before allocation can be agreed on. Then the most practical objective is to secure an allocation that is acceptable and, therefore, likely to induce compliance. There are few, if any, criteria for determining an equitable allocation. That these benefits be allocated in what is perceived to be an equitable fashion is to be highly desired, but satisfying this perception is less important than coming to agreement on an acceptable allocation.

These considerations also suggest the common interest of states involved (as high seas fishing states and coastal states) with regard to exploitation of straddling stocks. Each side can lose all benefits of a fishery unless it works out an accommodation with the other. For some time, at least, the high seas fishing state can refuse to cooperate, can ignore efforts to regulate harvests, and thereby conceivably reduce the stock abundance to the point that the catch from it will not cover costs. On the other hand, the coastal state can fish the EEZ part of the stock without regard to the high seas fishery and thereby minimize or even eliminate a profitable fishery. In the end, the coastal fishing state will be required either to stop fishing entirely or to fish at a very low level, far lower than if the adversaries agreed to maintain the abundance by cooperation.

The biggest single difficulty is that even if current high seas and coastal states reach agreement, the accord will be at the mercy of new entrants under current law. The coastal and fishing states, acting together if possible, may prohibit the new entrant from fishing the stock concerned. Otherwise, an effective regime for straddling stocks is unlikely.

If international cooperation is unavailing in producing agreement among a coastal state (or states) and high seas fishing states on conservation of straddling stocks, the question is whether the coastal state should be permitted to employ unilateral action to achieve the necessary conservation measures. Seldom do coastal states need to take measures to avoid extinction of a stock; the common problem is to maintain profitable yields. This usually means reduction of effort. It seems unconscionable to place all the burden of reduction on the high seas fishing state if it is agreeable to observing coastal state measures that increase coastal benefits measurably. The preference should be for sharing the benefits of management when some of the stock is outside national jurisdiction and a high seas state has a legitimate historical claim to some of the overall stock abundance.[17] On the other hand, if the high seas fishing state will not accept any accommodation, the coastal state's conservation measures should prevail.

Common Interests in a New System of Governance

International cooperation is the sine qua non for high seas fishery management. The question is how the states concerned can cooperate to organize themselves to achieve agreed objectives in conservation and allocation. The more specific questions here are as follows: (1) How should states and others organize to acquire the information and theory necessary for supporting management measures? (2) How should they effectively prescribe management measures? and (3) How should states achieve effective enforcement of these measures?

Acquisition, Collection, Analysis, and Dissemination of Scientific and Other Information. The major need in a new governance system is for an objective and credible body of theory, data, and knowledge on which to base management measures. This includes not only scientific information but also social, economic, and legal data appropriate for addressing the questions of maximizing and distributing benefits. Experience in other fishery management contexts counsels that there be an independent scientific staff that is responsible for this function. Another desirable mechanism would be a wholly separate body for developing the scientific information or at least a component of the fishery body that is kept separate from the rest of the agency. The precise need is to avoid use of data filtered by political and economic interests of members.

Comprehensive Coverage of Regulations. A fishery body should be permitted to prescribe regulations that are effective for all participants in a fishery. Mere power to make recommendations that may or may not be accepted or that are subject to a discretionary objection procedure will not permit effective management.

Adequate Enforcement Methods. Enforcement methods must be adequately funded and permitted to operate directly against those in the fishery. Resources for adequate surveillance and observation are indispensable. The prime goal is to achieve effectiveness, credibility, and sufficient stature that potential violations are deterred and, if committed, are likely to be penalized. Enforcement issues need to be addressed when management regulations are being developed.

Feedback Loop. A final overall function is a feedback loop to keep the entire system under continuous appraisal and subject to timely amendment.

PREVIOUS DECISIONS: PRINCIPLES OF INTERNATIONAL LAW FOR STRADDLING STOCKS

The most important decisions about high seas fisheries are those providing for the principle of freedom of fishing on the high seas, the general obligations of high seas fishing states, and the specific obligations owed to coastal states. Equally important are decisions to establish international fishery agencies to provide for conservation and allocation of high seas fisheries. The following discussion briefly examines these various decisions.[18] The dominance of freedom of fishing may be undergoing erosion as conflicting claims become more widespread and intense.

Rights of High Seas Fishing States: Freedom of Fishing on the High Seas

Article 2 of the 1958 Convention on the High Seas[19] embodies the general understanding, also considered to be part of customary law of the sea, that the freedoms of the high seas include freedom of fishing in this area.[20] This freedom, like all others on the high seas, must be conducted with reasonable regard for the interests of others in their exercise of the same or other freedoms of the high seas. Freedom of fishing has traditionally extended to all types of fishing gear, without exception.

Article 2 of the 1958 Convention on Fishing and Conservation of the Living Resources of the High Seas contains a more qualified formulation of the rights of states to fish on the high seas, declaring:

1. All States have the right for their nationals to engage in fishing on the high seas, subject (a) to their treaty obligations, (b) to the interests and rights of coastal States as provided for in this Convention, (c) to the provisions contained in the following articles concerning conservation of the living resources of the high seas.
2. All States have the duty to adopt, or to cooperate with other States in adopting, such measures for their respective nationals as may be necessary for the conservation of the living resources of the high seas.[21]

Although article 2 is in a treaty that has not been widely adopted,[22] paragraph 2 at least is now considered as part of customary law.[23] The International Court of Justice, in the *Fisheries Jurisdiction* case (*United Kingdom v. Iceland*), declared that a high seas fishing state must take full account of necessary conservation measures in conducting its operations.[24]

The most recent multilateral agreement dealing with freedom of fishing on the high seas affirms it once again and also confirms that states generally are agreed on obligations that burden this right. Article 87 of the 1982 United Nations Convention on the Law of the Sea provides as follows:

1. The high seas are open to all States, whether coastal or land-locked. Freedom of the high seas is exercised under the conditions laid down by the Convention and by other rules of international law. It comprises, *inter alia*, both for coastal and land-locked States:
   . . . (e) freedom of fishing, subject to the conditions laid down in section 2. . . .[25]

Article 116 in section 2 of part VII of the 1982 Convention repeats article 1 of the 1958 Geneva Fishing Convention in declaring that "[a]ll States have the right for their nationals to engage in fishing on the high seas" but adds significant new conditions that are relevant, *inter alia*, to straddling stocks, including high seas pelagic drift nets, as noted later in this chapter.[26]

Experience under Freedom of Fishing on the High Seas. The level of difficulty attending conservation and allocation of benefits may be gauged by recalling that they have been addressed largely by repetition of the doctrine of freedom of fishing and the associated injunction that states should cooperate to conserve high seas fisheries. International agreements have long been employed to restrict fishing operations outside national jurisdiction but almost invariably have been unsuccessful in producing benefits for the states concerned. Because cooperation through explicit fishing agreements has been generally fruitless, the success of the principle of free fishing has been severely limited. A review of this experience is useful.

Conservation under freedom of fishing. Under the principle of freedom of fishing, how much of a resource should be taken was also left to the participants in the high seas fishery, each of whom was virtually forced to join in depleting the fishery as a matter of rational behavior in such a context.[27] Depletion of most marine fisheries simply lowers the abundance of the target fish population and perturbs the ecosystem in mostly unknown ways but does not irreversibly damage a target species or even a stock. In contrast, the effect of unrestricted harvesting of marine mammals can, in the long run, result in extinction of the species. Worry over this possibility explains some of the widespread concern over the largely unknown total catch of species of marine mammals in the drift net fisheries.

In the straddling stock fishery, as noted earlier, freedom of high seas fishing means that the stock concerned cannot be managed for an optimum yield because international agreement rarely yields effective regulatory authority over the entire stock. Typically, the adjacent coastal state must bear the entire burden of conservation, which in all likelihood cannot be attained. The state fishing on the high seas meanwhile may harvest as much as the stock will bear in the area.

Allocation under freedom of fishing. Under freedom of fishing, allocation decisions were made in accordance with the principle that whoever could harvest stocks could take them, even though this meant that no one had any assurance that harvesting would continue to generate enough return to pay the costs. The result in every unregulated fishery (and in many purportedly regulated fisheries) is that the fishery becomes a commercial disaster for many participants and certainly a loss for society. In terms of allocation of stocks that are also found within an exclusive economic zone (EEZ), the benefits (if any) of the fishery are enjoyed by the high seas fishery while the costs are incurred by those who are subject to regulation within the EEZ. The overall levels of catch and effort are beyond anyone's control because no single entity has authority over the entire stock.

Institutional mechanisms and freedom of fishing. The institutional mechanisms for purely high seas stocks and for straddling stocks under freedom of fishing must be created by international agreement among the states concerned. Only if these states are willing to cooperate in establishing a regional agency to exercise adequate authority over the fishery is there any hope of constructive regulation. But as noted earlier, agreement on institutional mechanisms has not been successful in the past. Contemporary problems of straddling stocks in an international regime in the Northwest Atlantic suggest similar difficulties.[28] States acting wholly independently, even if they seek to cooperate in an ad hoc fashion, will probably have a very difficult time with overall management.

Almost all existing international fishery management bodies have been failures when measured by the benefits produced and distributed. Few have had access to independent research staffs, and most are forced to rely on research results and analysis that are carefully filtered through politically biased national scientists. The result may be management measures that pay little respect to currently available knowledge. Only rarely have fishery bodies been permitted to exercise a competence to regulate directly, with the result that their recommended regulations are routinely defied. Enforcement is the largest problem of all, and no fishery agency on the international level has ever been allowed authority equal to the task.

Despite this record, it is evident that the principle of freedom of fishing on the high seas continues to protect this activity. It is also evident, as a result of recent decisions, including the 1982 Convention, that conditions burden the exercise of this right and need to be taken into account in relation to fishing of straddling stocks. Significant modifications of the principle of freedom of fishing on the high seas are, arguably, introduced by the provisions of the 1982 Convention on the Law of the Sea that provide for straddling stocks, as is noted further later in this chapter.

General Obligations of High Seas Fishing States

Under contemporary international law, a fishing state has two sets of obligations for its operations in high seas fisheries, one set of which may modify the other substantially in the particular circumstances of straddling stocks. The first consists of all the obligations that must be observed for any high seas fishing activity; the second are those that are due a coastal state when high seas fishing takes, directly or incidentally, stocks that bear sufficient relationship to the coastal state.

The list of general principles establishing obligations for high seas fishing is substantial and includes the following obligations:

1. To conserve the living resources of the high seas (including a state's obligation to take unilateral measures for its own nationals when a fishery is conducted by them alone).
2. To cooperate with other states in conserving living resources of the high seas.
3. To negotiate with other states over conservation measures.
4. To generate, contribute, and exchange scientific and other information (catch and effort statistics).
5. To employ the best scientific evidence available to establish needed conservation measures.
6. To refrain from discrimination in conservation.

A major source recognizing the authority of these obligations is the *Fisheries Jurisdiction* case (*United Kingdom v. Iceland*), decided by the International Court of Justice.[29] The subsequent incorporation of some of these principles in the 1982 Convention confirms their status in international law. Recent actions of the United Nations General Assembly may also be cited to establish the general expectation that high seas fishing states have the obligations listed here.[30]

Obligations of High Seas Fishing States to Coastal States

The 1982 Convention on the Law of the Sea establishes obligations for states fishing on the high seas for certain living resources shared with the coastal state regardless of gear type. Article 63(2) requires a high seas fishing state and a coastal state to seek agreement where the stocks being fished on the high seas are also found within the exclusive economic zone. The obligation is to seek agreement on conservation (but not allocation) of the shared stock in the adjacent area (i.e., the high seas). As noted earlier, however, agreement on allocation is probably a prerequisite for agreed-on conservation measures, since the two issues are difficult to disentangle.

Article 64 refers to another specific instance of shared stocks: highly migratory species that are fished within as well as outside the exclusive economic zone. Under article 64, the high seas fishing state and the coastal state in a region are to cooperate "with a view to ensuring conservation and promoting the objective of optimum utilization of such species throughout the region."[31] The last three words contemplate that conservation measures are applicable on the high seas and also in areas subject to national jurisdiction, although the measures need not be identical.

Article 66 concerns anadromous species and provides, *inter alia*, that the state of origin has primary interest in and responsibility for such stocks and, subject to one condition, prohibits fishing for these species beyond the EEZ. The state of origin is to establish regulatory measures for fishing within its EEZ and for permissible fishing on the high seas beyond, including total allowable catches. The high seas fishing state and the state of origin shall maintain consultations "with a view to achieving agreement on terms and conditions of such fishing giving due regard to the conservation requirements and the needs of the State of origin in respect of these stocks."[32]

The obligations of the high seas fishing states under these several articles in part V of the 1982 Convention differ from those previously discussed because they are owed to specific states rather than to the general community of states. Furthermore, these articles, coupled with those concerning the high seas, provide that the high seas fishing state is not competent to decide alone on conservation measures for high seas fishing under the circumstances set out in these articles. A coastal state is a necessary associate of the high seas fishing state when it takes species subject to the coastal state's rights, duties, and interests.

These various articles concerning straddling stocks might be considered to be supplemental in nature because if the high seas states and the coastal states cooperate and negotiate successfully, that is, adopt and implement an effective regime of conservation measures, there is no need for resort to any other principles of international law to resolve conservation difficulties. But if these states are unable, by acting together, to take measures necessary for conservation of living resources on the high seas, coastal states might invoke the principles contained in the 1982 Convention to justify unilateral imposition of conservation measures for harvesting stocks on the high seas. At present, this seems unlikely, but it is not beyond the realm of possibility. Accordingly, the following discussion examines the purport of the articles in greater detail in relation to possible coastal action to achieve conservation of the high seas stocks involved.

Article 63(2) and Article 116. Article 63(2) establishes that the high seas fishing state has the duty to seek agreement with the coastal state; article 116 provides that "the right . . . to engage in fishing on the high seas is subject to . . . (b) the rights and duties as well as the interests of coastal states provided for, *inter alia*, in Article 63, paragraph 2, and Articles 64 to 67."[33] Article 87 declares that the freedom of fishing on the high seas is subject to several provisions, including articles 116 through 120. Thus, article 116 goes beyond requiring action to seek agreement and declares that the right to fish on the high seas is subject to the rights and duties of coastal states with respect to straddling stocks.

Because article 56 establishes the sovereign rights of the coastal state over the living resources of the EEZ, article 116 means that the right to fish on the high seas is subject to the sovereign rights, as well as the interests, of coastal states as provided in the articles of part V of the 1982 Convention. Accordingly, the 1982 Convention might be interpreted to provide that high seas fishing on stocks that also occur within a coastal state's EEZ is subject to the sovereign right of that coastal state.

The question that remains to be answered regarding article 116 is how the apparently superior right of the coastal state might be implemented in the specific context of straddling stocks, where high seas fishing takes stocks also occurring within the EEZ of the coastal state. Assuming the states concerned have not been able to conclude an agreement on conservation in the high seas, one interpretation of article 116 is that the coastal state would be considered authorized to establish conservation measures applicable to the stock as a whole, including the high seas portion, and to demand compliance with those measures by high seas fishing states. Refusal of the high seas fishing states to comply with genuine conservation measures would constitute a violation of the treaty (were it in force) and, otherwise, of its customary international law obligation to join in conserving the living resources of the high seas.[34]

It is more difficult to find a plausible basis in the 1982 Convention for enforcement of coastal measures that, in the interpretation offered here, apply to the high seas fisheries. Nothing in the treaty directly authorizes enforcement action on the high seas by a coastal state, and article 66 assumes there is no competence of this kind.[35] Certainly, this has been the traditional customary international law standard.

In at least two instances, however, the coastal state has a basis for gaining authority to enforce on the high seas. In the case of anadromous stocks, a state of origin might make its capacity to enforce its regulations on the high seas (which are definitely authorized by article 66) a condition of its concurrence in the continuation of high seas fishing, which is also required by article 66 for states that suffer economic dislocation.

The second instance concerns enforcement measures against fishing gear on the high seas, as distinct from the fishing vessel. Thus, appropriate enforcement of the customary law obligation of the fishing state to cease and desist from certain action could take the form of disruption, for example, of drift net gear or longlines sufficiently to deter violations of coastal state conservation measures on the high seas. There would be no boarding or arrest of a foreign fishing vessel on the high seas and therefore no interference with a vessel nor any claim to exercise jurisdiction over one. The sole jurisdiction would be to achieve compliance with conservation measures authorized to be prescribed to affect the gear being used.

In accordance with the 1982 Convention, objections to this enforcement effort could be resolved by submission of the dispute to a third party, whose decision would be binding on those concerned. Under the treaty, the coastal state and the high seas fishing state would be bound to submit to such a settlement procedure. Even apart from the treaty, the coastal state taking this course of action should consider itself bound to submit to such settlement. Conservation measures should remain in effect pending a decision.

Since the 1982 Convention is not in force, state practice becomes especially important. That practice does not yet establish any accepted or uniform approach to the problem. In specific contexts of straddling stocks, states have cooperated to exchange information, and in one drift net fishery (salmon in the North Pacific), agreements have been reached to resolve controversies in some circumstances. Nor does that practice yet confirm the initial point, that article 116 does establish a superior right that must be recognized and deferred to by high seas fishing states.

Recent action by the Northwest Atlantic Fisheries Organization (NAFO) may be an indication of a change in state practice. Canada and the European Economic Community successfully proposed a resolution in NAFO reflecting the view that states not party to the NAFO agreement should not fish for straddling stocks in the high seas contrary to NAFO management measures in the area. Although this proposal refers to regulatory measures established by a regional fishery organization for high seas fishing by its own members, the measures themselves must be consistent with coastal measures (Canada) within the EEZ.[36] The principle recognized in this resolution by inference is that noncoastal states should not fish for straddling stocks on the high seas contrary to a coastal state's conservation measures in its EEZ.

Thus far, despite the failure to resolve some well-known straddling stock problems by international cooperation, coastal states have not sought to take unilateral action to deal with conservation problems arising from unrestricted fishing on the high seas, whether by drift net or other gear. Such a pattern of inaction, however, if continued over a long enough interval, could give rise to expectations about the relative rights involved. Thus, if coastal states do not take the initiative to demand that high seas fishing states observe particular conservation measures prescribed by the coastal state to regulate harvest of the stocks concerned (salmon, tuna, or straddling stocks), the inference may be drawn that under international law the superior right of the coastal state cannot be implemented by coastal state action to prescribe regulations for high seas fishing. Furthermore, assuming that coastal states did demand compliance with measures prescribed by them for application on the high seas but did not take steps to enforce them, failure to act might support the inference that such action was not consistent with international law.

Article 64. Among the principal stocks that appear to be affected by drift net fishing are albacore tuna, both the population in the South Pacific and that in the North Pacific. It is now generally agreed that the 1982 Convention articles on these tuna species reflect or embody existing customary international law. In the case of tuna, nearly all states in the world consider that this species falls within coastal state jurisdiction while present within the exclusive economic zone and that this position is consistent with article 64 of the 1982 Convention. As noted earlier, under article 64 coastal states and states fishing for tuna in a region are to cooperate to ensure conservation throughout the region, including the high seas.

Although tuna species are dealt with as a discrete problem in article 64 of the treaty, there can be little doubt that they can be regarded as a straddling stock, since they are available both within and outside national jurisdiction. Harvesting in one area affects stock abundance and availability in the other. The considerations affecting this problem differ only in degree from those involving other species.[37]

Article 65: Marine Mammals. Under the 1982 Convention, marine mammals are subject to coastal states' sovereign rights within the exclusive economic zone in the same sense as is any other living resource of the zone. However, in accordance with article 65, the other provisions of the treaty regarding coastal states' obligations to provide access to a surplus of such species do not apply to marine mammals. Although there may be mammals available for exploitation in the EEZ, because there is little or no local take allowed[38] there is no obligation to permit foreign harvesting. It is well known that some marine mammals found within coastal state jurisdiction are also found on the high seas and are often also taken in fishing operations there. The analysis regarding stocks subject to article 116 and its provision for the superior right of the coastal state also appears to apply here. Article 120 makes article 65 applicable to conservation and management of marine mammals on the high seas.

In at least one instance, the marine mammal problem is aggravated because the species concerned has a special relationship with land areas subject to coastal jurisdiction and is now regarded as depleted by the United States. This is the northern fur seal, which bears its young on island rookeries in the North Pacific that are territorial possessions of some bordering states. The rookeries are located on islands of the United States and the former Soviet Union, with the largest populations being found in the United States. These animals were formerly dealt with by the North Pacific Fur Seal Convention, which prohibited their taking by pelagic fishing on the high seas, but that treaty has been allowed to lapse because of environmental opposition to it within the United States. In a sense, these mammals have a similar life history as anadromous species, with the pronounced difference that parent fur seals survive the reproductive phase. Although these species are not now the target of high seas pelagic fishing, it is reported that Japan has declared that it may have to "undertake procedures by which to resume pelagic sealing."[39]

Article 66: Anadromous Species. Because of their importance to the states in the North Pacific, salmon fisheries and the provisions of article 66 of the 1982 Convention merit specific comment. This species is especially important because it is vulnerable to high seas drift net fishing, both as target species and as incidental catch. Much of the concern about drift net fishing in the North Pacific arises from the incidental catch of salmon in the squid drift net fisheries of Japan, South Korea, and Taiwan.[40] Japan has targeted salmon with drift net gear for many years, and the practice has been regulated by agreement with the United States and Canada through the International North Pacific Fisheries Convention[41] (INPFC) and in a bilateral agreement with the former Soviet Union.[42] Now, fisheries by other nations using drift nets believed to be targeting salmon also excite opposition.[43]

Article 66 provides that in the absence of agreement with the state of origin of anadromous species there shall be no fishing for such species on the high seas, that is, beyond an exclusive economic zone. States that suffer from economic dislocation as a result of this prohibition (essentially Japan) may conduct such fisheries after consultations with the state of origin that achieve agreement on the terms and conditions of such fishing. (Again, however, if no agreement is reached, no fishing is authorized.) After such consultations, the host state establishes the total allowable catch for such species to ensure conservation on the high seas. The terms of the agreement between fishing state and host state include measures to renew anadromous species, particularly by expenditures for that purpose by the fishing state. The fishing state reaching this agreement is to be given special consideration by the state of origin in harvesting stocks originating in its rivers. To be effective, an agreement for high seas fishing must include provisions for enforcement of regulations in harvesting stocks on the high seas. Since such an agreement is a requirement for any high seas fishing at all, it is apparent that there would normally be no high seas fishing without provision for enforcement.

An initial question is whether article 66 applies to the incidental catch of salmon by drift nets on the high seas. Article 66 confines "fisheries for anadromous stocks" to areas landward of the outer limits of exclusive economic zones "except in cases where this provision would result in economic dislocation for a State other than the State of origin."[44] Does "fishing for" mean only target fishing on the high seas, or does it also include incidental catch?

As a practical matter, the difference between target catch and by-catch does not appear to be significant. Concerns about conservation and the high seas take of salmon are not diminished by how the fish are taken. It is enough to know that fishing on the high seas for one species is expected also to take another species in appreciable quantity.[45] Any such by-catch of salmon on the high seas diminishes the number of fish returning to the state of origin and affects the conservation of the stock as well as the catch of the state of origin. High seas fishing states manifest their awareness of concerns over incidental catch by their legislation, which forbids retaining salmon caught on the high seas and which forbids landing such fish.

The implication of article 66 is that it applies to all catches of salmon on the high seas. The state of origin may establish the total allowable catch of such fish, which in the normal management procedure must include the by-catch if coastal management is to be effective. It is the coastal state that has authority to establish regulations, and only the coastal state can know the conservation needs of particular salmon fisheries. If a significant portion of the catch is beyond regulatory control, the coastal state either cannot act to take adequate conservation measures or can take only ineffective measures. Accordingly, article 66 should be interpreted to extend to direct fishing and also to fish expected to be taken by harvesting in a particular fishery.

The key question concerns (1) a state that, not having fished these stocks before, harvests high seas stocks of anadromous species in violation of the prohibition or (2) a state that, though suffering economic dislocation, fishes without the agreement of the state of origin and in disregard of regulations that provide, *inter alia*, for a high seas catch of zero. In either case, such fishing would be in violation of the 1982 Convention prohibition on high seas fishing and may be subject to sanctions for this violation by the state of origin. In such circumstances, where the high seas fishing state acts in violation of the treaty, it cannot in good faith rely on the treaty provision concerning prior agreement to enforcement measures as a means of escaping the consequences of violation of the same treaty. To countenance such sleight of hand would make a mockery of international law.

The foregoing assumes that the 1982 Convention enumerates the obligations for the parties concerned. Otherwise, the law applicable is customary international law regarding high seas salmon fishing. The pattern of state practice in the North Pacific appears to be well established. All traditional high seas fishing states and coastal states in the region, including Japan, claim jurisdiction on the high seas over anadromous stocks originating in their rivers. All states whose vessels fish on the high seas with drift nets for species other than salmon forbid the retention of salmon by their vessels. Thus, the high seas fishing states recognize the authority of the state of origin to exercise its fishery management jurisdiction over the taking of salmon on the high seas. To take salmon in disregard of state of origin regulations, including that establishing the total allowable catch, would be a violation of customary international law. As noted earlier, appropriate enforcement measures would extend to prevention of the fishing.

Other General Principles Relevant to Straddling Stocks

Some urge that the provisions on semienclosed seas in the 1982 Convention provide a basis for the coastal states adjacent to a straddling stock fishery in such a sea to adopt a management regime with which any other state must comply. This argument is seldom explained fully. The basic contention is that under article 123 of the treaty the surrounding states should cooperate to exercise their rights, and to achieve this they are authorized "to coordinate the management, conservation, exploration and exploitation of the living resources of the sea." It is not clear how "coordination" is transformed into authority to direct, supervise, or control.

PROPOSED PRINCIPLES APPLICABLE TO HIGH SEAS FISHING, WITH PARTICULAR REFERENCE TO STRADDLING STOCKS AND DRIFT NETS

A considerable number of proposed arrangements have been suggested for dealing with straddling stock problems and the drift net controversy. Some of these rely on general principles, and some simply advance suggested remedies without identifying any justification in legal or other principle. These approaches are identified in the following discussion. It is important to remember that the issues are related because high seas drift net fishing also takes a quantity of straddling stocks.

Straddling Stocks

The major principles proposed for dealing with straddling stocks include (1) the primacy of coastal state conservation measures, (2) the dominant role to be played by adjacent states in a conservation and allocation regime, (3) the right to coordinate management, conservation, exploration, and exploitation of stocks in semienclosed seas, (4) the need for consistency with coastal zone action, and (5) the prohibition of all fishing. Each of these is described in the following sections.

Primacy of Coastal State Conservation and Allocation Measures. This refers to the position that in the absence of agreement on conservation measures applicable to straddling stocks on the high seas, the 1982 Convention establishes that such fishing is subject to coastal states' rights, duties, and interests, including the sovereign rights of the coastal state over stocks within the EEZ. Because fishing in disregard of coastal state conservation measures for a straddling stock effectively diminishes or eliminates the state's sovereign rights over the stock, the high seas fishing state or states must observe coastal state measures when fishing on the high seas. This proposal is conditioned on acceptance by the coastal states concerned of binding dispute settlement in which high seas fishing states can challenge the accuracy and validity of the scientific basis for the proposed measures.[46]

A difficulty with this, or any other, proposal that gives priority to coastal state interests and measures is that the coastal state measures almost inevitably must deal with distribution of the benefits of a conservation regime. The problem this causes is that in contrast to the scientific basis for conservation measures, there are no agreed-on criteria for reviewing the validity of allocations of benefits. The benefits may be differently conceived or defined by the states involved, and their distribution involves judgments that are highly subjective in nature.

The Dominant Role of Coastal States in the Regime for Straddling Stocks. The coastal state's dominant role is reflected in the proposals that have been discussed in the United States and the former Soviet Union for creation of a Bering Sea Fishery Authority.[47] In the drafting stage at least, the proposed organization appears designed to reflect the preferences of only these two coastal states in the Bering Sea, although the states fishing in the high seas area would at least have a forum for their views. Although the drafts are not identical, they do concur in providing for exclusive control by the two coastal states. The two would have the ultimate authority to determine the allowable catch and other conservation measures as well. Nothing in the 1982 Convention suggests a basis in legal principle for this arrangement other than that the stocks in the area subject to the agreement are straddling stocks, the only two coastal states are the United States and the former USSR, and the high seas harvests detrimentally affect coastal state efforts at conservation and management.

Another illustration of the assumed dominance of coastal states over adjacent high seas straddling stocks is to be seen in S. 396, a resolution adopted by the United States Senate in March 1988.[48] The Senate proposed in this resolution that the United States and the Soviet Union join together to declare a moratorium on fishing by all states in the central Bering high seas area, which would end only when a multilateral agreement was concluded with a majority of the concerned states. The effect of this approach would be to preclude any fishing in the "donut area" except that which was agreeable to the coastal states concerned. The draft treaty on the Bering Sea Fisheries Authority appears to implement the Senate resolution.

The Right of Adjacent Coastal States to Coordinate Management and Conservation Measures for High Seas Fishing of Straddling Stocks. The coastal states' right to coordinate has been urged as a means of recognizing the dominance of these states in decisions on conservation and allocation. The Soviet draft of the Bering Sea Fisheries Authority notes that this is an obligation of the two states under their bilateral agreement of May 31, 1988. One possible source of authority to coordinate conservation and management measures in the Bering Sea is article 123 of the 1982 Convention, which addresses the "[c]ooperation of states bordering semi-enclosed seas." However, the authority to coordinate derived from article 123 does not include any exclusive right to establish conservation measures or to enforce such measures, but this appears to be provided in the proposed drafts by both the United States and the former Soviet Union.

Consistency with Coastal State Conservation Measures. Consistency with coastal state measures is provided for in the existing international fishery body, the Northwest Atlantic Fisheries Organization, which was established to address the straddling stock question.[49] In a recent action, NAFO's council extended this principle to nonmember nations of NAFO, calling on members to take actions to ensure that these states act consistently with NAFO regulations in the area beyond 200 miles.[50]

It needs to be noted again that consistency applies here in a conservation regime that has distinct distributional implications. Thus, Canada establishes an annual allowable catch that reflects a desire to maintain population abundance at a level that is greater than would sustain the maximum sustainable catch. The effect of this decision is to reduce the total allowable catch and thereby to withhold from allocation a portion of the stock. Consistency with the coastal state measures means that less is made available to high seas fishing states.[51] This was a major part of the difficulty experienced by NAFO in recent years. Although the NAFO arrangement defers to the interests of the coastal state, the trade-off is that high seas states are given quotas in the area concerned. The alternative is unrestricted fishing, which, over time, may diminish the catch by all states involved.

As will be noted later, the consistency principle could have additional effects when the coastal state has regulatory measures affecting the by-catch of marine mammals. The drastic impact of this on high seas fishing raises the question of whether consistency should be observed in all instances.

Prohibition of Any Harvest of Straddling Stocks. This alternative is a radical one that may be advanced as a component of the consistency principle. This occurs in the following way. The drift net fishery in the North Pacific is known to result in the take of marine mammals of various species. One of these is the northern fur seal, which inhabits islands of the United States (among others) in the Bering Sea during the land phase of its life. Under United States law, a fishery that takes fur seals must obtain a permit under the Marine Mammal Protection Act. If the United States is unable to determine the optimum population of fur seals, as required by the act, it cannot issue a permit to take these animals as part of the incidental catch in a fishery targeting other stocks.[52] Therefore, if fishing states must take action consistent with coastal state conservation measures, these states would be unable to conduct the target fishery that takes fur seals.

Drift Nets

Several proposals address the controversy over drift nets operating on the high seas. These include a complete ban on drift net fishing (Pacific Island nations treaty;[53] United States Congress 1990[54]), a conditional moratorium established by an international body (United Nations Resolution 44/225),[55] and a proposed interpretation of this resolution by the United States.[56] After a brief description of the South Pacific agreement, the following addresses the UN resolution and especially the U.S. policy statement.

Drift Nets in the South Pacific. The South Pacific states are clearly on record regarding their general views about the obligations of high seas fishing states concerning the harvest of tuna by drift nets. The Tarawa Declaration of July 11, 1989, by the South Pacific Forum states, in relevant part:

> [R]ecalling the relevant provisions of the 1982 Convention on the Law of the Sea, and in particular articles 63, 64, 87, 116, 117, 118, and 119; recognizing that the use of driftnets as presently employed in the Southern Pacific Albacore Tuna Fishery is not consistent with international legal requirements in relation to rights and obligations of high seas fisheries conservation and management and environmental principles; resolves for the sake of this and succeeding generations of Pacific peoples to seek the establishment of a regime for the management of albacore tuna in the South Pacific that would ban driftnet fishing from the region; such a ban might then be a first step to a comprehensive ban on such fishing.[57]

This appears to mean that the specific drift net fishing on the high seas for South Pacific albacore tuna is inconsistent with principles of international law, derived from the various 1982 Convention articles identified in the Tarawa Declaration, but the suggested principles are not elaborated on in this statement. The subsequently adopted Convention for the Prohibition of Fishing with Long Driftnets in the South Pacific (South Pacific Driftnet Convention or Wellington Convention), discussed later, also does not further identify principles supporting coastal state authority; violation of the obligation to conserve high seas fish may have been meant.

States fishing for tuna on the high seas with drift nets would appear to be subject also to the "rights, duties, and interests" of the coastal state within whose waters those tuna also occur. As a straddling stock, tuna may differ from coastal species in the sense that tuna may be caught on the high seas in areas much farther removed from the coastal areas in which they also occur, making the relationship in stocks and fishing activities more difficult to establish. Assuming that the relationship is established, however, the legal relationship of dominant right would otherwise seem the same.

The South Pacific Driftnet Convention does not assert any jurisdiction by South Pacific Forum states over high seas drift net fishing by other states on the high seas, although it provides for measures within coastal jurisdiction that might discourage such fishing. These states obviously oppose this fishing and demand its termination, but other than the possible implications of the general language in the Tarawa Declaration they have not suggested that they have jurisdiction over the fishing itself on the high seas. The convention and associated protocols are limited to proposed agreements by the states of the region, other states adjacent to it, and states fishing in it to prohibit their nationals or registered vessels from conducting a drift net fishery for tuna in the region defined in the treaty. Except pursuant to the agreement among the parties, it does not claim jurisdiction directly to prohibit that fishing and to enforce that prohibition.[58] There seems to be no other evidence of such a claim to jurisdiction.

United Nations Resolution and U.S. Policy Statement. The United Nations drift net resolution[59] considered alone is simply a recommendation by the General Assembly addressing an issue of high seas fishing, setting out principles and procedures that, if interpreted uniformly, might resolve differences among states. The U.S. policy statement on the resolution, however, seeks to transform a UN recommendation into an immediately binding legal principle.[60]

The U.S. policy statement has special importance because (1) it appears to declare U.S. policy that is not limited to high seas drift nets but could extend to other fisheries; (2) propositions in the policy statement are inconsistent with the 1982 Convention; and (3) the policy statement is inconsistent with the United Nations resolution itself. The following observations bear on these points.

Scope of policy pronouncement. The policy statement in form is addressed to the implementation of the UN Driftnet Resolution, which the United States joined in cosponsoring, but the statement suggests that its contents are expressions of "sound management and conservation principles" that apply generally to fishing activities in the high seas.[61] The high seas drift net fisheries are simply a "special example" of the need for such management. Presumably, these are principles that would be applicable under the relevant provisions of the 1982 Convention.

If the United States has identified sound principles, they should also be applicable both to high seas fisheries other than drift nets and, for that matter, to fisheries conducted within the exclusive economic zone. At the very least, the principles would apply to high seas fisheries of every type, not merely those using drift nets. It would be very difficult to establish that general principles that are sound for high seas fishing by drift nets are not also sound for conservation of other fisheries that are believed to have comparable effects.

Alleged principles applicable to high seas fishing. In application to high seas fishing, the policy statement creates or advocates two new standards: (1) the fishing must not impose "unacceptable impacts" on target and nontarget species; and (2) a particular fishery cannot proceed unless it is shown that it can be conducted without unacceptable impacts. This latter standard is especially important because the policy specifically declares that "the absence of suitably reliable data for impacts assessment will not justify continued large-scale pelagic driftnet fisheries beyond the June 30, 1992 date for the moratoria."[62] Moreover, in this instance, the showing must be accomplished as a joint effort by "all concerned parties of the international community,"[63] not by a single state or party.

Apparently, unacceptable impacts are not necessarily violations of conservation measures or standards. It may be noted that the UN Driftnet Resolution itself mentions "unacceptable impacts" as a consideration separate from conservation. The U.S. policy statement that a drift net fishery is unacceptable if the harvest (mortality) is not sustainable does not necessarily identify a violation of conservation principles in the 1982 Convention. That treaty includes a provision (article 61) that would permit a coastal state to legitimate a yield from a fishery subject to its jurisdiction that may be unsustainable at least for some period. Similarly, the principle applicable to a high seas fishery (article 119) does not require catches to be maintained at a sustainable level at all times. To the extent that the notion of "unacceptable impacts" is a new imperative to be observed by high seas fisheries, it is inconsistent with the 1982 Convention, as well as extremely subjective.

A most important point to note is that the UN resolution and the U.S. policy statement place the burden of proof on "concerned parties of the international community," which must be discharged before vessels can continue or initiate fishing on the high seas. The U.S. statement specifically declares that if the absence of data prevents a showing that no unacceptable impacts occur, then the fishing must cease or must not be initiated.[64] Furthermore, the showing must be positive and agreed on; if it is disputed or cannot be made, the fishery must terminate, even if no unacceptable impacts are shown. This mechanism seems almost to ensure termination of fishing, since an opponent of drift net fishing may withhold approval of an alleged showing of no impact.

Nothing in the 1982 Convention sets out conditions such as these, and, as noted later, this also appears to be inconsistent with the UN resolution itself. If this position were an acceptable interpretation of general principles of the law of the sea, few lawful high seas fisheries would be conducted without extremely costly and complex prior international investigations to show that there no unacceptable impacts could be alleged by concerned members of the international community. As noted earlier, it is not a small point that "unacceptable impacts" are left undefined in the UN resolution and that they do not necessarily involve conservation principles or even questions of allocation of benefits.

One could argue by reference to the 1982 Convention that article 116, which subjects high seas fishing to coastal state rights, duties, and interests in specific instances, could be used to justify imposing this "unacceptable impacts" doctrine that the United States now advocates. The difficulty is that to the United States, this doctrine means that high seas fishing must terminate unless "concerned parties" can carry the burden of proving that no unacceptable impacts will transpire. According to the U.S. position, where data are inadequate, the foreign fishing must stop. But the 1982 Convention appears to require cooperative efforts among states to deal with the problem, not unilateral direction to cease fishing entirely or to refrain from beginning.

This conclusion is reinforced by the analysis offered by the United States of the doctrine of unacceptable impacts. These impacts are measured by all of the relevant fishing that affects a stock, which presumably includes the target and incidental catches that occur within areas of national jurisdiction in the region involved. In the case of anadromous fisheries, the relevant fisheries and most of the mortality occur within the internal waters of adjacent coastal states. For other species, such as pollock in the Bering Sea, the preponderant mortality apparently occurs within the exclusive economic zones of the United States and the former Soviet Union. If there is a connection between pollock fishing and the decline of the Steller's sea lion, for example, then presumably the mortality on the latter from the donut hole fishery would be combined with that within 200 miles to determine whether the impacts will be "unacceptable." More to the point, if the known mortality within areas of national jurisdiction is calculated but there are no data for the high seas component of the fishery, it may prove to be impossible to show that no unacceptable impact occurs from either fishery. The fishery would then have to terminate according to the United States's theory of unacceptable impact. The 1982 Convention calls for cooperation, not compliance with unilateral fiat.

The provisions of the 1982 Convention do address the question of incidental catches on the high seas, in article 119(1)(b). Literally interpreted, this article might be construed to support the position now advanced by the United States. In a provision identical to that addressed to fishing within the exclusive economic zone, the fishing states are "to take into consideration the effects on species associated with or dependent upon harvested species with a view to maintaining or restoring populations of such associated or dependent species above levels at which their reproduction may become seriously threatened."[65] Although the phrase "to take into consideration . . . with a view to . . ." places but a slight legal burden on fishing states, the preliminary information gathering to enable this consideration could require enormous work.

In contrast to the United States's policy statement that information required before a fishery can be initiated or continued must show that no unacceptable impacts will occur, the treaty requires only that states employ the best scientific evidence available. Although this evidence may be inadequate to show no unacceptable impact, the treaty permits the state to act in reliance on it to adopt conservation measures and to continue fishing. Actually, according to the UN resolution, scientific evidence may not even be relevant to "unacceptable impacts," since such impacts may not be scientific in character.[66]

If these sound conservation and management principles require termination of the use of drift nets on the high seas, it is difficult to understand why the reasoning involved and similar considerations would not apply to other gear on the high seas, such as tuna purse seines or longlines, or to this and all other gear taking fish within 200 miles. At the very least, if purse seines have effects on dolphins, as is known to be the case, their use might be considered to be unacceptable to other states when the fishing states are unable to establish that purse seining does not lead to declines in the dolphin population. Again, the fishery would be required to terminate because of the lack of information. The 1982 Convention makes no such provision.

For fisheries within 200 miles, presumably the coastal state has sovereign rights that it may use to control their harvest. But these fisheries unquestionably have effects on marine mammals that also are found on the high seas. If high seas fisheries can be stopped because of unacceptable impacts, defined in biological terms, the same principle should be considered to burden the coastal state whose vessels also inflict mortality on the same species within national jurisdiction.

There are two separate elements in the specific situation of coastal state fisheries within national jurisdiction that affect high seas stocks. One involves the coastal fishery where there is no accompanying high seas fishery. In such a context, it is only the coastal fishery mortality that raises a question of unacceptable impact on the stock. In the absence of adequate coastal state knowledge of the abundance of the stock, the same principle of unacceptable impact would argue for termination of this fishery. This specific situation does not seem to be addressed in the U.S. statement.

But the United States's policy statement does address this issue when the high seas fishery inflicts some mortality on a stock in addition to that of the coastal fishery. Without explanation or explicit statement of reasons, the U.S. statement provides that the fisheries within national jurisdiction have priority. The high seas fishery must carry the entire burden of the joint impact of mortality on the species affected. This is not wholly without foundation, but in the absence of some effort to negotiate with the foreign fishing state, it seems an unreasonable application of coastal authority under the 1982 Convention. This is, in fact, the problem of straddling stocks. This position of the United States, which would justify unilateral termination of the high seas fishery without first attempting cooperative actions, seems incompatible with the provisions of the 1982 Convention.[67]

Finally, it is not compatible with this treaty to conclude that compliance with the recommendation in the UN resolution must be considered to determine the lawfulness of a high seas fishery. The U.S. policy statement proclaims that unless a joint assessment of a specific drift net fishery "concludes there is no reasonable expectation of unacceptable impacts by that fishery, the conditions of relief from the moratorium recommended in U.N. Resolution 44/225 are not met. In this event, such a pelagic driftnet fishery cannot operate legitimately in areas beyond the EEZ of any nation after June 30, 1992."[68] The 1982 Convention contains different conditions of legitimacy. The treaty provisions should not be displaced by a unilateral interpretation of a United Nations recommendation adopted by the General Assembly. This way lies the chaos that the treaty was intended to avoid.

U.S. policy inconsistent with UN Resolution 44/225. The U.S. policy statement also is not consistent with the UN resolution, especially paragraph 4(a). The U.S. summary of this paragraph distorts the resolution and attributes to it undertakings that it does not require. The United States asserts that 4(a) requires an assessment of a specific drift net fishery that shows that no unacceptable impacts of the fishery will occur. This is not what the resolution provides. It calls for a "statistically sound analysis" to be the basis for effective conservation and management measures to prevent "unacceptable impact of such fishing practices on that region and to ensure the conservation of the living marine resources of that region."[69] What is required is a "statistically sound analysis" on which management measures are based to prevent unacceptable practices, and so forth. If this is done, the resolution is satisfied. This is not the same as requiring an assessment showing that no unacceptable impacts occur. A statistically sound analysis may not demonstrate that there are no unacceptable impacts because the data are insufficient, but it does not follow that the fishery must terminate. The analysis may still be sufficient to indicate needed conservation measures, and if these are instituted, the recommendation is satisfied.

There is an enormous difference between these two obligations. The one provided by the UN resolution can be satisfied, while that urged by the United States cannot be because it calls for a prognosis that an unidentified condition will not occur. Even if the prospective condition can be identified, a prognosis might not be possible because data may not be available to support the specific conclusion or to enable the specific assessment to be made. The former obligation, the one actually demanded by the resolution—an analysis showing that the conservation measures are needed—can be accomplished on the basis of the best available scientific data, even though those data might be insufficient to conclude that there are no possible impacts. A sound statistical analysis may be made of data that are insufficient for the purpose of ensuring that no specific impacts will actually occur. In other words, the United States seeks to transform the obligation from one that permits a judgment based on available data (which may not be very good) to one that is impossible to discharge because there are no data or because data are inadequate. Certainly, any alleged principle of this kind is incompatible with the 1982 Convention, which in respect to high seas fishing calls for cooperation to conserve fisheries by measures that take into account the best scientific evidence available. If the United States's alleged principle of jointly agreed assessments were actually required by the law of the sea, the effect would be to halt virtually all fishing now undertaken beyond the 200-mile EEZ, and if it were applied inside 200 miles, it would have similar impacts.

CONCLUSION

The preceding discussion identifies newly created conditions on freedom of fishing on the high seas whose implementation would permit some progress to be made in bringing such fishing under management control. Thus far, however, the states concerned have been unwilling or unable to advance or adopt interpretations of the available principles that would support appropriate action. Nonetheless, coastal states having straddling stock difficulties have proposed that they alone should prescribe the conservation measures with which fishing states should comply on the high seas, but they refrain from identifying and explaining the legal principles that support this result or suggest a course of conduct. Despite this reticence, the somewhat abstract provisions and language of parts V and VII of the 1982 Convention provide a basis for action. Article 116 subjects high seas fishing to the rights, duties, and interests of coastal states in some circumstances and thereby raises a serious problem of interpretation. If, as some assert, a coastal state has no authority whatsoever under the treaty beyond the exclusive economic zone, how are such rights, duties, and interests to be given effect in the face of a refusal to recognize them? To insist on the ancient freedom of fishing doctrine simply ignores the problem and dismisses the treaty provisions as irrelevant.

An appropriate interpretation of the treaty would facilitate protection of such rights, duties, and interests and would also preserve high seas fishing. Where necessary, this would entail some degree of coastal authority beyond 200 miles while also maintaining the substantial rights of the high seas fishing state. The latter rights might be qualified and conditioned but not eliminated.

In several contexts of high seas fishing, the lack of adequate mechanisms for decision making has resulted in the unilateral employment of economic sanctions by one state, the United States, which presently has the market strength to mount such sanctions for its own objectives. Perceived problems involving high seas fishing and current international organizations include those concerning whales, tuna and dolphins, salmon, and drift nets. Yet it is not at all clear that the United States's sanctioning efforts regarding such problems (all of which have been high seas disputes) have actually achieved the aims set for them or will achieve them. True, there have been embargoes and refusals of permits to fish, but these have not removed the problem, and the difficulties persist. Moreover, the reactive measures may either be counterproductive in other international relations or have undesirable side effects. In some instances, the sanctioning effort has been considered to be a violation of international trade agreements and has led to controversies over this impact.[70] In no instance known to this author have the unilateral sanctions led to greater understanding between the fishing state and the objecting state such that the problem was substantially removed once and for all.

Among the difficulties of national sanctions such as those employed by the United States is that they may well support standards that are idiosyncratic or provincial. The United States's experience documents this point in regard to its position on tuna and, more recently, drift nets. The 1990 tuna amendment to the Magnuson Act finally brings the United States into step with the rest of the world regarding coastal state jurisdiction over these species. But we now also have the spectacle, derived from political forces in Congress, of the legislative branch seeking to direct the executive branch to negotiate a worldwide ban on any use of drift nets on the high seas, which will be sanctioned by the embargo provisions of the Fishermen's Protective Act. Apart from the arrogance and ignorance afflicting this posture as a matter of substantive policy,[71] the effect is to mandate a policy for the United States that is in conflict with its position on the same subject at the United Nations and with the UN resolution itself. In view of this contradiction, it may be hoped that the executive branch will simply ignore the recommendation and the threat of further economic sanctions. Certainly, the United States should take no action to interfere with an activity that complies with the UN resolution.

Remedies for the many high seas problems do not seem likely to arise from any new set of general principles applicable to everything happening on the high seas as a whole. The more hopeful alternative is to deal with the many problem areas on a regional or other multilateral basis, fashioning the decision-making system, the objectives to be sought, and the criteria of decision for each context. In this regard, what are some criteria for acceptable solutions?

1. As for any fishery within national jurisdiction, on the international level a single management unit should be fashioned that can effectively handle the conservation and allocation tasks in approximately the same manner as if the fishery were within national waters. That is, effective authority must be exercised by a single entity, whether national or international. It is conceivable that in some instances this might actually be accomplished by a group of states coordinating their conservation activities and measures even without a formal and separate organization, but this would be very difficult. Even if uniform regulatory measures are adopted by such states, enforcement on a national basis may be ineffective and in the end may destroy the credibility of the regulatory scheme. Even the first essential step of management, that of building a solid information base, is difficult to accomplish on a decentralized basis, and the use of the product of such a system is often contaminated by political considerations inimical to sound and effective management. In the end, a separate fishery management body seems the only effective solution. The virtually unbroken record of failure by the existing and previous fishery bodies is not encouraging, but if nations genuinely cooperate to realize maximum net benefits, the gains should be enough to encourage effective management.

2. Where international agreement is unobtainable after good faith negotiations, a required management unit will probably need to be established by the coastal state acting unilaterally in regard to fisheries in an appropriate geographic context. This unilateral move might be accomplished by relying on interpretations of the 1982 Convention's provisions in support, coupled with resort to dispute settlement institutions as a means of removing or at least mitigating any arbitrariness. This latter way of proceeding offers some hope that the high seas fishing state will not be completely excluded from the fishery. The availability of adequate dispute settlement arrangements is critical to solving high seas fishery problems.

3. A third criterion of acceptability is that shared stocks actually be shared. Straddling stocks are not stocks that are wholly within national jurisdiction, as are exclusive economic zone stocks, over which the coastal state has largely discretionary authority. Amicable arrangements for their regulation may be feasible only if the high seas fishing state retains benefit from the fishery in which

it engages. If acceptance of a coastal state regulation simply means complete loss of fishing opportunity, the opposition of the high seas state is ensured.

This point needs to be emphasized because the "conservation" measures involved will probably always be influenced by and aimed at maintaining or improving the coastal state's economic benefits. If the coastal state sets an allowable catch for the single stock, it will do so either with an eye on stock abundance and its relationship to net benefits of coastal fishermen or with an eye on the overall foreign effort and how that will affect coastal benefits. But with this in mind, it is necessary that benefits be shared rather than enjoyed by only one participant. The outcome must be an all-win, no-lose result rather than one that is completely one-sided. The precise sharing should be negotiated.

In this context, it makes sense (as it does not in other contexts) to postulate that the aim of regulation is to permit the largest sustainable catches from a particular fishery. On the international level, provision for larger sustainable catches might be optimum because it is more likely than provision for lesser catches to permit individual states to achieve their particular national purposes. This recommendation is not wholly realistic, however, because of the severe difficulties in satisfying states whose goals relate to different species in a multi-species fishery. Negotiating these trade-offs will be a daunting task.

4. For straddling stocks, a major principle is consistency with coastal state management measures for the target fishery. Where the coastal state has management measures in force, the adjacent high seas state should comply with them, so long as they do not discriminate. This is a common recommendation by states with straddling stocks, and it has a good deal of merit.

Consistency also presents a large problem, however, because if this recommendation is applied also to by-catches, it could mean the termination of the target fishery on the high seas. For example, the U.S. policy statement on drift nets calls in effect for application of the standards of the U.S. Marine Mammal Protection Act on the high seas. It would follow that where high seas fishing takes marine mammals for which inadequate information exists to determine an estimate of population abundance, it would be impossible to show no unacceptable impact and therefore the high seas fishery must terminate. It is an understatement to say that this outcome is not likely to be widely acceptable. An alternative approach could incorporate the existing provisions of the 1982 Convention on the by-catch problem, for these do not completely prohibit all by-catch of marine mammals, but they do call for avoiding a level of harvest that might seriously threaten their reproduction.

5. Reduction of high seas fishing effort, that is, of capital investment, must be a high priority because this is a pervasive problem, afflicting almost all fisheries.[72] Coastal state measures to conserve high seas resources are not enough by themselves. Successful efforts to limit overcapitalization in one fishery are likely

to shift the increase in effort to other fisheries, where the added effort is also a waste of resources and dissipates the benefits of the fishery. This goal perhaps should be the highest priority. The implications of it are significant, including the implication that it can no longer be acceptable for individual fishing states to provide subsidies for their flag state vessel fishing activities. Another is that continued use of flags of convenience in order to escape fishing regulations must be made unacceptable.

6. Proposals for international fishery bodies to deal with high seas fishery problems should include all of the species and populations that require regulation, including by-catch. It is not enough to establish regional or multilateral regimes to deal with some of the target species that are considered needful of regulation. In most instances, there are by-catch problems that are significant, and these need also to be resolved.

NOTES

1. A number of situations illustrate this claim, including the dispute about fishing in the high seas enclave called the "donut" in the central Bering Sea, the ongoing controversy in the Northwest Atlantic over fishing outside the Canadian 200-mile fishing zone, Mexican claims to tuna in the high seas of the eastern tropical Pacific Ocean, the fishery in the Southwest Atlantic off the coast of Argentina, the jack mackerel fishery in the Southeast Pacific outside the Chilean 200-mile zone, and fishing for orange roughy near New Zealand. *See generally* E. Miles & W. Burke, *Pressure on the United Nations Convention on the Law of the Sea of 1982 Arising from New Fisheries Conflicts: The Problem of Straddling Stocks*, 20 OCEAN DEV. & INT'L L. 343, 344-49 (1989).
2. U.P.I. Wire Serv., Nov. 20, 1990.
3. Miles & Burke, *supra* note 1, at 344-45; *see also infra* note 36 and accompanying text.
4. Miles & Burke, *supra* note 1, at 346-47.
5. Pub. L. No. 101-627 §§ 101(b)(1), 103(a)-(c). Although the effect of this new authority over tuna in the U.S. EEZ was delayed until January 1, 1992, the president's statement on signing the amendments states that "[a]s a matter of international law, effective immediately the United States will recognize similar assertions by coastal state[s] regarding their exclusive economic zones." 26 WEEKLY COMP. PRES. DOC. 1932 (Dec. 3, 1990). *See also* Treaty on Fisheries Between the Governments of Certain Pacific Island States and the Government of the United States of America, Apr. 2, 1987, 26 I.L.M. 1048 (1987) [hereinafter FFA States–U.S. Fisheries Treaty].
6. One factor accounting for this development is that apparently only one or two U.S. flag tuna-fishing vessels still operate in the eastern tropical Pacific. All of the others have been transferred to foreign flags. Accordingly, the reason for the tuna exclusion in the U.S. law no longer exists. The United States had by agreement recognized the jurisdiction of the Pacific Islands in their EEZs. See FFA States–U.S. Fisheries Treaty, *supra* note 5.

7. ˮThe former Soviet Union has placed a total ban on salmon fishing in Soviet territorial waters in the North Pacific by 1992. Kyodo News Serv., June 8, 1990. *See also United States Urged to Join Soviets in Bering Sea Fishing Ban*, Reuters Wire Serv., Oct. 10, 1990. The U.S.-Soviet drafts of salmon agreements for the North Pacific that are currently under examination would achieve this objective.

8. "[S]ignificant yields [of Antarctic minke whales] can be sustained, probably indefinitely and at least for decades." Gulland, *Commercial Whaling—the Past, and Has It a Future?*, 20 MAMMAL REV. 3, 11 (1990). For a discussion of these issues, see *id.* and Holt, *Whale Mining, Whale Saving*, 9 MARINE POL'Y 192, 212-13 (1985).

9. *See generally* Hall & Boyer, *Incidental Mortality of Dolphins in the Tuna Purse-Seine Fishery in the Eastern Pacific Ocean During 1988*, 40 REP. INT'L WHALING COMM'N 461 (1990); Hofman, *Cetacean Entanglement in Fishing Gear*, 20 MAMMAL REV. 53 (1990); Coe, Holts, & Butler, *The "Tuna-Porpoise" Problem: NMFS Dolphin Mortality Reduction Research, 1970-81*, 46 MARINE FISHERIES REV. 18 (1984).

10. *See, e.g.*, Suzuki, *Description of Japanese Pelagic Driftnet Fisheries and Related Information*, in *Report of the Expert Consultation on Large-Scale Pelagic Driftnet Fishing*, U.N. Food and Agriculture Organization (Fisheries Rep. No. 434, Rome 1990), App. H, U.N. Doc. FIPL/R433. For a description and assessment of drift net fishing by ocean, *see Report of the Secretary-General to the United Nations General Assembly, Large-Scale Pelagic Driftnet Fishing and Its Impact on the Living Marine Resources of the World's Oceans and Seas*, U.N. Doc. A/45/663 (Oct. 26, 1990).

11. *See Obscure Measure Blocks Marine Conservation Efforts*, U.P.I. Wire Serv., Dec. 11, 1990.

12. *See U.S. Drafting Catch Limits As Sharks Grow Scarce*, N.Y. Times, Mar. 11, 1990, § 12 at 1, col. 1.

13. As noted *infra*, the United States policy regarding drift nets is cast in terms that could embrace other gear.

14. There is also some concern over exploitation of Antarctic krill as affecting dependent species. *See* Beddington, Basson, & Gulland, *The Practical Implications of the Eco-System Approach in CCAML*, 10 INT'L CHALLENGES 17 (1990).

15. *See* James Carr & Matthew Gianni, *High Seas Fisheries, Large-Scale Drift Nets, and the Law of the Sea* (chapter 18 in this book).

16. Serge Garcia & J. Majkowski, *State of High Seas Resources*, in THE LAW OF THE SEA IN THE 1990S: A FRAMEWORK FOR FURTHER INTERNATIONAL COOPERATION 175 (Tadeo Kuribayashi & Edward L. Miles eds. 1992).

17. Even within national jurisdiction, prior foreign fishing is recognized in the 1982 Convention on the Law of the Sea as a factor bearing on decisions about access and allocation, although the coastal state retains ultimate authority over whether to give weight to this factor and how much. United Nations Convention on the Law of the Sea, Dec. 10, 1982, U.N. Doc. A/CONF.62/122, 21 I.L.M. 1261 (1982), arts. 62(3), 69(4), 70(5), in 17 *Official Records, Third United Nations Conference on the Law of the Sea* 139 (1984) [hereinafter 1982 Convention].

18. The principles of international law mentioned in this discussion are also applicable to fishing conducted with drift net gear. As is evident from recent investigations into

the impact of the gear, fish and mammals that are straddling stocks are also inter-
cepted by drift nets on the high seas.

19. Convention on the High Seas, *entered into force* Sept. 30, 1962, 450 U.N.T.S. 82, 123
    U.S.T. 2312, T.I.A.S. No. 5200.

20. The preamble to the 1958 Convention on the High Seas notes also that the
    principles in this treaty were adopted as "generally declaratory of established
    principles of international law." 2 OFFICIAL RECORDS, UNITED NATIONS
    CONFERENCE ON THE LAW OF THE SEA 135, U.N. Doc. A/Conf.13/L.53
    (1958).

21. *Id.* at 139.

22. As of January 1990, thirty-six states were reported to have ratified or acceded to this
    treaty, none since the adoption of the 1982 Convention. *Multilateral Treaties Deposited
    with the Secretary-General, Status as at Dec. 31, 1988*, U.N. Doc. ST/LEG/Ser.B/7
    (1989).

23. This was not the interpretation in 1968. Denmark ratified in that year and declared
    that it did not consider itself bound by the last sentence of article 2.

24. Fisheries Jurisdiction case (United Kingdom v. Iceland), 1974 I.C.J. 3, 31, para. 72
    (Merits).

25. The convention is not yet in force, but it has been signed by virtually all states in the
    world and some other entities, a total of 159 political units. The United States, the
    United Kingdom, and the Federal Republic of Germany are the major nonsignato-
    ries. *See* OFFICE OF THE SPECIAL REPRESENTATIVE OF THE SECRETARY-GENERAL
    FOR THE LAW OF THE SEA, THE LAW OF THE SEA: STATUS OF THE UNITED
    NATIONS CONVENTION ON THE LAW OF THE SEA (1985). The principal articles in
    the convention concerning fisheries are widely considered now to be part of
    customary international law, especially those concerning fisheries in the exclusive
    economic zone.

26. *See infra* notes 33–37 and accompanying text.

27. Clark, *Bioeconomics of the Ocean*, 31 BIOSCIENCE 231 (1981).

28. The NAFO controversy is examined in detail in OCEAN INSTITUTE OF CANADA,
    MANAGING FISHERY RESOURCES BEYOND 200 MILES: CANADA'S OPTIONS TO
    PROTECT NORTHWEST ATLANTIC STRADDLING STOCKS (1990). *See also* B. Ap-
    plebaum, *The Straddling Stocks Problem: The Northwest Atlantic Situation, International
    Law, and Options for Coastal State Action*, in IMPLEMENTATION OF THE LAW OF THE
    SEA CONVENTION THROUGH INTERNATIONAL INSTITUTIONS (Alfred H. A.
    Soons ed. 1990).

29. 1974 I.C.J. 3.

30. Large–Scale Pelagic Driftnet Fishing and Its Impact on the Living Marine Resources
    of the World's Oceans and Seas, G.A. Res. 44/225, U.N. GAOR Supp. (No. 49),
    U.N. Doc. A/44/746/Add.7 (1989) [hereinafter UN Driftnet Resolution], *reprinted
    in* 20 ENVTL. POL'Y & L. 36 (1990). For a more extended discussion of these
    principles, *see* William T. Burke, The International Law of the Sea Concerning
    Coastal State Authority over Driftnets on the High Seas (paper prepared for United
    Nations Food and Agriculture Organization, Feb. 1990, in press). *See also* Douglas

M. Johnston, *The Driftnetting Problem in the Pacific Ocean: Legal Considerations and Diplomatic Options*, 21 OCEAN DEV. & INT'L L. 5 (1990).

31. 1982 Convention, *supra* note 17.

32. *Id.* art. 66(3)(a).

33. *Id.* art. 116.

34. *See* Miles & Burke, *supra* note 1, at 350–53.

35. 1982 Convention, *supra* note 17, art. 66(3)(d). *See also* art. 66(5), which states: "The State of origin of anadromous stocks and other States fishing these stocks shall make arrangements for the implementation of the provisions of this article, where appropriate, through regional organizations."

36. See *infra* note 53.

37. Kunio Yonezawa, *Some Thoughts on the Straddling Stock Problem in the Pacific Ocean*, in THE LAW OF THE SEA IN THE 1990S: A FRAMEWORK FOR FURTHER INTERNATIONAL COOPERATION 175 (Tadeo Kuribayashi & Edward L. Miles eds. 1992).

38. A prohibition against local and foreign harvesting of a stock might also be authorized by article 61 of the 1982 Convention, *supra* note 17, which contains broad language permitting the coastal state to regulate yields in accordance with environmental and economic factors. In addition, the coastal state is authorized to protect associated or dependent species. Such protection may require a complete cessation of fishing that takes such species.

39. Zilanov & Vylegzhanin, *Termination of the Interim Convention on Conservation of North Pacific Fur Seals* (1957), in SOVIET Y.B. MAR. L. 79, 86 (1989).

40. The first detailed observations of this fishery, still not considered wholly representative of the whole fishery, nonetheless report that the by-catch of salmon observed to be taken is almost infinitesimal. CANADIAN DEPARTMENT OF FISHERIES AND OCEANS, FISHERIES AGENCY OF JAPAN, U.S. NATIONAL MARINE FISHERIES SERVICE, & U.S. FISH AND WILDLIFE SERVICE, FINAL REPORT OF SQUID AND BYCATCH OBSERVATIONS IN THE JAPANESE DRIFTNET FISHERY FOR NEON FLYING SQUID (*OMMASTREPHES BARTRAMI*), JUNE–DECEMBER 1989 OBSERVER PROGRAM (June 30, 1990) [hereinafter FINAL REPORT]. As the Japanese have not failed to point out, the harvest of salmon in Alaska in 1989 was the largest on record.

41. International North Pacific Fisheries Convention (INPFC), *signed* May 9, 1952, 205 U.N.T.S. 65, *amended in* 1962, 14 U.S.T. 953; *amended in* 1978, 30 U.S.T. 1095; *amended in* 1986, 86 D.S.B. 73–74 (June 1986). For a discussion of the INPFC, *see* Sathre, *The International North Pacific Fisheries Commission: A Thirty-Year Effort to Manage High Seas Salmon and Some Suggestions for the Future*, Anadromous Fish Law Memo No. 29, May 1985.

42. For a discussion of the Japan–USSR agreement, *see* E. MILES ET. AL., THE MANAGEMENT OF MARINE REGIONS: THE NORTH PACIFIC 88, 127 (1982). For the outcome of more recent negotiations, *see* U.S. NATIONAL MARINE FISHERIES SERVICE, JAPAN–USSR SALMON FISHERY COOPERATIVE AGREEMENTS, 1985–87 (Apr. 2, 1987).

43. The United States threatened sanctions against both Korea and Taiwan for drift net fishing activities in the North Pacific. Agreement between these countries was completed in late 1989.

The U.S.-Korea agreement is in the arrangements concluded on September 8, 1989, by an exchange of letters between the United States and Korea: Agreement Regarding the High Seas Squid Driftnet Fisheries in the North Pacific Ocean, with Record of Discussions and Exchange of Letters, *effected by exchange of notes at* Washington Sept. 13 and 26, 1989, *entered into force* Sept. 26, 1989, *reprinted in* 29 I.L.M. 464 (1990) (Korea); Letter to the Speaker of the House and the President of the Senate Reporting on Korean and Taiwanese Driftnet Fishing, *reprinted in* 25 WEEKLY COMP. PRES. DOC. 1278 (Aug. 28, 1989); *White House Fact Sheet on Environmental Initiatives, reprinted in* 25 WEEKLY COMP. PRES. DOC. 1392 (Sept. 18, 1989).

44. 1982 Convention, *supra* note 17, art. 66(3)(a).

45. But there may well be disagreement about what is appreciable. A given number of fish or birds may appear to be absolutely large but may be relatively small. Thus, 1 million fish sounds like a great many, but the number loses impact if it is considered part of a population of several hundred million. The recent joint report by Japan, Canada, and the United States on observations of the Japanese high seas squid drift net fishery showed that the salmon by-catch was almost nonexistent. The total of observed salmon in the by-catch was seventy-nine; in most months, no salmon were taken. See FINAL REPORT, table 3, *supra* note 40. It may be, however, that the area covered by the observations was not one where salmon would be expected to be found.

46. For a discussion of this argument, *see* Miles & Burke, *supra* note 1, at 352-54.

47. Drafts of these proposals are in the author's files.

48. S. Res. 396, 100th Cong., 2d Sess., 135 CONG. REC. S2617 (daily ed. Mar. 21, 1988).

49. Article XI of the NAFO Convention provides:

> 2. The Commission may adopt proposals for joint action by the Contracting Parties designed to achieve the optimum utilization of the fishery resources of the Regulatory Area. In considering such proposals, the Commission shall take into account any relevant information or advice provided to it by the Scientific Council.
>
> 3. In the exercise of its functions under paragraph 2, the Commission shall seek to ensure consistency between:
>
> (a) any proposal that applies to a stock or group of stocks occurring within the Regulatory Area and within an area under the fisheries jurisdiction of a coastal State, or any proposal that would have an effect through species interrelationships on a stock or group of stocks occurring in whole or in part within an area under the fisheries jurisdiction of a coastal State, and
>
> (b) any measures or decisions taken by the coastal state for the management and conservation of that stock or group of stocks with respect to fishing activities conducted within the area under its fisheries jurisdiction.

The "Regulatory Area" under NAFO is the Convention Area beyond the fishery jurisdiction of coastal states.

50. NAFO Resolution on Non-NAFO Fishing Activities, NAFO/GC Doc. 90/8, 12th Annual Meeting (Sept. 1990). For text, *see* annex II in W. Burke & F. Christy,

Jr., *Options for the Management of Indian Ocean Tuna*, U.N. Food and Agriculture Organization (Fisheries Technical Paper No. 315, 1990). *See also* B. Applebaum, *The Straddling Stocks Problem: The Northwest Atlantic Situation, International Law, and Options for Coastal State Action*, in IMPLEMENTATION OF THE LAW OF THE SEA CONVENTION THROUGH INTERNATIONAL INSTITUTIONS 282 (Alfred H. A. Soons ed. 1990).

51. *Id.* at 11–12.

52. Kokechik v. Secretary of Commerce, 839 F.2d 795 (1988), *cert. denied* 488 U.S. 1004 (1989); U.S. Policy Statement on Large-Scale Pelagic Driftnets 4 (submitted to the United Nations Office of Ocean Affairs and the Law of the Sea, July 1990) [hereinafter U.S. Policy Statement].

53. Convention for the Prohibition of Fishing with Long Driftnets in the South Pacific, *reprinted in* 14 L. SEA BULL. 38 (Dec. 1989).

54. Pub. L. No. 101-627 (1990). Section 107 provides for an embargo on importation of fish caught by drift nets, and elsewhere the 1990 amendments to the Magnuson Act call for a permanent ban on drift nets.

55. UN Driftnet Resolution, *supra* note 30.

56. U.S. Policy Statement, *supra* note 52.

57. Tarawa Declaration of July 11, 1989, *reprinted in* 14 L. SEA BULL. 24 (Dec. 1989).

58. Although the area to be regulated under the convention contains a significant portion of high seas, jurisdiction over prohibition of drift net fishing remains within the member states' individual fisheries jurisdiction. The only possible extension of enforcement jurisdiction is contained in article (3)(1)(b).

59. UN Driftnet Resolution, *supra* note 30.

60. After referring to a failure to meet the requirements of the joint assessment under the drift net resolution, the United States policy statement declares: "In this event, such a pelagic driftnet fishery cannot operate legitimately in areas beyond the EEZ of any nation after June 30, 1992." U.S. Policy Statement, *supra* note 52, at 6. This leap from a description of a United Nations resolution to illegality is not explained.

61. *Id.* at 2.

62. *Id.* at 3.

63. *Id.*

64. *Id.*

65. 1982 Convention, *supra* note 17, art. 119(1)(b).

66. The UN resolution called for evaluation on the basis of "sound statistical analysis," although the need for more scientific information is mentioned in paragraphs 2 and 3. *See* UN Driftnet Resolution, *supra* note 30, at para. 4(a).

67. Article 297(3)(a) requires states to settle fishing disputes in accordance with the compulsory procedures contained in the 1982 Convention, *supra* note 17.

68. U.S. Policy Statement, *supra* note 52, at 6.

69. UN Driftnet Resolution, *supra* note 30, at para. 4(a).

70. *See* the GATT tuna-dolphin discussion in Melinda P. Chandler, *Recent Developments in the Use of International Trade Restrictions as a Conservation Measure for Marine Resources* (chapter 21 in this book).

71. This provision was adopted despite the fact that the studies mandated by Congress to determine the extent of the problem resulting from high seas drift netting were not complete. The amendment itself refers to a "pressing need for detailed and reliable information" on species taken in the by-catch. Presumably, the "pressing need" is for data that will support the policy of completely banning high seas drift netting.

72. *See* Claudia J. Carr, *The Legacy and Challenge of International Aid in Marine Resource Development* (chapter 22 in this book).

18 High Seas Fisheries, Large-Scale Drift Nets, and the Law of the Sea

James Carr and Matthew Gianni

GLOBAL FISHERIES ARE increasingly threatened with overexploitation, waste, and in some cases collapse. As competition for dwindling fish stocks intensifies, more powerful and efficient vessels and methods, including large-scale drift nets, have been developed and deployed in a race to exploit remaining fish stocks. This burgeoning fishing capacity is outpacing the abilities of scientists, managers, politicians, and the industries themselves to control the overexploitation now common in every region of the world's oceans. "The relative failure of international management to establish sustainable fisheries in many areas . . . is clearly demonstrated by the dwindling resources, excessive catching capacity, uncontrolled transfers of fishing effort between resources and oceans, and depletion of many highly valuable resources," according to the United Nations Food and Agriculture Organization.[1]

The freedom to fish on the high seas provides open access to marine resources. As competition has intensified for highly prized species of tuna, billfish, salmon, and squid, high seas drift net fishing has proliferated.[2] Fleets deploying huge lengths of nylon netting, often as much as 50 to 60 kilometers per boat per night, can be a highly effective means of intercepting fish in relatively unknown and vast expanses of water, even when stocks are dispersed. Yet this method of fishing results in a high incidental catch of nontarget species of fish and other wildlife.[3] Virtually all observer reports and experimental drift net fisheries lead to the conclusion that high by-catch is unavoidable. Further, concerns are now being expressed by coastal states, including many developing countries, that highly migratory stocks important for potential future fisheries may be irreversibly depleted by the operation of these immensely effective fleets.[4]

The 1982 United Nations Convention on the Law of the Sea has established general provisions for cooperation and conservation in high seas fisheries and specific regimes for highly migratory species, straddling stocks, and some endangered species.[5] These provisions have found expression in United Nations General Assembly Resolution 44/225 on high seas drift nets, adopted by consensus in 1989, which called for a provisional cessation to this indiscriminate and wasteful high seas fishing method.[6] This was an important step toward international cooperation and the long-term health of the ecosystem on which we depend.[7]

UN Resolution 44/225 states that "large-scale pelagic driftnet fishing . . . can be a highly indiscriminate and wasteful fishing method that is widely considered to threaten the effective conservation of living marine resources, such as highly migratory and anadromous species of fish, birds and marine mammals."[8] The resolution calls for a cessation of large-scale drift net fishing in the South Pacific region by July 1, 1991, an immediate halt to the expansion of large-scale drift net fishing on the high seas in all regions of the world's oceans, and a global moratorium on large-scale drift net fishing on the high seas by June 30, 1992, unless and until "effective conservation and management measures" based on statistically sound analysis can be implemented that prevent unacceptable impacts and ensure the conservation of living marine resources of the high seas.

Subsequent to passage of the resolution, the debate continued and expanded as evidence became increasingly available with respect to the impacts of high seas drift net fishing. In 1990, the United Nations adopted Resolution 45/197, a reaffirmation of Resolution 44/225. International scientific reviews on the impact of large-scale drift net fishing, as called for in both the 1989 and 1990 resolutions, were conducted for the northern and South Pacific regions. These reviews established that it could not be demonstrated that the impacts of the drift net fisheries were not "unacceptable." In December 1991, the UN General Assembly affirmed that the conditions for allowing continuation of the drift net fisheries, as set out in the previous resolutions, had not been met and unanimously adopted Resolution 46/215. This resolution calls for a global moratorium on all large-scale drift net fishing on the high seas by December 31, 1992. It remains to be seen whether this moratorium will be effectively implemented and enforced. Full implementation of the moratorium is critical to the success of future negotiations and agreements on high seas fisheries in general.

IS EFFECTIVE MANAGEMENT OF HIGH SEAS DRIFT NET FISHING POSSIBLE?

Proponents of high seas drift net fishing claim that large-scale pelagic drift nets are no different from any other fishing gear that may result in some negative impacts on the marine environment.[9] They also have claimed that high seas drift netting can be managed as an ecologically sound fishing method and that such fishing must be allowed to continue in order to determine the effects of drift netting on the marine ecosystem.

These views are based on the premises that there is no qualitative difference between large-scale drift net fisheries on the high seas and small-scale coastal gill nets and that high seas drift nets can be utilized in conjunction with conventional measures such as area or seasonal closures and modifications in design and deployment of the gear to avoid "unacceptable" impacts.[10]

Small-Scale versus Large-Scale Drift Nets

Small-scale coastal gill nets are among the most widely used gear type in the world. Although numerous examples exist of this method being responsible for significant incidental takes of marine animals, small-scale gill net fishing can be, if conducted in an appropriate fashion, one of the more selective forms of fishing. In any case, several important distinctions exist between large-scale drift nets used on the high seas and small-scale gill nets used in coastal areas within EEZs, especially concerning the potential to prevent negative impacts on the ecosystem through the use of regulatory measures.

Large-scale pelagic drift nets on the high seas are generally used in areas where stocks are widely dispersed and where the migratory patterns, distribution, and abundance of populations of targeted species are poorly understood. Large quantities of netting must be deployed on the high seas to ensure adequate catches. Because information on the distribution and abundance of target populations is rarely available, even less is likely to be known of nontarget species of fish, birds, and marine mammals on the vast expanses of the high seas, and therefore they cannot be easily avoided as by-catch.

On the other hand, small-scale gear is often deployed in areas where the distribution of target species is relatively concentrated and better known. Also better known are the locations and migration routes of potential by-catch species, including nontarget species of fish, whales, dolphins, seals, and other marine animals. This information affords much greater opportunity to avoid potentially harmful incidental takes.

High seas drift net fisheries also use greater lengths of net to recoup the significant capital expenditures and fuel costs involved in high seas fishing operations. High by-catches on nontarget species of fish, mammals, and birds resulting from this intensive and lucrative fishing method are likely to be viewed merely as irrelevant externalities.

Small-scale coastal gill net fisheries generally operate close to shore, and delivery to markets is consequently much faster. This proximity vastly reduces the costs to the fishers. Shorter lengths of net are therefore economically viable.

Escapement and predation on the catches also are likely to be far less, as is the by-catch of birds and marine mammals attracted to the prey in the nets. Further, with the use of small nets, the quality of the catch is likely to be higher because of less "soak time."

Large-scale drift nets are more apt to be lost or cut loose by vessels under adverse conditions. Lost nets are known to pose a hazard to navigation.[11] Furthermore, vessels passing over these nets are likely to sever sections, adding to the problem of lost gear, referred to as "ghost nets." The Japanese government has estimated conservatively that 0.05 percent of the nets set every night by Japanese vessels are not recovered, which means that 10 miles of netting is lost *each night* in the North Pacific from Asian vessels alone.[12] These estimates do not take into consideration the amount of net that may be discarded at sea during the course of fishing operations as the netting material becomes progressively more damaged and less effective. Japanese sources report that as much as one-third of the net material becomes unusable in an average trip by a squid drift net vessel.[13]

The information available on the effects of lost or discarded large-scale drift net gear certainly gives cause for concern.[14] Contrary to the claims of proponents of high seas drift net fishing that these nets simply ball up and sink, and therefore do not pose a threat of "ghost fishing,"[15] many scientists believe that lost or discarded nets may continue "passively" to ensnare and kill marine creatures indefinitely.[16]

Small-scale gear, in contrast, can be much more effectively tended and monitored. In coastal gill net fisheries, vessels often will remain attached to the nets while fishing. Small-scale gear can more readily be retrieved under adverse weather conditions or in the presence of migrating marine mammals and birds, approaching vessels, or other potential obstacles. Therefore, lost netting and consequent ghost fishing is likely to be much less of a problem in small-scale gill net fisheries than it is with high seas drift netting.

According to the 1990 report of the secretary general to the UN General Assembly, as much as 40 percent of the catch is discarded or lost as "drop-out" when large-scale pelagic drift nets are retrieved.[17] "High grading," the practice of discarding all but the largest or most valuable fish in a catch, is commonly done in high seas drift net fisheries to maximize the limited hold space of the vessels.

By contrast, in coastal fisheries much of the nontarget catch may be retained on board to be sold to local markets or sold "over the side" to vessels that deliver to the markets. Thus, at least the potential exists for reducing much of the waste associated with discards. There is also, of course, much greater opportunity for enforcement of any regulations against discarding within national waters, as opposed to the high seas. Moreover, many of the fish and other marine animals that survive an encounter with high seas drift nets suffer considerable damage in terms of skin and scale loss, effects that have been documented in the South Pacific. In addition to drop-out of dead fish and wildlife, this may represent a considerable source of eventual mortality.[18]

Collecting data and monitoring activities of fleets is potentially easier for fisheries within national jurisdictions as opposed to international waters. Reliable data are not readily available for catches on the high seas, and data collection is complicated by the mobility of high seas fleets and the difficulties of independently monitoring vessels for the accuracy of reported catches (especially in the case of by-catch).[19]

Endangered and Threatened Species

Large-scale drift net fishing raises serious concerns for threatened or endangered species of marine life.[20] Such concerns have played a major role in decisions by a number of governments to ban the practice within exclusive economic zones (EEZs) and by flag vessels on the high seas.

Regarding threatened or endangered species, the United Nations Food and Agriculture Organization's "Report of the Expert Consultation on Large-Scale Pelagic Driftnet Fishing" has stated that

> the conservation of all components of the ecosystem is an important consideration. This is not only because we know little of the role of each species in the ecosystem, but it is thought to be important in preserving stability of the system itself.[21]

Observer data gathered in the Japanese North Pacific squid fishery during 1990 show that a number of species are seriously threatened by the use of large-scale drift nets in the region.[22] For instance, leatherback turtles, considered endangered and listed in appendix I of the Convention on International Trade in Endangered Species of Wild Fauna and Flora (CITES), have a high mortality rate in that fishery, a situation that is "likely to have a serious impact on population numbers."[23] Loggerhead and green turtles are also known to be caught incidentally in the North Pacific high seas drift net fleets; the loggerhead turtle is listed as "vulnerable" and the green turtle as "endangered" by the International Union for Conservation of Nature and Natural Resources (IUCN).[24]

The negative impact of high seas drift net fishing on species of large whales is likely to be substantial but probably can never be determined with any degree of accuracy. Scientists have expressed concern for a number of whale species, including blue, sei, humpback, Bryde's, and minke whales, any of which may be taken in the large-scale drift net fisheries, according to the international "Scientific Review of North Pacific High Seas Driftnet Fisheries."[25] The review particularly called attention to the situation of both the blue whale and the eastern Pacific population of the northern right whales, which are "near extinction" (the population is estimated to be no larger than fifty animals).

Many of the cetacean species entangled or susceptible to entanglement in the North Pacific and other drift net fisheries are listed in appendix I of CITES, and the remaining cetacean species are listed in appendix II. There is no reason to doubt that the same holds true for high seas drift net fisheries in other ocean regions. The question is: Is it worth risking species extinction in order to continue the practice of high seas drift net fishing when alternative methods of fishing are available or can be developed to harvest the same fish stocks more selectively?

Populations of slow-growing and slow-reproducing species that are threatened or endangered not only must be prevented from declining but also must be allowed to recover. It could be argued that any mortality or significant risk of mortality to an endangered species itself constitutes an unacceptable impact. Yet attempts to assess the direct mortality of vulnerable species in high seas drift net fisheries is extremely difficult given the very limited amount of data reported by drift net fleets. Assessing indirect impacts is more difficult still. Little is known about feeding and migration habits of many of these species, and ascertaining, for example, the impact from forage reduction as a result of catch or by-catch of prey species in high seas drift net fisheries is, at present, virtually an impossible task.

Management Options

Another argument offered by proponents of high seas drift nets is that "high seas driftnet fishing, like other fishing methods, can be managed under appropriate conservation and management methods."[26] Implicit in this statement is the notion that high seas drift net fishing can be managed to reduce by-catch through conventional measures such as gear modifications, restrictions on fishing effort, and area and seasonal closures.

Results of experiments using modified large-scale drift net gear to determine whether a reduction of by-catch is possible are not encouraging. These experiments have been limited, generally focusing only on reducing the mortality of birds and mammals and not the full range of species caught in significant quantities. Further, they have shown little promise in reducing by-catch under commercially viable conditions.[27]

Also unencouraging are the results of experiments conducted employing various acoustic methods to avoid incidental takes of cetaceans. Experiments that fitted drift nets with air-filled plastic tubes and metallic beaded chains did not result in reduction of incidental marine mammal takes.[28] After reviewing the results of such experiments, the FAO concluded in 1990 that "[t]he use of either active or passive sonic devices has not so far been shown to reduce cetacean mortality."[29]

Adjusting the hanging ratio (tension) of the nets to reduce the incidental capture of cetaceans would probably also prove ineffective. Large cetaceans would probably continue to become entangled and tow sections of netting away from the area of deployment. Furthermore, large-scale drift nets are often loosely hung in order to increase the catch of high-value species of billfish, a fact noted by the South Pacific Forum in its review of drift net fishing.[30] Enforcing hanging ratio regulations would obviously be extremely difficult, particularly if altering the hanging ratio to avoid, for instance, mammal by-catch resulted in reduction of target catches.[31]

Based on a summary of data taken from Japanese research cruises, marine mammals were caught in virtually all mesh sizes used. The Japanese large-mesh fishery for tuna and billfish is regulated by a 150-millimeter mesh size. Presumably, this is to minimize the catch of undersized target species. However, larger meshes appear to be associated with higher mammal mortality, according to observer data.[32] It is ironic that efforts to conserve stocks of tuna and billfish by establishing minimum mesh size regulations may actually increase the by-catch of marine mammals.[33]

In fact, reducing the mortality of especially vulnerable species in large-scale high seas drift net fisheries to a level that may help ensure survival most probably would render the fisheries uneconomical. The FAO's Driftnet Report states:

> It was further noted that in the case of multiple species being caught by an unselective gear, the rate of harvest must be restricted to that rate which allows populations of the slowest growing and reproducing species to survive. This will be lower than the allowable catch possible on some commercially valuable components with gear adapted to single species capture.[34]

Applying conventional measures such as area and seasonal closures or reducing effort through lessening the total amount of net allowed per vessel to a more "manageable" level of length probably would also make the fishery economically untenable. In addition to the purely economic constraints on these options, virtually no means of enforcing such restrictions on the high seas exists, nor have credible means been proposed. Furthermore, given the lack of data on high seas migration patterns of threatened species or the distribution and abundance of nontarget species in general, there is presently no way to ascertain what constitutes an "acceptable" level of fishing effort. In this respect, reducing the total number of vessels allowed to fish with drift nets would pose similar problems. Given the by-catch associated with large-scale drift net fishing, it is virtually impossible to determine how many vessels could safely fish with drift nets and still ensure the conservation of all affected species. This may not be the case with many of the alternatives to high seas drift netting, which are known to provide higher-quality fish, and thus higher prices, because of the absence of net marks on the fish and greater freshness due to more timely retrieval. This in turn can provide equal or higher economic yield for fewer fish caught, thus placing less pressure on the resources while avoiding the significant problems associated with substantial by-catch.[35]

Proponents of large-scale drift net fishing also argue that drift netting should be allowed to continue for the purpose of collecting information on the status of target and associated species.[36] However, the South Pacific Forum concluded in its submission to the UN General Assembly that "[t]his argument . . . is completely at odds with the precautionary approach . . . and with general principles of modern fisheries management and development which require that exploitation of living marine resources be sustainable in the long term, result in the maximum benefit for all concerned parties, and not endanger the sustainability of other stocks or species caught."[37]

The Problem of Enforcement

Lack of fisheries enforcement on the high seas is one the most important distinctions between coastal gill net fishing and high seas drift net fishing. Even if a conservation and management regime were devised to address high seas drift net fishery problems, there would be no means to ensure that such a regime could ever be effectively implemented and enforced.[38]

Underscoring the difficulties with policing the high seas drift net fisheries has been the continued discovery of "pirate" fishing. Taiwanese authorities have acknowledged that "clandestine operations" are working in conjunction with large fish brokers and canneries selling salmon caught outside zones agreed on by North Pacific governments. Estimates of the illegal catches were as high as 30,000 tons.[39] A joint Canadian-U.S. surveillance program reported in 1991 sightings of several dozen Asian drift net vessels a considerable distance north of the internationally agreed boundary for squid fishing (set at 40 degrees north). The U.S. National Marine Fisheries Service reported that in mid-July 1990, seventeen ships from Korea and twenty-one ships from Taiwan were in areas off limits under the agreement. Some of the vessels were as much as 120 kilometers north of the boundary, and they were known to be setting and retrieving their nets.[40]

In another incident illustrating the vast enforcement problems associated with high seas drift netting, the Soviets seized a fleet of ten supposedly North Korean drift net vessels on May 20, 1990, only to find that more than 140 of the 200 crew taken into custody were in fact Japanese.[41] Apparently, Japanese boats had been painted with the flag of North Korea. Furthermore, the catch of several thousand tons of illegally caught salmon was destined for Tokyo fish markets, probably through transfer from the North Korean–flagged fishing boats to Japanese vessels. This embarrassing "poaching" incident is in clear violation of the existing agreements on controls designed to protect salmon stocks in the North Pacific. Countries may try to police these violations, but costs are excessive even for a developed country and are not realistic for most countries in areas such as the Indian Ocean and the South Atlantic.[42]

Coastal fisheries are potentially far more manageable than high seas fisheries from the standpoint of government regulation and enforcement. Further, coastal fishers are directly dependent on long–term sustained production from local fish stocks and are more likely to adhere to regulations. By contrast, capital–intensive distant-water fleets have tended to disregard the future of the resource in favor of short–term profits. After stocks are depleted, these fleets have the option to move on to less exploited species or areas. This is the now–familiar pattern known as "pulse fishing" that has been a major reason for the overexploitation prevalent in world fisheries over the course of the past few decades.

In conclusion, there is virtually no institutional framework in place to regulate high seas drift net fisheries. Efforts to control high seas fisheries are generally made in the context of international conventions and fishery commissions, usually by consensus. However, most of these organizations do not have the leverage or even the most basic data necessary to provide for conservation needs.[43] To date, the only high seas drift net fisheries to be brought under any form of international agreement are the squid and salmon drift net fisheries in the North Pacific. These fisheries, however, are subject to restrictions only with respect to their catch or by-catch of salmon, and even these restrictions are ineffective or extremely difficult to enforce. The take of the dozens of other species in these fisheries goes completely unregulated. Although recently some high seas drift net fisheries in the North Pacific have become subject to limited observer programs,[44] these programs do not by any means constitute a conservation and management regime. They will, for the foreseeable future, serve only to continue to document the highly indiscriminate and wasteful nature of these fisheries.

It is impossible to assess the status of species taken incidentally in high seas drift net fisheries or to determine the real impact of drift net fishing on populations because of insufficient information regarding the distribution, migratory patterns, and abundance of these species. Even less is known about the potential impact on species associated with or dependent on those incidentally caught species. Because an inordinately complex amount of research needs to be done to assess the full effects of drift net fishing on high seas ecosystems, the UN Driftnet Report says that "prudence would dictate that the management measures taken should be precautionary in nature."[45]

A precautionary approach would require a solid base of ecological information and a regulatory regime that allows for sufficient monitoring of the fishery to detect damage to the viability of populations of target and nontarget species. Such programs would include adequate time-series data on growth rates, recruitment, spawning areas and seasons, predator-prey relations, migration and distribution, natural mortality, catch and effort data, and fishing mortality of both target and nontarget species. In addition, the regime must also include adequate predictive models for assessing ecosystem dynamics over time in relation to the effects of fishing mortality. Although one cannot expect all of the above criteria to be fully realized before any fishing can occur, virtually none of them is met in the high seas drift net fishery.

Furthermore, the cost of gathering and evaluating such necessary data is prohibitive. Development, implementation, and enforcement costs for a regime to minimize the negative impacts of the fishery will far outweigh the short-term economic gain provided by using this technology, as opposed to other, more selective gear. Indeed, the South Pacific Forum's review of drift net fishing arrived at the following conclusion:

Attempts to develop management regimes for high seas drift net fishing are likely to be confounded by the fact that the costs of acquiring the information necessary to form the basis for rational and sustainable management for all fish and wildlife species affected by the fishery may be well in excess of the net value of sustainable production from the target species.[46]

The issue of high seas drift nets and the serious problems they are known to cause has been under international scrutiny for many years. It is clearly the consensus of the international community, as embodied in the UN Driftnet Resolutions, that high seas drift net fishing must end. The proponents of high seas drift net fishing have not demonstrated that the risks associated with this type of fishing are insignificant or that the fisheries can be managed to reduce such risks until the impact is adequately assessed. This is true with respect to the conservation not only of species caught as by-catch but also, in some cases, perhaps even of target species themselves. In fact, all evidence, from both observer programs and the available regional scientific reviews, convincingly demonstrates the insurmountable practical difficulties in managing this method of fishing. Given the minimum requirements of a fishery management regime, based on principles of comprehensive data and credible analytical methods, it is inconceivable that this large-scale and demonstrably indiscriminate method of fishing could regain the approval of the international community in the foreseeable future.

THE UN DRIFTNET RESOLUTIONS AND THE UNITED NATIONS CONVENTION ON THE LAW OF THE SEA

International efforts to restrict or eliminate high seas drift net fishing are considered to be inconsistent with customary international law and practice.[47] It has been argued that measures taken by nations to prohibit drift net fishing in the international waters of the South Pacific, unilateral measures taken by the United States with respect to high seas drift net fishing in the North Pacific, and, ultimately, the UN Driftnet Resolution unfairly impinge on the freedom to fish on the high seas as elaborated in article 87 and articles 116 through 120 of the 1982 Convention.

Articles 117, 118, and 197, however, establish the obligation of states to cooperate in the conservation of the living marine resources of the high seas and, in general, in the preservation and protection of the marine environment. The UN Driftnet Resolution is fully consistent with these provisions of the 1982 Convention. The United Nations General Assembly provided a forum for all interested nations to negotiate on the subject of high seas drift net fishing, and this process resulted in a series of resolutions to which *all* member nations have unanimously agreed. The United Nations General Assembly reaffirmed its commitment to the recommendations contained in Resolution 44/225 with the adoption, again by consensus, of UN General Assembly Resolution 45/197 in 1990 and Resolution 46/215 in 1991. As such, the UN Driftnet Resolutions fully express the 1982 Convention's principles on cooperation among states on the issue of high seas fisheries and the protection and preservation of the marine environment.

The criteria by which conservation measures are to be established in high seas fisheries are found in article 119(1) of the 1982 Convention:

In determining the allowable catch and establishing other conservation measures for the living marine resources in the high seas, States shall:
(a) take measures which are designed, on the best scientific evidence available to the States concerned, to maintain or restore populations of harvested species at levels which can produce the maximum sustainable yield, as qualified by relevant environmental and economic factors, including the special requirements of developing States, and taking into account fishing patterns, the interdependence of stocks and any

generally recommended international minimum standards, whether subregional, regional or global;

(b) take into consideration the effects on species associated with or dependent upon harvested species with a view to maintaining or restoring populations of such associated or dependent species above levels at which their reproduction may become seriously threatened.

The provisions and recommendations of the UN Driftnet Resolutions are consistent with the high seas conservation criteria elaborated in article 119. Under the 1982 Convention, conservation measures must be taken on the "best scientific evidence available." The UN Driftnet Resolutions reflect this same requirement. Repeated observations of the by-catch in high seas drift net fisheries constitute the best scientific evidence available, and these observations have indicated that high seas drift net fishing is or may be having an impact on numerous species "above levels at which their reproduction may become seriously threatened." It is now clear from the international scientific reviews of the impacts of drift net fishing in both the North Pacific and the South Pacific, and the review of the impacts on cetaceans in other regions by the International Whaling Commission (IWC), that many populations of slow-reproducing species of marine mammals, seabirds, and sea turtles are either known to be threatened or likely to be negatively impacted by the continued operations of high seas drift net fleets. These include species internationally recognized to be in danger of extinction.

The 1982 Convention has established a general framework for cooperation among states and the duty to take such measures as may be necessary for the conservation of living resources of the high seas.[48] The UN Driftnet Resolutions recognize that high seas drift net fishing "can be a highly indiscriminate and wasteful fishing method that is widely considered to threaten the effective conservation of living marine resources," and in 1991 the United Nations recommended that global moratoriums be imposed on high seas drift net fishing by December 31, 1992. There is nothing in the 1982 Convention that precludes the adoption of such an approach to fisheries conservation on the high seas. To the contrary, unless and until conservation can be ensured, articles 116 through 120, in particular article 119, would appear to support a moratorium on the practice.

UN Resolution 44/225 refers to the conservation of living marine resources and the prevention of unacceptable impacts. The concept of conservation as it applies to fisheries often refers only to the sustainability of a particular fishery. We would argue that conservation should be considered from a broad environmental perspective—the maintenance or restoration of the health and abundance of all species either directly or indirectly affected by the fishery and the overall biodiversity of marine ecosystems, not merely the sustainability of catches from a stock of fish of commercial importance or the economic sustainability of a fishery itself. Furthermore, conservation in high seas fisheries must be viewed in light of fisheries for (and the general conservation of) the same, associated, or dependent species within EEZs. Again, a broad definition of conservation in the context of high seas fisheries would appear to be generally consistent with article 119 of the 1982 Convention.

The UN Driftnet Resolutions do not define the term "unacceptable" as it pertains to the impacts of high seas drift net fishing. The FAO has suggested (with respect to fisheries in general) that "[a]ny human activity may irreversibly change some components of the marine environment. The 'concept of acceptable' impacts as associated with the concept of sustainability implies that reversibility is ensured."[49] Article 119(1)(b) of the 1982 Convention goes further and refers to maintaining or restoring populations of affected species "above levels at which their reproduction may become seriously threatened."

Any definition of unacceptable impacts should be based on a number of criteria, including (1) any level of mortality that significantly impedes the recovery of threatened or endangered species, (2) the mortality of substantial numbers of any population of marine animals whose status or role in the ecosystem is unknown, (3) any impact that may threaten the overall biodiversity of marine ecosystems, and (4) any impact that threatens the viability of established coastal fisheries and dependent communities and cultures, especially coastal fisheries that serve to contribute to the food security of a given area or region.

Many nations are concerned over the actual or potential impact of high seas drift net fishing on fisheries within national zones.[50] The FAO "Report on Environment and Sustainability in Fisheries" recently stated that "[m]ost high-seas resources have a phase of their life-cycle inside two hundred mile limits (EEZs). . . . The conservation of those species being exploited on the high seas is a matter of serious international concern."[51] Developing states are particularly concerned with the issue of food security.[52] The economics of high seas drift net fisheries dictate that great quantities of food fish are wasted, whether as discard of low-value by-catch species or as discard of damaged fish. All nations should have the right to set standards collectively to govern the harvest and utilization of high seas resources when high seas fisheries resources are considered to be a common property resource and when high seas fisheries directly or indirectly affect fisheries within EEZs.

The permanent representative of Barbados expressed to the Second Committee of the UN General Assembly the sentiments of many nations regarding efforts to eliminate the practice of high seas drift net fishing: "Barbados remains convinced that until this method of fishing is completely banned no region will be safe from its destructive effects: the consequences for small states, dependent on the seas for their very existence, would be particularly disastrous should we fail."[53]

GLOBAL STANDARDS FOR THE CONDUCT OF FISHERIES ON THE HIGH SEAS

Historically, most fisheries management has been reactive.[54] A fishery often is allowed to develop or expand relatively unmonitored and virtually unmanaged until it has reached a state of crisis. The crisis may be one of overcapitalization and excess capacity as a result of the unrestricted or subsidized growth of the fishery; intensive competition and conflicts among fishers; declines in catches; or negative impacts on nontarget species. Only then is any real effort made either to assess the status of affected species or to impose regulatory measures to reduce fishing effort or the impact of the fishery on other endangered or threatened species.

Rather than allowing unrestricted growth in a fishery until a crisis occurs, a precautionary approach would allow for the rational and sustainable development of a fishery. This view was expressed by recommendations of a conference convened in La Jolla, California, in 1990 by the International Whaling Commission (IWC) to address the entanglement of cetaceans in passive fishing gear: "The development of any new fisheries, or expansion of existing fisheries, should only be countenanced *after* a rigorous multidisciplinary environmental impact assessment that includes the potential effects on target and nontarget species, including cetaceans."[55] Both the precautionary approach and placement of the burden of proof on the proponents of the fishery are embodied in the UN Driftnet Resolution.

It is obvious that fisheries in international waters need to be brought under much greater international control. Although a set of "generally recognized international minimum standards" would be useful, they may not be sufficient. Article 119(1)(a) of the 1982 Convention requires only that nations take such standards into account; it does not oblige states to abide by them. The issue of high seas fisheries was the subject of intense debate in the negotiations leading up to the United Nations Conference on Environment and Development (UNCED). A number of nations advanced a detailed set of proposals for regulation of fisheries on the high seas, referred to as the Santiago Declaration. However, it was not possible to reach consensus on an international approach to high seas fisheries in particular because of reservations on the part of the European Community. Instead, it was agreed at UNCED that the United Nations would convene an intergovernmental conference on high seas fisheries.

What is ultimately needed is an internationally binding regime for the management of fisheries on the high seas, jointly negotiated by all nations, to ensure the conservation of all species and the biodiversity and integrity of open ocean ecosystems and to prohibit environmentally harmful fishing practices such as the use of large-scale drift nets.

CONCLUSION

The UN Driftnet Resolutions are arguably the most important development in international cooperation for fisheries conservation since the conclusion of the law of the sea negotiations in 1982. The UN General Assembly called for scientific evidence that high seas drift net fishing could be conducted in such a way as to ensure the long-term conservation of marine species. The overwhelming conclusion from the scientific observations and assessments conducted to date has been that conservation cannot be ensured. The UN General Assembly thus agreed, in 1991, to a moratorium on large-scale high seas drift net fishing by the end of 1992. The United Nations reaffirmed the commitment to a moratorium in agenda 21 of the UN Conference on Environment and Development.

Large-scale drift net fishing is the most obvious example of the failure to control fisheries adequately on the high seas. The current trend of over-capitalized fleets moving to international waters to exercise their "freedom" to fish on the high seas, by whatever means, fundamentally contradicts the goal of ensuring the long-term integrity and productivity of the marine ecosystem, in international waters and within EEZs. Rapidly expanding and inadequately regulated fisheries are now being recognized as the best-documented cause of decline of ocean ecosystems.

The challenge that the international community has put forward on the issue of high seas drift net fisheries represents a positive sign for future international cooperation in the increasingly crucial issue of marine conservation. If the challenge is not met with full and effective implementation of a moratorium on this blatantly destructive fishing method, there is little hope that more comprehensive agreements for fisheries on the high seas can be expected or implemented in the near future.

The UN Driftnet Resolutions strongly express international cooperation in demanding real evidence from a few countries that their vast enterprises are not compromising the future of the ecosystem on which we all depend. It is a sign that the spirit of the best provisions of the 1982 Convention, so painstakingly negotiated over many years, may be further realized and strengthened. The ancient concept of freedom of the seas can and must be transformed into a modern and progressive ideal of rational international cooperation to ensure that irreversible ecological harm to the oceans does not become our contribution to our common future.

NOTES

1. *Review of the State of World Fishery Resources*, Marine Resources Service, U.N. Food and Agriculture Organization (Fisheries Circular No. 710, Rev. 7, Rome 1990).
2. *See Report of the Secretary-General to the United Nations General Assembly, Large-Scale Pelagic Driftnet Fishing and Its Impact on the Living Marine Resources of the World's Oceans and Seas*, U.N. Doc. A/45/663 (Oct. 26, 1990) [hereinafter *UN Driftnet Report*]. Large-scale drift nets are many times longer than gill nets in coastal waters to produce higher returns on the huge investment of distant-water fishing fleets and to harvest more of the dispersed, and relatively unknown, stocks of fish beyond continental shelves. Drift nets on the high seas of the North Pacific, for example, are 40 to 60 kilometers in total length per vessel in the squid fishery. In contrast, coastal drifting gill nets deployed throughout the world are very much smaller. They are generally only a few hundred meters in length, although a few coastal fisheries, notably in the Indian Ocean, may use as much as 10 kilometers of netting. Many nations have banned drift nets larger than 2.5 kilometers from their exclusive economic zones.
3. *See* the following reports on the impacts of high seas drift nets: INTERNATIONAL NORTH PACIFIC FISHERIES COMMISSION, FINAL REPORT OF 1990 OBSERVATIONS OF THE JAPANESE HIGH SEAS DRIFTNET FISHERIES IN THE NORTH PACIFIC OCEAN (JOINT REPORT BY THE NATIONAL SECTIONS OF CANADA, JAPAN, AND THE UNITED STATES) [hereinafter 1990 NORTH PACIFIC DRIFTNET OBSERVER REPORT]; SOUTH PACIFIC FORUM, REVIEW OF DRIFTNET FISHING IN THE SOUTH PACIFIC OCEAN—ISSUES AND IMPACTS (SUBMITTED TO THE SECRETARY-GENERAL OF THE UNITED NATIONS BY MEMBERS OF THE SOUTH PACIFIC FORUM IN CONSULTATION WITH THE FORUM FISHERIES AGENCY AS REQUESTED IN OPERATIVE PARAGRAPH 4(A) OF UN RESOLUTION 45/197) (1991) [hereinafter FORUM REPORT]; CANADIAN DEPARTMENT OF FISHERIES AND OCEANS, SCIENTIFIC REVIEW OF NORTH PACIFIC HIGH SEAS DRIFTNET FISHERIES (REPORT FOR PRESENTATION TO THE UNITED NATIONS PURSUANT TO RESOLUTIONS 44/225 AND 45/197) (1991) [hereinafter CANADIAN REVIEW]; CANADIAN DEPARTMENT OF FISHERIES AND OCEANS, FISHERIES AGENCY OF JAPAN, U.S. NATIONAL MARINE FISHERIES SERVICE, & U.S. FISH AND WILDLIFE SERVICE, FINAL REPORT OF SQUID AND BYCATCH OBSERVATIONS IN THE JAPANESE DRIFTNET FISHERY FOR NEON FLYING SQUID (*OMMASTREPHES BARTRAMI*), JUNE-DECEMBER 1989 OBSERVER PROGRAM (JUNE 30, 1990); U.S. NATIONAL MARINE FISHERIES SERVICE, FINAL ENVIRONMENTAL IMPACT STATEMENT AND ECONOMIC IMPACT ANALYSIS ON THE INCIDENTAL TAKE OF DALL'S PORPOISE IN THE JAPANESE SALMON FISHERY (Washington, DC 1987);

Report of the Expert Consultation on Large-Scale Pelagic Driftnet Fishing, U.N. Food and Agriculture Organization (Fisheries Rep. No. 434, Rome 1990) [hereinafter FAO *Driftnet Report*]; SCIENTIFIC COMMITTEE OF THE INTERNATIONAL WHALING COMMISSION, REPORT OF THE SUB-COMMITTEE ON SMALL CETACEANS (Annex G.IWC/44/4) (May 1991) [hereinafter IWC]; Northridge, Driftnet Fisheries and Their Impacts on Nontarget Species: A Worldwide Review (paper presented to the International Whaling Commission Workshop on Mortality of Cetaceans in Gill Net and Trap Fisheries, La Jolla, Cal., Oct. 1990); M. Earle & M. Hagler, Mortality of Cetaceans in Gillnet and Trap Fisheries (paper submitted to the International Whaling Commission Workshop on Mortality of Cetaceans in Gill Net and Trap Fisheries, La Jolla, Cal., Oct. 1990); T. COFFEY & ASSOCIATES LTD., A PRELIMINARY ASSESSMENT OF THE IMPACT OF DRIFTNET FISHING ON OCEANIC ORGANISMS: TASMAN SEA, SOUTH PACIFIC (E.I.A./SDF (G.P.03/1990), Jan. 1990); T. H. Woodley & M. Earle, Observations on the French Albacore Driftnet Fishery of the Northeast Atlantic (preliminary report prepared for Greenpeace International, July 1991).

4. *See, e.g.*, FORUM REPORT, *supra* note 3 (concluding that "the non-selective impacts of driftnets with regard to size and species caught" indicated that high seas drift nets should be prohibited and stating that "[i]n the South Pacific, driftnet fishing is considered to have posed an appreciable threat to the viability of established longline and troll fisheries and to have the potential to reduce the albacore resource to unacceptably low levels where [the] driftnet fishing method might have become the 'only economically viable fishing method' ").

5. United Nations Convention on the Law of the Sea, Dec. 10, 1982, U.N. Doc. A/CONF.62/122, 21 I.L.M. 1261 (1982) [hereinafter 1982 Convention].

6. Large-Scale Pelagic Driftnet Fishing and Its Impact on the Living Marine Resources of the World's Oceans and Seas, G.A. Res. 44/225, U.N. GAOR Supp. (No. 49), U.N. Doc. A/44/746/Add.7 (1989) [hereinafter UN Driftnet Resolution].

7. The UN Driftnet Resolution was preceded by several intergovernmental actions on the issue. The South Pacific Forum produced the Tarawa Declaration in July 1989, calling for a ban on drift net fishing in the region. The Joint Assembly of the African, Caribbean and Pacific States and the European Economic Community (ACP/EEC) adopted a resolution in September 1989 urging all member states to ban drift nets in their waters and to work toward a global ban. The Langkawi Declaration, adopted by the Commonwealth Heads of Government in Kuala Lumpur in October 1989, supported the Tarawa Declaration and agreed to seek to ban pelagic drift net fishing. In November 1989, the Organization of Eastern Caribbean States adopted the Castries Declaration, outlawing the use of drift nets in the region. The Wellington Convention, which prohibits the use of large drift nets in the South Pacific, was opened for signature in November 1989 (*reprinted in* 29 I.L.M. 1449). Some of the most vehement protests have come from coastal developing countries concerned about depletion of stocks of highly migratory species, which would endanger the future of their marine ecosystems and possibly their future livelihoods from the sea.

8. UN Driftnet Resolution, *supra* note 6.

9. *See* Kazuo Sumi, *The International Legal Issues Concerning the Use of Drift Nets, with Special Emphasis on Japanese Practices and Responses* (chapter 19 of this book).

10. Government of Japan, The View of the Government of Japan Concerning High Seas Driftnet Fishing in Conjunction with the U.N. Resolution 44/225 Concerning "Large-Scale Driftnet Fishing" (paper submitted to the UN Office of Ocean Affairs and the Law of the Sea, Sept. 12, 1990) [hereinafter Government of Japan]. *See also* JAPAN FISHERIES ASSOCIATION, THE HIGH SEAS DRIFTNET ISSUE: WHAT IS KNOWN AND WHAT IS NOT KNOWN (Mar. 1991).

11. *FAO Driftnet Report, supra* note 3.

12. D. K. Conner & R. O'Dell, *The Tightening of Marine Plastics Pollution* 30 ENV'T 1 (1988).

13. J. Sproul, *Driftnets: A Burning Issue in Japan*, PAC. FISHING, Sept. 1990.

14. D. W. Laist, *Overviews of the Biological Effects of Lost and Discarded Plastic Debris in the Marine Environment*, 18 MARINE POLLUTION BULL., June 1987, at 6B.

15. Government of Japan, *supra* note 10.

16. *UN Driftnet Report, supra* note 2. *See also* FORUM REPORT, *supra* note 3.

17. *UN Driftnet Report, supra* note 2.

18. *Id. See also Fewer Damaged Tuna*, FISHING NEWS INT'L, Aug. 1991 (noting that a decline in the number of drift net-damaged fish caught by the albacore troll fleet east of New Zealand coincided with the cessation of Japanese drift netting and a reduced Taiwanese fleet in the South Pacific).

19. *See Environment and Sustainability in Fisheries*, U.N. Food and Agriculture Organization, U.N. Doc. COFI/91/3 (1991) [hereinafter *FAO Sustainability Report*]. *See also FAO Driftnet Report, supra* note 3.

20. *UN Driftnet Report, supra* note 2.

21. *FAO Driftnet Report, supra* note 3.

22. S. Northridge, Driftnet Fisheries and Their Impact on Nontarget Species: A Worldwide Review (paper presented to the International Whaling Commission Workshop on Mortality of Cetaceans in Gill Net and Trap Fisheries, La Jolla, Cal., Oct. 1990). *See also* CANADIAN REVIEW, *supra* note 3, *and* 1990 NORTH PACIFIC DRIFTNET OBSERVER REPORT, *supra* note 3.

23. CANADIAN REVIEW, *supra* note 3.

24. *Id.*

25. *Id.*

26. JAPAN FISHERIES ASSOCIATION, THE HIGH SEAS DRIFTNET ISSUE: WHAT IS KNOWN AND WHAT IS NOT KNOWN (Mar. 1991).

27. T. Murray, *Report of the Expert Consultation on Large-Scale Pelagic Driftnet Fishing*, U.N. Food and Agriculture Organization (Fisheries Rep. No. 434, Rome 1990), App. I (noting how a reduction in by-catch was achieved by adjusting net depth). Despite the improvement, however, vessels on the high seas might deploy more nets in an effort to make up for a loss of target species.

28. *Id.* (according to studies reported in 1987 by Hembree and Harwood).

29. *FAO Driftnet Report, supra* note 3.

30. FORUM REPORT, *supra* note 3.

31. *Id.* (concluding that "no trials involving modifications to commercial-scale driftnets have proved successful in reducing bycatch without significantly affecting the target fish catch").

32. Northridge, *supra* note 22.

33. FORUM REPORT, *supra* note 3 (noting that "unlike normal gillnets, changing the mesh size of driftnets does not necessarily improve selectivity").

34. *FAO Driftnet Report, supra* note 3.

35. Japan has been conducting research utilizing modified jigging gear and midwater trawls for squid as alternatives to large-scale drift netting. Korea is also testing jigging as a means of reducing by-catch (the method is used with success in other areas of the world). The results of the research from both countries were expected to be available during 1991 and 1992, according to representatives from both countries present at the Scientific Review of North Pacific Driftnet Fisheries held in Canada in June 1991. *See* CANADIAN REVIEW, *supra* note 3.

36. *Japan Works to Beat U.N. Gill Net Ban,* FISHING NEWS INT'L, June 1990.

37. FORUM REPORT, *supra* note 3.

38. *See The Regulation of Driftnet Fishing on the High Seas: Legal Issues,* U.N. Food and Agriculture Organization (Legislative Study No. 47, Rome 1991).

39. Douglas M. Johnston, *The Driftnetting Problem in the Pacific Ocean: Legal Considerations and Diplomatic Options,* 21 OCEAN DEV. & INT'L L. 5 (1990).

40. D. Charles, *Satellite Nets Illegal Fishing Ships,* NEW SCIENTIST, Aug. 24, 1991.

41. Seattle Post-Intelligencer, May 30, 1990 (from a N.Y. Times article).

42. Information presented at the International Symposium on High Seas Driftnet Fishing, Sidney, British Columbia, 1991.

43. *FAO Sustainability Report, supra* note 19.

44. Drift net agreements concluded between the United States and Japan, Taiwan, and Korea pursuant to the 1987 Driftnet Impact Monitoring and Enforcement Act, Pub. L. No. 100-220, §§ 4004-09, 101 Stat. 1477, have resulted in observer programs and data-reporting requirements of varying scope.

45. *UN Driftnet Report, supra* note 2.

46. FORUM REPORT, *supra* note 3.

47. For a discussion, see William T. Burke, *The Law of the Sea Concerning Coastal State Authority Over Driftnets on the High Seas, in The Regulation of Driftnet Fishing on the High Seas: Legal Issues,* U.N. Food and Agriculture Organization (Legislative Study No. 47, Rome 1991), and Kazuo Sumi, *The International Legal Issues Concerning the Use of Drift Nets, with Special Emphasis on Japanese Practices and Responses* (chapter 19 in this book). *See also* Johnston, *supra* note 39.

48. The authors would like to express some reservation with the use of the term "resources" in reference to the conservation of marine species. The term "resource" implies a component of the marine environment that has been targeted for human utilization. We would not necessarily view other components of the marine environment as "resources," since this implies that their sole reason for existence is their direct benefit to humanity. Rather, all species contribute to the health and stability

of marine ecosystems, and the conservation of exploited species should be viewed in the overall context of the conservation of the marine environment.

49. *FAO Sustainability Report, supra* note 19.

50. Articles 63(2) and 64 of the 1982 Convention, *supra* note 5, are not entirely clear on the degree to which coastal states may exercise authority over fisheries on the high seas. In particular, the issue of coastal state authority over fisheries on straddling stocks in adjacent high seas areas (article 63(2)) was a subject of contentious debate during the final years of the Third UN Conference on the Law of the Sea negotiations and was considered by many nations to have been inadequately resolved. (For a discussion, see William T. Burke, *Fishing in the Bering Sea Donut: Straddling Stocks and the New International Law of Fisheries*, 16 ECOLOGY L.Q. 285 (1989).)

Nevertheless, the concern of (coastal) states in the South Pacific appears to have convinced the Japanese government to suspend drift net fishing in the South Pacific beginning with the 1990–1991 season:

> It is reported that Japan has made this decision, taking into account the fact that the South Pacific island countries' economies rely on fisheries resources, that these countries wish to develop their economies based upon albacore fishing and that they, therefore, have a grave concern about driftnet fishing in the region. By this decision, Japan hopes to further promote good relations with South Pacific island countries.

C. Mizukami, *Fisheries Problems in the South Pacific Region*, 15 MARINE POL'Y 111 (1991).

51. *FAO Sustainability Report, supra* note 19.

52. A recent publication of the International Center for Ocean Development estimated that more than a billion people in Asia alone depend on fish and seafood as their major source of animal protein and that approximately 60 percent of people in developing countries derive 40 percent or more of their animal protein from fish. International Center for Ocean Development (ICOD), *Viewpoint: The World Food Situation and the Importance of Fish*, 1 ICOD INFO. 2 (1988).

53. Statement by H.E. Mr. Besley Maycock, Permanent Representative of Barbados to the United Nations, to the Second Committee, General Debate, Oct. 9, 1990.

54. For a discussion of global minimum standards as applicable to the management of high seas fisheries, see E. Hey, *The Provisions of the United Nations Law of the Sea Convention on Fisheries Resources and Current International Fisheries Management Need*, in *The Regulation of Driftnet Fishing on the High Seas: Legal Issues*, U.N. Food and Agriculture Organization (Legislative Study No. 47, Rome 1991). *See also* Johnston, *supra* note 39.

55. International Whaling Commission Workshop on Mortality of Cetaceans in Gill Net and Trap Fisheries, La Jolla, Cal., Oct. 1990.

19 The International Legal Issues Concerning the Use of Drift Nets, with Special Emphasis on Japanese Practices and Responses

Kazuo Sumi

IN SPITE OF THE USE OF DRIFT NET FISHING for more than a century, environmentalists recently have harshly criticized the practice of this fishing method, particularly on the high seas. The concerns have also led to conflicts between pelagic drift net fishing nations and coastal fishing nations.

In 1989, during the forty-fourth session of the United Nations General Assembly, the United States presented a draft resolution, cosponsored by seventeen countries, including New Zealand, Australia, Canada, and Vanuatu, calling for an "immediate ban" or "moratorium" on drift net fishing.[1] Japan opposed the proposed moratorium and submitted a counterproposal, suggesting that any regulatory measures be based on scientific data and analysis.[2]

As a result of a compromise between the United States and Japan, on December 22, 1989, the UN General Assembly adopted Resolution 44/225, titled "Large-Scale Pelagic Driftnet Fishing and Its Impact on the Living Marine Resources of the World's Oceans and Seas."[3] The UN Driftnet Resolution provides that a moratorium on "all large-scale pelagic driftnet fishing" will be implemented by June 30, 1992, unless "effective conservation and management measures . . . based on statistically sound analysis to be jointly made by concerned parties of the international community" are taken.

But the problem has not been resolved by adoption of this resolution. There remains a difference of opinion between countries calling for a complete and immediate ban on drift net fishing and countries calling for conclusive scientific evidence on the effects of drift net fishing, based in part, at least, on differences in perceptions by countries with a "fish-eating culture" and countries with a "meat-eating culture." This chapter addresses the specific objections to drift net fishing, with particular emphasis on the practices and position of Japan.

WHAT ARE THE REAL PROBLEMS?

Although the use of large-scale drift nets on the high seas has been widely denounced, critics do not address the broader suite of issues associated with this fishing method. The following sections describe the two problems currently ignored by those who oppose the use of large-scale drift nets.

High Seas and Coastal Drift Net Fisheries

Environmentalists denouncing the use of drift nets have directed their attacks at large-scale high seas fishing. They have not expressed concern over the use of small-scale coastal drift nets.[4] The sole basis for making this distinction is that drift nets used on the high seas are larger than those used in coastal waters. From the viewpoint of conservation of living marine resources, however, is discrimination between drift nets on the high seas and those in coastal waters justifiable?

Living marine resources, including marine mammals, seabirds, and turtles, are far more abundant in coastal waters than on the high seas. Furthermore, although the length of net deployed per fishing vessel is shorter in coastal drift net fisheries, the number of fishing vessels is far greater in coastal waters than on the high seas. The use of drift nets may actually be more destructive in coastal waters than on the high seas.[5]

Despite suggestions that coastal fisheries may be better managed, it is doubtful whether drift nets in coastal waters have been better managed than those on the high seas and whether more information and data concerning coastal fisheries have been accumulated than those concerning the high seas.[6]

It seems that the UN Driftnet Resolution has assumed that although large-scale high seas drift nets have adverse impacts on marine resources, smaller-scale coastal drift nets do not, though entanglement problems may also occur in coastal drift net fisheries.

The By-Catch Issue

In preambular paragraph 2 of the UN Driftnet Resolution, drift net fishing is denounced as "a highly indiscriminate and wasteful fishing method." Is this denunciation based on valid reasoning?

All types of fishing gear take species other than those targeted. As long as fishing activities are carried out in a natural milieu, their impact on nontarget species is more or less inevitable. In the trawl fishery, for example, both target and nontarget species are caught without discrimination. At present, no persuasive evidence exists showing drift net fishing to be more environmentally destructive than any other commercial fishing method. Comparative studies about the effects of various types of fishing gear should precede political decisions limiting their use.

The National Oceanic and Atmospheric Administration (NOAA) of the U.S. Department of Commerce has condemned high seas drift netting.[7] Its condemnation, however, contains no reference to incidental take by coastal drift netting. According to NOAA, "approximately 1,000 marine mammals were entangled or killed in the Prince William Sound/Copper River Delta salmon drift gillnet fishery during the 1978 season," and an "estimated 335 harbor seals and 45 California sea lions were killed annually incidental to gillnetting in the Columbia River, Willapa Bay and Grays Harbor fisheries."[8] The agency contradicts itself, demanding a total ban on high seas drift net fishing on the grounds of bycatch while keeping silent about the use of drift nets in coastal waters.

The U.S. government also seeks to transform a moratorium recommendation embodied in the UN Driftnet Resolution into a binding legal principle. The U.S. policy statement on the UN resolution reads:

> Unless joint assessment by all concerned members of the international community of scientifically sound data from a specific large scale pelagic driftnet fishery concludes there is no reasonable expectation of unacceptable impacts by that fishery, the conditions of relief from the moratorium recommended in UNGA 44/225 are not met.[9]

The notion of "unacceptable impacts" will inevitably bring an unnecessary interpretative problem into high seas fisheries, particularly since the question of whether the impacts will be unacceptable must be determined by "all concerned members of the international community," not by a single state. In addition, the doctrine of unacceptable impacts will have a significant impact on coastal fisheries. If this doctrine is applied to coastal fishing, a number of fishers who use small-scale drift nets or other gear will be deprived of a means of living.

Salmon-Fishing Issues

Because salmon originate from and spawn within national territory but also migrate to the high seas during their life cycle, they pose a unique fisheries management problem. Although international legal mechanisms to prevent high seas harvesting of these national resources exist, they have proven difficult to implement and enforce.

Incidental Catch of Salmon in U.S. Waters. Under the framework of the 1952 International Convention for the High Seas Fisheries of the North Pacific Ocean,[10] the Japanese high seas salmon drift net fishery has been regulated through negotiation within the commission established under the International North Pacific Fisheries Convention (INPFC). The Japanese high seas salmon fishery in the North Pacific has also been subject to a bilateral agreement between Japan and the former Soviet Union.[11] Recently, however, the anti–drift net campaign has challenged these long-standing cooperative legal frameworks.

The INPFC was renegotiated in 1978 following establishment of the United States's 200-mile fishery zone under the Magnuson Act.[12] As a result, Japanese fishers were given permission to conduct a salmon drift net fishery in the U.S. 200-mile zone and were granted a three-year exemption from the Marine Mammal Protection Act of 1972 (MMPA).[13]

In 1981, under the MMPA, a general permit for three years authorizing the annual incidental taking within the U.S. 200-mile zone of marine mammals— 5,500 Dall's porpoises, 450 northern fur seals, and 25 Steller's sea lions—was issued to the Federation of Japan Salmon Fisheries Cooperative Association (Japanese Salmon Federation). The general permit for the Japanese fishery was extended until 1986.

On July 18, 1986, the Japanese Salmon Federation sought a general permit for a five-year period authorizing an incidental take of the same level of marine mammals as in the preceding years. On May 22, 1987, the U.S. Department of Commerce issued a general permit to the Japanese salmon drift net fleet for the incidental take of as many as 6,039 Dall's porpoises over a three-year period. Because that permit contained no reference to a quota for northern fur seals, the Japanese Salmon Federation filed a lawsuit challenging it.

On the other side, Alaskan fishing groups and environmental organizations, including the Sierra Club Legal Defense Fund and Greenpeace, challenged the permit on the ground that the fleet cannot operate in U.S. waters without killing northern fur seals, a species declared depleted and protected by the MMPA.

On June 12, 1987, the court upheld a preliminary injunction prohibiting the Japanese fleet from operating within the U.S. 200-mile zone on the ground that no permit had been issued for the taking of northern fur seals.[14] On appeal, the United States Supreme Court denied review.[15] As a result, Japanese fishers are now unable to fish in the U.S. 200-mile zone.

The High Seas Salmon Interception Problem. Critics also condemn drift net fishing on the ground that it intercepts salmon on the high seas and invoke the concept of "national interest" as justification for their arguments. For example, the U.S. government argues that:

[a]lthough the Japanese salmon fisheries have been sanctioned under the bilateral agreement and the INPFC, we remain convinced that high seas driftnet fishing for immature salmon is inefficient and indiscriminate.[16]

The U.S. government has negotiated with the former Soviet government regarding the interception issue under the framework of the Intergovernmental Consultative Committee.[17] As a result, on February 9, 1989, both governments signed a memorandum of understanding on anadromous species in the North Pacific, agreeing to cooperate on high seas enforcement activities and the exchange of information on "illegal" high seas salmon fishing.[18] A U.S.-Soviet draft of a proposed convention has been provided to Japan and Canada for consideration, the ultimate goal of which is to end high seas salmon fishing.

Salmon "Piracy." Another serious problem related to high seas drift net fishing in the North Pacific is the accidental catch of salmon by squid drift nets. A widespread view among American fishers and environmental groups, and even in the U.S. government, is that Asian squid drift net fishers are "pirating" American-origin salmon and "decimating" the stocks.[19]

During testimony before the U.S. Senate, Kate Troll spoke of "salmon piracy":

> Under well established international law, high seas salmon belong to the country of origin. Squid ships from Taiwan, Korea, and Japan are stealing them on the high seas. This constitutes piracy, pure and simple.[20]

Supporting these arguments, the U.S. government has said:

> Over 1,000 driftnet vessels from Japan, Korea, and Taiwan fish in the North Pacific, mainly for squid and tuna/billfish. The United States is especially concerned with the interception of valuable U.S.-origin salmon and steelhead trout by these fisheries. Some driftnet vessels are also illegally targeting salmon which is then smuggled to world markets where it competes with legitimate product.[21]

With regard to these views, two questions occur: first, whether the norm of piracy under international law applies, and second, whether taking of salmon on the high seas is illegal.

The first question relates to the definition of piracy. Under traditional international law, piracy is defined as "every act of illegal violence committed on the open sea by the crew of a private vessel against another vessel," and pirates are treated as *hostis humani generis*—the enemy of all mankind.[22] Therefore, "all maritime States are authorized by general international law to capture on the open sea individuals who are guilty of piratical acts in order to punish them."[23] However, under the definition of piracy under the 1982 Convention, the law of piracy is unequivocally not applicable to the case of salmon interception and has no relevance to high seas drift net fishing.[24]

With regard to the second question, the U.S. claim over anadromous species such as salmon is based on an arbitrary interpretation of the law of the sea. The U.S. Fishery Conservation and Management Act of 1976 declares that the United States exercises "exclusive fishery management authority" over all anadromous species throughout the migratory range of each such species beyond the fishery conservation zone.[25] But this provision contravenes article 66 of the 1982 Convention, which provides that jurisdiction of the state of origin over anadromous stocks is not exclusive.

A fundamental principle of international law is that a state cannot invoke national law to evade obligations under international law. If a state's domestic legislation is inconsistent with international law, that state logically should rectify the irregularity. But instead of revising its domestic legislation, the United States has tried to impose restrictions on other countries' exercise of legitimate rights on the high seas.

Furthermore, since North Pacific Asian-origin salmon intermingle with U.S.-origin salmon, it is not an easy task to discriminate between them. While claiming exclusive jurisdiction over anadromous species beyond its 200-mile zone, the United States irrationally has refused to allow Japanese fishers to take Asian-origin salmon within the U.S. 200-mile zone.

Lawfulness of Unilateral Measures

With the clear objective of regulating drift net fishing activities of Asian countries in the North Pacific, the U.S. Congress passed the Driftnet Impact Monitoring, Assessment, and Control Act of 1987.[26] This legislation requires that the secretary of commerce, through the secretary of state and in consultation with the secretary of the interior, negotiate cooperative agreements with those countries conducting high seas drift net fishing. Specifically, the act calls for adequate monitoring and assessment programs concerning the deployment of scientific observers on drift net vessels[27] and similar agreements regarding effective enforcement of laws, regulations, and treaties governing high seas drift net fishing.[28] If these negotiations do not produce agreements with the designated countries within eighteen months after the date of enactment, the secretary of commerce must certify such fact to the president of the United States. Such certification may trigger the Pelly Amendment to the Fishermen's Protective Act,[29] authorizing the president to ban imports of fish products from the country concerned.

Under threat of such an embargo by the United States, Japan was compelled to conclude an agreement on June 23, 1989.[30] Through this agreement, Japan undertook several measures that allowed U.S. and Canadian research and observation of Japanese fishing activities on Japanese vessels.[31] Subsequently, Taiwan and Korea concluded similar agreements with the United States on North Pacific high seas drift net fishing.[32]

Under international law, these agreements are of dubious legal validity because they were brought about by the threat of sanctions under the Pelly Amendment. A treaty is void *ab initio* when concluded under coercion. Article 52 of the Vienna Convention on the Law of Treaties[33] provides that a "treaty is void if its conclusion has been procured by the threat or use of force in violation of the principles of international law embodied in the Charter of the United Nations."[34] Although analysts have concluded that the word "force" refers only to armed force, the *travaux préparatoires* of this article show that its meaning is open-ended.[35] In addition, in the Vienna Convention on the Law of Treaties, the prohibition of "the threat or use of force" is regarded as a typical example of *jus cogens*, that is, a peremptory norm of general international law from which no derogation is permitted (article 53).[36]

Because the threat of economic sanctions under the Pelly Amendment and Magnuson Act coerced the consent of Japan, Taiwan, and Korea to those agreements, they should be considered null. The concerned Asian nations have not yet made this claim, however.

Moreover, strong dissatisfaction with these agreements persists among U.S. fishing groups and environmental organizations. Specifically, the Pacific Seafood Processors Association, the Bering Sea Fishermen's Association, and the United Fishermen of Alaska have called for the U.S. government to develop a new international "nation of origin" convention, with at least Japan, Russia, and Canada as prospective members; the convention would establish "salmon zones" identifying where U.S. salmon are likely to be beyond the 200-mile zone(s), require satellite transponders on vessels fishing in salmon zones, and expand coverage under the Pelly Amendment to include all products, "not just those of marine origin."[37]

In response to these pressures, the U.S. government has asked that the existing agreement between the United States and Japan be rejected and that a new agreement be negotiated requiring all Japanese fishing vessels to carry position-indicating satellite transponders, requiring placement of more U.S. observers on Japanese vessels, and requiring a ban on squid-fishing activities in the North Pacific. Under these circumstances, the Japanese government decided in 1990 to install satellite transmitter equipment on all fishing vessels engaging in North Pacific squid drift netting and to increase the number of foreign observers on the squid drift net fishing vessels.

The real intention of the United States is to stop squid drift net fishing itself. But under general international law, fishing activities on the high seas are governed by the "flag state" principle. Accordingly, any steps taken on the high seas require agreement among all countries involved. If a country were free to coerce other countries to abstain from the exercise of legitimate rights by the threat of sanctions under domestic legislation, it would vitiate the fundamental concept of treaty making itself.

Incidental Catch in Squid Fishing

With regard to the North Pacific squid fishery, drift net fishing has been denounced as environmentally destructive by the anti-drift net campaign. Greenpeace criticizes it as follows:

> High seas driftnet fisheries are taking a dangerous toll on marine resources in the North Pacific. The large driftnet fleets of Japan, Taiwan and the Republic of Korea are slaughtering tens of thousands of porpoises and dolphins, other marine mammals, and hundreds of thousands of marine birds during fishing operations annually. Driftnets are also dangerously depleting fisheries resources—including salmon species of North American origin. If ocean wildlife is to be preserved and fish stocks kept at sustainable yield levels, this practice of strip-mining of the seas must be phased-out.[38]

In response to such criticism, the Japanese government has taken the afore-mentioned measures to reduce accidental takes of nontarget species by squid drift net fishing vessels. In addition, the Japanese government has undertaken scientific research programs and foreign observer programs to collect data and information concerning incidental takes. Distrust still prevails, however, among American fishers and environmental groups.[39]

Although by-catch data and information currently available are seriously deficient, this is an accusation without scientific basis. Such a biased view is counterproductive and is less likely to lead to a positive and lasting solution to the problem.

A joint U.S.-Canada-Japan report was recently issued on squid by-catch observations in the Japanese squid drift net fishery for the 1989 season from June through December.[40] During the 1,402 reported retrievals, only 79 salmonid were accidentally caught, while 3,119,061 flying squid were taken. The accidental capture of marine mammals included 208 northern fur seals; 914 dolphins, including 141 Dall's porpoises; and 22 marine turtles. With respect to seabirds, a total of 8,534 shearwaters were accidentally captured. These figures should be viewed in relation to the total population size of the marine mammals and birds concerned.

"Ecological" Approach

Western critics believe that ecologically sound fishing should be selective, and they condemn drift net fishing on the ground of its nonselectivity. For example, Greenpeace insists that "[u]sing driftnets makes sound, ecological management of fisheries resources impossible."[41] The government of New Zealand similarly notes that large-scale drift net fishing is "an environmentally unsound and unsustainable activity which threatens effective conservation and management of living marine resources" and that "[d]riftnets are a relatively indiscriminate or unselective method of fishing."[42]

Others doubt, however, that selective methods of fishing are ecologically sound. Fishing conducted in the natural environment with a complicated food web leads inevitably to by-catch of species other than targeted fish. Therefore, all species caught should be used to the maximum extent.

In addition, the concept of ecosystem balance is not static but dynamic. To cite one example, in the Antarctic, as a result of past overexploitation of some whale species by Western countries, including Japan, the present ecosystem balance is quite different from the original one. Although the large whales, such as blue whales and fin whales, have been reduced in number, the smaller whales, such as sei and minke, have increased beyond previous numbers. Since the food of all the baleen whales is the same, namely, krill, there is scientific opinion that taking of minke whales will accelerate the recovery of the larger whales. Although there is debate among scientists over this opinion, it should not be discounted.

Similar competitive interactions may occur among target and nontarget species in other sea areas. Therefore, selective catch of specific fishes with commercial value may invite an ecological change at a local, regional, or global level.

Removal of a few selected species may result in the remaining undesirable fish becoming dominant in the specific fishing ground; the undesirable species may then suppress the recovery of the more desirable species. When a full cross-section of species is taken so as to avoid overfishing of each species, the original ecosystem balance is likely to be restored. Accordingly, fishing that removes a full cross-section of marine life in a specific fishing ground may be ecologically healthier than fishing that removes only a few selected species.

Moreover, environmental groups' estimates of incidental takes[43] are exaggerated. The research results of the 1989 cooperative observer program among Japan, Canada, and the United States demonstrate that the by-catch proportion is less than expected. Tens of millions of dolphins and hundreds of millions, perhaps billions, of seabirds live in and around the Pacific. Available food supplies and environmental conditions have a greater effect on the population size of these animals than does entrapment by drift nets.

Efforts should, of course, be made to avoid unnecessary by-catch. It is not rational, however, to claim that there should be no incidental take of even a single animal. The key question is the measurement of the population stock from which a marine mammal is taken and the determination of the allowable level of accidental take so that the marine mammal stock will not be disadvantaged.

The "Ghost Fishing" Issue

Anti-drift net campaigners claim that lost or discarded plastic drift nets will continue to fish permanently. Greenpeace says that "[s]ince the plastic netting is non-biodegradable it can continue to ghost fish, entangling and killing these creatures indefinitely."[44] On the other hand, in defending the drift net fishery, the Japanese government contends that

> [e]ven if the nets were failed to be retrieved, it is reported that they will lose their ability as fishing gear quite rapidly. According to the results of independent experiments conducted separately by Japanese and the U.S. scientists, lost nets will lose [their] length exponentially as time goes by to be a solid mass in, at most, two weeks.[45]

Although countries disagree about the fate and impact of ghost nets, the loss or disposal of netting material poses some threat to marine life and the environment, and it should be discouraged.

This problem is subject to some regulation under international law. The major multilateral treaties concerning drift net regulation include the 1972 Convention on the Prevention of Marine Pollution by Dumping of Wastes and Other Matter (London Dumping Convention, or LDC)[46] and the International Convention for the Prevention of Pollution from Ships (MARPOL 73/78).[47]

Under article 4 of the London Dumping Convention, "[t]he dumping of wastes or other matter listed in Annex 1 is prohibited."[48] Annex 1 lists "[p]ersistent plastics and other persistent synthetic materials, for example, netting and ropes, which may float or may remain in suspension in the sea in such a manner as to interfere materially with fishing, navigation or other legitimate uses of the sea."[49] Accordingly, under this convention the deliberate disposal of drift nets at sea is prohibited, and it is the responsibility of the flag state to ensure compliance.

Annex V of MARPOL 73/78 concerns regulations for the prevention of pollution by garbage from ships. The disposal into the sea of "all plastics, including but not limited to synthetic ropes, synthetic fishing nets and plastic garbage bags" is prohibited.[50] However, an exemption is allowed for "the accidental loss of synthetic fishing nets or synthetic material incidental to the repair of such nets, provided that all reasonable precautions have been taken to prevent such loss."[51]

In June 1983, Japan accepted annex V of MARPOL 73/78 and now prohibits its fishing vessels from discarding nets in all areas of the sea, both inside and outside national jurisdiction.

A new convention imposing on fishing vessels obligations for gear marking and retrieval is the solution to the accidental loss of drift nets. Japan has already adopted a system of mandatory marking of fishing gear for identification. Adoption of this system should be considered at the international level. With regard to the retrieval problem, an international obligation to retrieve accidentally lost nets should also be established. Within this framework, if lost nets are found by others, the offending vessel should pay a penalty to the appropriate international authority. The system might also reward anyone who finds and retrieves lost fishing gear. Efforts should also be made to improve net materials, making them more biodegradable so that lost gear does not continue to "ghost fish."

TOWARD A SOLUTION

Although controversy over the use of large-scale drift nets may persist, both sides can begin to work toward solving high seas fisheries problems. There is a critical need for scientific data and the use of risk assessment methodology. These tools will enhance international and regional efforts to manage and conserve living marine resources.

The Need for Scientific Research

The current drift net controversy is characterized by a paucity of data on the impact of this fishing gear on targeted fish and on the accidental take of untargeted fish, marine mammals, and seabirds.

The UN Driftnet Resolution recognizes the need for scientific research on this matter. Preambular paragraph 6 of the resolution states that "any regulatory measure to be taken for the conservation and management of living marine resources should take account of the best available scientific data and analysis." In addition, operative paragraph 3 recommends that "all interested members of the international community . . . by 30 June 1991, review the best available scientific data on the impact of large-scale pelagic driftnet fishing." However, operative paragraph 4(a) concerning moratoriums refers only to "statistically sound analysis," which is not the same as "the best available scientific data." The former may be inconsistent with the latter and may require less scientific evidence.

Under the resolution, conservation and management measures need *not* be based on the best available scientific data. Accordingly, a moratorium is to take effect even if there is no good scientific evidence. The assumption underlying this resolution is that drift nets have adverse effects on marine ecology unless otherwise shown. This approach is counterproductive and is likely to lead to unnecessary conflict.

The same is true of the doctrine of "unacceptable impacts" in the U.S. policy statement on the UN resolution.[52] Introduction of this approach will bring confusion into sound conservation and management principles. Furthermore, its effects will not be limited to the drift net fishery issue. The adoption of a standard of "unacceptable impacts" will have far more significant effects on fishing as a whole on the high seas and in coastal waters.

Environmental Impact Assessment

The adoption of new or large-scale technology may have significant unintended or unforeseen effects on an ecosystem's balance. Therefore, it is essential to carry out impact evaluation and risk assessment before introducing such technology. This assessment should be made at local, regional, and international levels. On the high seas, an appropriate mechanism also must be devised for assessing impacts of drift net fishing technology on marine ecosystems.

In 1973, Norway submitted "Draft Articles on the Protection of the Marine Environment Against Pollution" to the United Nations Sea-Bed Committee.[53] Article 15 referred to environmental impact statements. Senator Claiborne Pell submitted a similar resolution to the United States Senate on August 25, 1976.[54] The preambular paragraph of the resolution states that the United States government should seek the agreement of other governments to a proposed treaty requiring the preparation of an international environmental impact statement for any major project, action, or continuing activity that may be reasonably expected to have a significant adverse effect on the physical environment or environmental interests of another nation or of a global commons area.

Until now, the international community has not succeeded in adopting an international environmental impact assessment system.[55] Nevertheless, part XII of the 1982 Convention reflects the idea of an environmental impact assessment in an embryonic form. Article 192 enunciates the general obligation to protect and preserve the marine environment, and articles 204, 205, and 206 make reference to monitoring and environmental assessment. These related provisions call for states to assess and monitor the effects of their activities on the marine environment.

From a long-term perspective, institutional mechanisms to make assessments in a more integrated and systematic way will be devised and developed at the regional or international level. Ideally, a single comprehensive agency should be established to direct the diverse functions involved in effectively managing the marine environment and resources.

Resource and Management Mechanism

To date, the ancient doctrine of the freedom of the high seas has safeguarded high seas fishing. Fishing activities on the high seas are under "flag state" jurisdiction. International law leaves it to the countries concerned to determine the nature of the conservation measures, if any, employed in the fishery and imposes no restriction on fishing gear. Fishing with drift nets on the high seas is a lawful exercise of this freedom. Accordingly, until recently no question has been raised about the lawfulness of this fishing activity.

Nevertheless, high seas fishing is not permitted on a *laissez-faire* basis. As seen already, article 117 of the 1982 Convention confirms the general principle that fishing countries have the duty to take appropriate conservation measures on the high seas, either for their own nationals alone or in cooperation with other countries for their nationals together.

Therefore, countries fishing on the high seas may be bound by specific commitments undertaken by agreement with other countries on a bilateral or multilateral basis. As far as the drift net fishery is concerned, the only fishing operation carried out under the auspices of a multilateral convention is the Japanese salmon drift net fishery operating under the International Convention for the High Seas Fisheries of the North Pacific Ocean.[56] The North Pacific squid and billfish fisheries and all other high seas drift net fisheries, such as those in the Atlantic and Indian oceans, are not regulated by any international fisheries convention.

Although a multilateral approach is preferable to a patchwork of bilateral treaties, it is not reasonable to confine the coverage of any proposed management convention to some specific fish species, such as salmon. As discussed earlier, since all living creatures constitute parts of a complex food chain, it is not advisable to deal with a few selected species.

In addition, since fishery resources do not respect political boundaries and most high seas stocks are transboundary with adjacent EEZs, measures implemented solely in the high seas area can never lead to viable long-term conservation of the North Pacific resources. Therefore, a solution to the problem should be sought in establishment of a comprehensive management mechanism on resource use at the level of the entire North Pacific region, including the areas within the EEZs of coastal states.

In view of the increased impacts of human activities on the marine environment, an innovative idea should be introduced into ocean management, in lieu of the historical concept of the "freedom of the high seas," to establish comprehensive resource management mechanisms at the regional and global levels. Furthermore, to achieve proper management of complex multispecies resources on the high seas, all of the elements related to the marine ecosystem should be covered by the management authority.

Cooperative research activities are a basis for joint conservation measures. Accordingly, agreement should be made on scientific research programs in each high seas region. In this context, the Food and Agriculture Organization should play a role in both initiating such research programs and establishing regional management bodies and serving as a clearinghouse for interregional exchange of data about conservation and management of living marine resources.

CONCLUSION

Recent increased attention to high seas drift net fishing is not derived from scientific knowledge about the impacts of this gear on living marine resources, as such knowledge does not yet exist. It emerges for the most part from emotional and political considerations. If drift net technology is ecologically destructive, it should logically be prohibited, not only on the high seas but also in coastal waters.

Nevertheless, existing fishing methods and trading patterns of fish products must be reconsidered in the light of growing environmental concern. First of all, reconsideration should be given to large-sized fishing vessels and gear. Since high seas drift nets may reach about 50 kilometers in length, such a fishing method cannot be regarded as normal. Modern technology has made possible a much more intensive use of the sea. Fishing vessels and gear technology have become larger in scale in the pursuit of economic efficiency and cost reduction. The new fishing method enhances the efficiency of catch efforts but also encourages ecologically insensitive fishing. Therefore, what is needed is to change or improve existing fishing techniques and gear and to find more appropriate technologies.

A second serious problem is by-catch of undesirable or noncommercial species. A significant proportion of fish taken by drift nets is dumped overboard. Such practices must be discontinued so that the maximum use of fishery resources is made.

Fishing fleets must also minimize the incidental catch of marine mammals, seabirds, and turtles. Modification of net material, reduction of gear length, or subsurface fishing might reduce the undesirable impacts on these animals.

Third, more attention should be paid to the involvement of transnational corporations in the trade of fish products. Environmentalists' denunciation of Japan for smuggling illegally caught salmon to world markets is off the point. Such unfair practices are chargeable to Japanese trading companies who invest in fishing companies in foreign countries, such as Korea and Taiwan. Unfortunately, such trading practices are outside the jurisdiction of the Japanese Fisheries Agency. Therefore, what is really needed is a way to regulate irresponsible trading practices of transnational corporations.

In conclusion, if it is scientifically demonstrated that drift net fishing is ecologically destructive, such a fishing method should be prohibited without exception. According to the scientific knowledge available at present, however, drift nets should be managed with only gear restrictions and modifications, as with any other fishing gear.

At the same time, a new management policy, conserving and managing marine resources as the common heritage of humanity, should be formulated and adopted by the international community. In this context, it is high time to reconsider the traditional concept of freedom of the high seas. As preambular paragraph 8 of the UN Driftnet Resolution rightly states, "all members of the international community have a duty to co-operate globally and regionally in the conservation and management of living resources on the high seas, and a duty to take, or to co-operate with others in taking, such measures for their nationals as may be necessary for the conservation of those resources."

In light of today's understanding of global environmental interdependence, review of the existing legal regimes of the seas, including the freedom of the high seas and the 200-mile EEZ, should be initiated by the international community. For that purpose, it might be advisable to hold a Fourth United Nations Conference on the Law of the Sea.

NOTES

1. U.N. Doc. A/C.2/44/L.30/Rev. 1 (1989).
2. U.N. Doc. A/C.2/44/L.28 (1989).
3. G.A. Res. 44/225, 29 I.L.M. 1555 (1990) [hereinafter UN Resolution 44/225].
4. *Id.* at 1.
5. The Australian government, however, has suggested that it is more likely that EEZ-based drift net fisheries will be monitored or regulated than that high seas fishing will be regulated. Australian Government, Australia's Comments on United Nations Resolution 44/225 on Driftnet Fishing 1 (1990).
6. UN Resolution 44/225, *supra* note 3, at 4.
7. *NOAA Releases Report on Driftnet Use*, U.S. Dep't Com. News, July 6, 1990, at 1.

8. U.S. Department of Commerce, National Oceanic and Atmospheric Administration, Notice of Final List of Fisheries, 54 Fed. Reg. 16,072 (Apr. 20, 1989).

9. U.S. State Department, U.S. Actions Concerning Large-Scale Pelagic Driftnets 6 (paper submitted to the United Nations Office of Ocean Affairs and the Law of the Sea, July 1990) [hereinafter U.S. Policy Statement].

10. International Convention for the High Seas Fisheries of the North Pacific Region, May 9, 1952, 205 U.N.T.S. 65, *amended in* 1962, 14 U.S.T. 953; *amended in* 1978, 30 U.S.T. 1095; *amended in* 1986, 86 D.S.B. 73-74 (June 1986) [hereinafter INPFC Convention].

11. This chapter refers to the Soviet Union as it existed before its breakup and subsequent reformation into a commonwealth of states. The question remains as to whether any, all, or some of the states will continue to honor and enforce this agreement.

12. INPFC Convention, *supra* note 10; Magnuson Fishery Conservation and Management Act, 16 U.S.C. §§ 1802(6), 1811 (1988).

13. 16 U.S.C. §§ 1361-1407 (1988) [hereinafter MMPA].

14. Federation of Japan Salmon Fisheries Cooperative Association v. Baldridge, 679 F. Supp. 37 (D.D.C. 1987).

15. Verity v. Center for Environmental Education, 488 U.S. 1004 (1989).

16. U.S. Policy Statement, *supra* note 9, at 3-4.

17. The U.S.-U.S.S.R. Intergovernmental Consultative Committee on Fisheries was established by the Agreement on Mutual Fisheries Relations, with annexes, May 31, 1988, United States-U.S.S.R. *See* William T. Burke, *Anadromous Species and the New International Law of the Sea*, 22 OCEAN DEV. & INT'L L. 95, 126 (1991).

18. See Burke, *supra* note 17, at 126.

19. *See* GREENPEACE, NORTH PACIFIC HIGH SEAS DRIFTNET FISHERIES 7 (Feb. 1989).

20. Kate Troll, *Testimony Before the U.S. Senate Committee on Commerce, Science and Transportation and National Ocean Policy Study* 3 (May 17, 1989).

21. U.S. Policy Statement, *supra* note 9, at 1.

22. HANS KELSEN, GENERAL THEORY OF LAW AND STATE 343-44 (1961).

23. *Id.*

24. *See* article 101 of the United Nations Convention on the Law of the Sea, Dec. 10, 1982, U.N. Doc. A/CONF.62/122, 21 I.L.M. 1261 (1982). Article 101 defines piracy as

   (a) any illegal acts of violence or detention, or any act of depredation, committed for private ends by the crew or the passengers of a private ship or a private aircraft, and directed:
   (i) on the high seas, against another ship or aircraft, or against persons or property on board such ship or aircraft;
   (ii) against a ship, aircraft, persons or property in a place outside the jurisdiction of any State;
   (b) any act of voluntary participation in the operation of a ship or of an aircraft with knowledge of facts making it a pirate ship or aircraft;

(c) any act of inciting or of intentionally facilitating an act described in subparagraph (a) or (b).

25. 16 U.S.C. § 1811(b)(1) (1988).

26. Driftnet Impact Monitoring, Assessment and Control Act of 1987, Pub. L. No. 100-220, §§ 4004-09, 101 Stat. 1477.

27. *Id.* at § 4004, 101 Stat. 1478 (1989).

28. *Id.* at § 4006, 101 Stat. 1479 (1989).

29. 22 U.S.C. § 1978(a) (1988).

30. This agreement was adopted pursuant to the 1987 Driftnet Impact Monitoring, Assessment and Control Act, *supra* note 26.

31. *See* CANADIAN DEPARTMENT OF FISHERIES AND OCEANS, FISHERIES AGENCY OF JAPAN, U.S. NATIONAL MARINE FISHERIES SERVICE, & U.S. FISH AND WILDLIFE SERVICE, FINAL REPORT OF SQUID AND BYCATCH OBSERVATIONS IN THE JAPANESE DRIFTNET FISHERY FOR NEON FLYING SQUID (*OMMASTREPHES BARTRAMI*), JUNE-DECEMBER 1989 OBSERVER PROGRAM (June 30, 1990).

32. Agreement Between the Republic of Korea and the United States Regarding the High Seas Squid Driftnet Fisheries in the North Pacific Ocean, effected by Exchange of Notes, Sept. 13 and 26, 1989; Letter to the Speaker of the House and President of the Senate Reporting on Korean and Taiwanese Driftnet Fishing, *reprinted in* 25 WEEKLY COMP. PRES. DOC., Aug. 28, 1989, at 1278.

33. May 23, 1969, 115 U.N.T.S. 331, 8 I.L.M. 679 (1969).

34. The Charter of the United Nations, article 2(4), provides that "[a]ll Members shall refrain in their international relations from the threat or use of force against the territorial integrity or political independence of any state, or in any other manner inconsistent with the Purposes of the United Nations."

35. In the drafting process of this article, Sir Humphrey Waldock, special rapporteur, said that "the text was open-ended in the sense that any future interpretation of the law of the Charter would affect the rule embodied in article 36 [present article 52]." *Summary Records of the 840th Meeting [1966]*, 1:1 Y.B. INT'L L. COMM'N 120, PARA. 100, U.N. DOC. A/CN.4/183 and Add.1-4, A/CN.4/L.107 (1966).

36. According to the explanation of the International Law Commission, "[t]he law of the Charter concerning the prohibition of the use of force in itself constitutes a conspicuous example of a rule in international law having the character of *jus cogens*." *Report of the International Law Commission to the General Assembly*, 2 Y.B. INT'L L. COMM'N 76, U.N. DOC. A/CN.4/186/1966/Add.1-7.

37. Henry V. E. Mitchell, *Statement Before the U.S. Senate Committee on Commerce, Science, and Transportation*, May 17, 1989, at 13.

38. GREENPEACE, PACIFIC CAMPAIGN: DRIFTNETS 1 (Spring 1989).

39. *See, e.g.,* Earthtrust, *Statement of Earthtrust to the United States Senate Committee on Commerce, Science and Transportation on the Magnuson Fishery Conservation and Management Act Reauthorization* 10, 22 (May 17, 1989).

40. CANADIAN DEPARTMENT OF FISHERIES AND OCEANS, FISHERIES AGENCY OF JAPAN, U.S. NATIONAL MARINE FISHERIES SERVICE, & U.S. FISH AND WILDLIFE SERVICE, FINAL REPORT OF SQUID AND BYCATCH OBSERVATIONS IN THE JAPA-

NESE DRIFTNET FISHERY FOR NEON FLYING SQUID (*OMMASTREPHES BAR-TRAMI*), JUNE-DECEMBER 1989 OBSERVER PROGRAM (June 30, 1990).

41. GREENPEACE, *supra* note 38.

42. Government of New Zealand, Statement on Driftnet Fishing 1, 11 (paper submitted to the United Nations Office of Ocean Affairs and the Law of the Sea, Sept. 1990).

43. GREENPEACE, *supra* note 38.

44. GREENPEACE AUSTRALIA, OCEAN ECOLOGY: DRIFTNETS 2 (1989). *See also* Government of New Zealand, *supra* note 42, at 14.

45. Government of Japan, The View of the Government of Japan Concerning High Seas Driftnet Fishing in Conjunction with the U.N. Resolution 44/225 Concerning "Large-Scale Driftnet Fishing" 12 (paper submitted to the UN Office of Ocean Affairs and the Law of the Sea, Sept. 12, 1990).

46. Convention on the Prevention of Marine Pollution by Dumping of Wastes and Other Matter, Dec. 29, 1972, 26 U.S.T. 2403, T.I.A.S. No. 8165, IMO Doc. LDC/19/INF.2 (May 28, 1985) [hereinafter LDC].

47. Convention for the Prevention of Pollution from Ships, Nov. 2, 1973, T.I.A.S. No. 10561, 12 I.L.M. 1319 (1973); 1978 Protocol, 17 I.L.M. 546 (1978) [hereinafter MARPOL 73/78].

48. LDC, *supra* note 46, art. 4.

49. *Id.* annex I.

50. MARPOL 73/78, *supra* note 47, regulation 5.

51. *Id.* regulation 6.

52. U.S. Policy Statement, *supra* note 9.

53. UN Doc. A/AC.138/SC.II/L.43.

54. S. Res. 521, 94th Cong., 2d Sess., 122 CONG. REC. 27,583-84 (1976).

55. *But see* the ECE Convention for environmental assessment in the transboundary context, signed recently.

56. *See supra* notes 10-25 and accompanying text.

20 Reconsidering Freedom of the High Seas: Protection of Living Marine Resources on the High Seas

Catherine Floit

IN THE HISTORY OF THE INTERNATIONAL LAW OF THE SEA, the concept of freedom of the seas has not been static. Rather, the concept has changed as the nature of the international community and its interests have changed.[1] As a result of developing technology and extraordinarily efficient fishing techniques, the "freedom of the high seas" principle has become inadequate to cope with the increasing pressures on the world's living marine resources.

This chapter discusses the problem of high seas drift net fishing and then argues that the use of this fishing technique is destructive and violates customary international law and provisions of the 1982 United Nations Convention on the Law of the Sea.[2]

HIGH SEAS DRIFT NETS

Before World War II, fishing nets were made of natural fibers, such as cotton, flax, or hemp. The nets were visually and acoustically detectable to marine mammals and sank or disintegrated within months if lost or discarded. After World War II, the United Nations Food and Agriculture Organization (FAO) underwrote development of modern fishing technology to help developing countries become more self-sufficient. As a result, monofilament drift nets were in widespread use by the mid-1970s. Today, the use of plastic nets[3] is virtually universal, with the exception of a few subsistence fisheries. These synthetic nets, manufactured for durability, may float for years and do not easily biodegrade or disintegrate.[4]

310

Drift nets range up to 40 miles in length and between 25 and 50 feet in depth. These fine monofilament or multifilament nylon mesh nets hang in the water like large curtains from floats on the surface; weighted lead lines keep the nets hanging straight. Drift nets may be anchored to fish in place or may be left to drift with the winds and currents. Their depth can be controlled by adjusting their buoyancy.[5] Fishing passively, the nets are transparent and almost invisible to the eye as well as acoustically invisible to the echo-locating frequencies of some cetaceans. Mesh size may vary, depending on the area and target species.[6] Marine life is vulnerable to high seas drift net fishing for a number of reasons. First, a high percentage of marine life is concentrated in the thin upper layer of the ocean. Second, many marine species, including marine mammals, feed on drift net target species such as squid. Third, migration routes often cross areas where drift net fishing is conducted.[7]

Unfortunately, scientific data on the impacts of drift nets on various species are limited and data on populations of affected species are lacking, in part because little field data collection has been done on most marine mammal species.[8] Typically, information is anecdotal.[9] Nevertheless, scientists recognize that the incidental take of drift nets is enormous and that the persistent plastic materials of lost or discarded nets increase the mortality of marine mammals, birds, nontarget fish, and turtles.[10]

LIMITS TO HIGH SEAS FREEDOMS: CUSTOMARY INTERNATIONAL LAW OBLIGATIONS

The modern attitude toward freedom of the high seas[11] is that the concept is flexible and adaptable to changing conditions and situations not contemplated when the principle became part of international law.[12] At any particular time, the concept represents a balance of interests that best serves the international community, and the concept constantly adjusts to new problems. The extent of "freedom of the seas" therefore depends on state practice rather than on any innate quality of the high seas.[13]

In determining what limits exist on the exercise of high seas freedoms—particularly the freedom to fish—we must first consider whether there is any rule of customary international law that prohibits a particular activity as a freedom of the seas. Typically, any high seas activity not expressly prohibited by international law is a permitted freedom of the seas.[14]

Although equality of access to the high seas is assumed,[15] the right of access and the right to fish are not equivalent to an unfettered right to exploit living resources. Rather, freedom of the seas is limited or qualified pursuant to various customary international law principles.[16] Since this freedom is exercised in the interests of all to enjoy, it is regulated; limitations on it are not meant to restrict high seas freedoms but are designed "to safeguard its exercise in the interests of the entire international community."[17]

Reasonable Use

The accommodation of interests, or the balancing of rights and duties, is a recurrent theme in the formulation of various jurisdictional zones in the law of the sea. The operative test is one of reasonableness; high seas freedoms are limited by the legal obligation that they be exercised reasonably.[18] One commentator notes that the usual high seas freedoms of fishing and navigation are permitted because they are, at worst, a momentary exclusivity. Those activities are prohibited if they are conducted unreasonably.[19]

During the meetings at which the 1958 conventions[20] were drafted, the International Law Commission gave support to the requirement of reasonable use,[21] and the codified definition of freedom of the seas in the 1958 Geneva Convention on the High Seas, which purports to codify customary international law, embodies reasonable use ideas. Article 2(2) provides that high seas freedoms, including the freedom to fish, "[s]hall be exercised by all States with reasonable regard to the interests of other States in their exercise of the freedom of the high seas."[22]

The 1982 Convention also embodies the concept of reasonable use. It recognizes that freedom of the high seas is subject to limiting rules of international law and the 1982 Convention itself. Article 87 declares that "[t]hese freedoms shall be exercised by all States with due regard for the interests of other States in their exercise of the freedom of the high seas, and also with due regard for the rights under this Convention with respect to activities in the Area."[23]

Abuse of Rights

Because the high seas are beyond the jurisdiction of any state, traditional freedoms are susceptible to abuse. Destructive fishing technologies arguably represent an abuse of right. In international law, there are prohibitions against abusing rights.[24] These prohibitions arise when one nation's exercise of its freedoms interferes with the exercise of the corresponding freedom by the other state, for "[n]o state may be vested with exclusive competence or unfettered liberty even as to its own resources when the interests of other states are implicated."[25]

There is no clear consensus on what constitutes an abuse of rights; the doctrine is defined both narrowly and broadly.[26] It is broadly defined as a "general civil law concept which permits action to be taken against an exercise of a right which is technically permitted under the law but which is unfair or malicious."[27] The International Court of Justice (ICJ) applied the abuse of rights doctrine to the *Anglo-Norwegian Fisheries* case in determining that Norway's delimitation of its fisheries zone was not contrary to international law.[28] The principle prohibiting the abuse of rights is codified in article 300 of the 1982 Convention: "States Parties . . . shall exercise the rights, jurisdiction and freedoms recognized in this Convention in a manner which would not constitute an abuse of right."

The test of whether an action constitutes an "abuse of rights" involves balancing the potentially conflicting interests of individual states against the wider general interest.[29] Individual rights may be adversarial to the benefits gained by society from limiting those rights. In determining whether an abuse of rights has occurred, one should analyze both the right and the limits of the right. Limiting abuse of rights is intended to "safeguard the general interests of the international community . . . freedom of the high seas has always contained an inherent danger of abuse, and indeed the international community has long since evolved rudimentary rules to insure that the high seas do not become a legal vacuum."[30] The principle provides an underlying logic for preventing harm to common natural resources from overexploitation by one or several states.

The concepts of "reasonable use" and "abuse of right" are important in another context. Persistent objectors to development of rules of customary international law may not be bound by those emerging rules.[31] This exception is nevertheless subject to the customary rule of reasonableness and cannot be invoked in bad faith. A state that acts in bad faith may have abused its rights to the detriment of other states, and it may be required to conform to the generally accepted standard.[32]

Duty to Conserve Living Resources of the High Seas

It is now generally agreed that the freedom of the high seas is subject to a duty to adopt measures to conserve living resources of the high seas:[33]

> Indeed, it would appear contrary to the very principle of freedom of the seas to permit actions which could interfere with the undisturbed and safe use of the sea by the community of nations and to destroy the living resources whose common enjoyment has always been one of the principal objects of the freedom of the high seas.[34]

The interference caused by high seas drift netting is potentially very broad; other states will be denied the opportunity to exercise resource rights if the resource is harmed or depleted.

Provisions for conservation on the high seas are found in article 2(2) of the 1958 Convention on Fishing and Conservation of the Living Resources of the High Seas, articles 87(1)(e) and 117 of the 1982 Convention; various other treaties, and United Nations resolutions and declarations. *Opinio juris* on this issue is evidenced by the number of multilateral conventions and bilateral agreements establishing quotas for fishing or measures for conservation and management.[35]

The obligation includes the responsibility to prevent harm to the shared resources of the high seas. In the *Fisheries Jurisdiction* case (*United Kingdom v. Iceland*), the ICJ held that although the high seas freedoms included fisheries, these freedoms were not absolute; "both States (Iceland and the U.K.) have an obligation to take full account of each other's rights and of any fishery conservation measures the necessity of which is shown to exist in those waters."[36] Judge Dillard's concurring opinion stated that the obligation to conserve living resources "may qualify as a norm of customary international law."[37] Moreover, the court stated that the "laissez-faire treatment of the living resources of the sea in the high seas has been replaced by a recognition of a duty to have due regard to the rights of other states and the needs of conservation for the benefit of all."

The duty to conserve extends to the fishing state's own nationals, alone or in cooperation with nationals of other states interested in the resource.[38] The obligation to cooperate with other states in establishing conservation measures is also customary international law.[39] The 1982 Convention specifically provides for such cooperation, and the *Fisheries Jurisdiction* case required the states "to examine together, in the light of scientific and other available information, the measures required for the conservation and development, and equitable exploitation, of those resources."[40] The duty to cooperate further includes the duties to negotiate conservation measures and to share scientific research.[41]

United Nations activities provide further evidence of the duty to conserve the living resources of the high seas. United Nations resolutions tend to embody policies of developing customary international law,[42] and they may function as sources of customary international law and treaty law. Several international documents may stand as quasi-legal documents urging states to cooperate to manage resources in a sensitive and rational manner.[43]

In 1972, the United Nations Stockholm Conference on the Human Environment adopted a Declaration of the Human Environment, containing preambular language and twenty-six principles. An action plan set forth in part II of the Stockholm Conference's report contains a number of principles relevant to resource conservation.[44] Principle 1 states that humans have "a solemn responsibility to protect and improve the environment for present and future generations." Other principles deal with the safeguarding of renewable resources, particularly wildlife. For example, Principle 2 provides: "The natural resources of the earth including the air, water, land, flora and fauna and especially representative samples of natural ecosystems must be safeguarded for the benefit of present and future generations through careful planning or management, as appropriate." Principle 3: "The capacity of the earth to produce vital renewable resources must be maintained and, wherever practicable, restored or improved." Principle 4: "Man has a special responsibility to safeguard and wisely manage the heritage of wildlife and its habitat which are now gravely imperilled by a combination of adverse factors. Nature conservation including wildlife must therefore receive importance in planning for economic development." The principles recognize that many factors threaten natural resources and that rational management requires an integrated approach.[45]

Recommendations 37 and 38 of the Stockholm Conference and the action plan call for protection of internationally significant ecosystems and for cooperation in that protection. Recommendations 32 and 50 urge cooperation to protect living resources; and recommendation 5 calls for cooperation in the management of shared water resources. The action plan calls for research, monitoring, and the exchange of information.

Furthermore, Principles 14, 17, and 18 adopted by the Stockholm Conference recommended that states alone and collectively establish adequate environmental assessment and monitoring programs. The United Nations Environment Program (UNEP) has subsequently indicated that states should (1) ensure that environmental effects are taken fully into account before decisions are made; (2) undertake environmental assessments for all significant activities so that impacts are known and considered; and (3) include a description of the affected environment and impacts.[46]

The Nairobi Declaration notes the continued validity of the Stockholm principles.[47] The United Nations Governing Council, with respect to oceans, notes trends toward overfishing and environmentally inappropriate exploitation of marine and coastal resources. Governments are asked to develop plans and procedures to manage marine resources.

On November 6, 1989, the United States and eleven other countries, including Canada, Australia, and New Zealand, introduced a resolution recommending an immediate ban on drift net fishing in the South Pacific and a moratorium on all high seas drift net fishing by June 30, 1992.[48] The General Assembly of the United Nations adopted this Driftnet Resolution,[49] noting that overexploitation of high seas resources tends to have adverse impacts on resources within EEZs. The UN Driftnet Resolution emphasizes the obligation of the international community to cooperate globally and regionally in the conservation and management of living resources on the high seas and to take such measures as are necessary to conserve the living resources of the high seas.

In the UN Driftnet Resolution, the United Nations General Assembly calls on nations engaged in large-scale pelagic drift net fishing to cooperate with the international community in collecting and sharing scientific data to assess the impacts of drift netting. The resolution recommends that the international community review the scientific data on impacts by June 30, 1991, and agree on regulatory and monitoring measures. The General Assembly recommends (1) moratoriums on large-scale pelagic drift netting on the high seas by June 30, 1992, unless jointly agreed-on effective conservation and management measures are taken to prevent unacceptable impacts; (2) cessation of the same activity in the South Pacific by July 1, 1991; and (3) cessation of further expansion in the North Pacific high seas and all other high seas. To some extent, the UN Driftnet Resolution embodies the new formulation of the precautionary principle,[50] since it calls for regulations even in the absence of complete scientific certainty.

Recent activities with regard to drift nets, including the 1989 UN Driftnet Resolution; agreements between the United States and Korea, Japan, and Taiwan pursuant to the Driftnet Impact Monitoring, Assessment, and Control Act of 1987;[51] the international Convention for the Prohibition of Fishing with Long Driftnets in the South Pacific, signed by the Pacific Island states;[52] and the Castries Declaration[53] provide ample evidence that states recognize the duty to conserve living resources of the high seas. Failure to conserve the high seas living resources is therefore a violation of customary law.

The Principle of Precautionary Action

The principle of precautionary action is an important principle of environmental policy that provides a conservative basis from which to solve future problems; conceptually, it may be applied to areas other than that of pollution control, in which it is most often cited. When possible damage to the marine environment or human health may be so serious that to wait for complete clarification of scientific uncertainties is irresponsible, the principle of precautionary action must be applied.[54]

The principle requires environmental quality objectives. Dr. Lothar Gundling has proposed the following understanding of the precautionary principle:

"Precautionary action" is a more stringent form of preventive environmental policy. It is more than repair of damage or prevention of risks. Precautionary action requires reduction and prevention of environmental impacts irrespective of the existence of risks. This, however, must not be understood in the sense that aspects of risks are not relevant; the crucial point is that environmental impacts are reduced or prevented even before the threshold of risks is reached. This means that precautionary action must be taken to ensure that the loading capacity of the environment is not exhausted, and it also requires action even if risks are not yet certain but only probable, or, even less, not excluded.[55]

The precautionary principle has been accepted at several forums.[56]

Although expression of the principle in treaties or declarations seems to be relatively recent, it may be an implied or underlying principle of international law.[57] The precautionary principle underlies the 1985 Vienna Convention for the Protection of the Ozone Layer and the 1987 Montreal Protocol on Substances That Deplete the Ozone Layer.[58] The principle appears to underlie the UN Driftnet Resolution, which also calls for curtailing of activity whether or not scientific evidence is available. The German Council of Experts on Environmental Matters has declared that the principle is required by international, supranational, and national law.[59]

Use of a "best available" standard for scientific evidence may tie in with use of the precautionary principle. Information may be difficult to obtain and may be subject to different interpretations. Research pertaining to the high seas may especially fall within this category. The requirement to conserve, however, allows the use of any evidence as support for institution of conservation measures.

This principle represents an emerging standard of customary international law.

Summary

The right to fish on the high seas is threatened by unregulated exploitation, which may deplete fish stocks. High seas drift net fishing is not a reasonable exercise of the freedom to fish, since the activity interferes with the rights of other states to fish. High seas drift net fishing fails to comply with the rule of reasonable use, that freedoms must be exercised with reasonable regard for the interests of other states and of the international community. High seas drift net fishing also fails to comply with the abuse of rights doctrine.[60] Finally, the precautionary principle requires that activities be regulated if there is a possibility of harm. Although in the past the party using the high seas did not need to establish positively a right to do so,[61] changing circumstances require a change in practice. Indeed, states now more often are on the defensive and in the position of having to justify their actions on the high seas.

OBLIGATIONS UNDER THE 1982 CONVENTION ON THE LAW OF THE SEA

The 1982 Convention sets forth an international legal regime for the high seas and the conservation of living marine resources. As discussed in this section, the use of high seas drift nets violates various provisions of this convention. Part VII[62] of the 1982 Convention, which applies to the high seas,[63] details the high seas freedoms in a nonexclusive list in article 87:

> 1. The high seas are open to all States, whether coastal or land-locked. Freedom of the high seas is exercised under the conditions laid down by this Convention and by other rules of international law. It comprises, *inter alia*, both for coastal and land-locked States
>   (a) freedom of navigation;
>   (b) freedom of overflight;
>   (c) freedom to lay submarine cables and pipelines, subject to Part VI (Continental Shelf);
>   (d) freedom to construct artificial islands and other installations permitted under international law, subject to Part VI;
>   (e) freedom of fishing, subject to the conditions laid down in section 2;
>   (f) freedom of scientific research, subject to Parts VI and XIII (Marine Scientific Research).

General Qualifications on the High Seas Freedom to Fish

The 1982 Convention qualifies the high seas freedoms with "the conditions laid down by this Convention and by other rules of international law. . . . These freedoms shall be exercised by all States with due regard for the interests of other States in their exercise of the freedom of the high seas."[64] The freedom to fish is thereby subject to customary international law and other treaties on the subject. Furthermore, the parties to the 1982 Convention are to "exercise the rights, jurisdiction and freedoms recognized in this Convention in a manner which would not constitute an abuse of right."[65]

The freedom of fishing is further qualified. Article 87(1)(e) states that the freedom to fish on the high seas is "subject to the conditions laid down in section 2," titled "Conservation and Management of the Living Resources of the High Seas," articles 116 through 120.[66] Article 116 provides that high seas fishing is subject to (1) treaty obligations; (2) the rights, duties, and interests of coastal states.provided in article 63(2) and articles 64 through 67; and (3) the provisions of section 2 (articles 116 through 120). In general, qualified freedoms are subject to regulation in a manner consistent with the restriction's purpose.[67]

Rights, Duties, and Interests of Coastal States in Living Marine Resources of the High Seas

The 1982 Convention exempts several types of fish from the traditional freedom of fishing on the high seas by classifying them separately. Article 63(2) applies to a stock or stocks of associated species that occur within an EEZ and on adjacent high seas; these are known as straddling stocks. Article 64 applies to highly migratory species, which are not defined in the 1982 Convention but are listed in annex I. Article 65 applies to marine mammals and permits coastal states or international organizations to regulate the exploitation of these species more strictly than does the 1982 Convention. Article 66 applies to anadromous species; article 67, to catadromous species. The 1982 Convention assigns primary interest in and responsibility for these species and stocks to coastal states but emphasizes cooperation and agreement among all states concerned.

Provisions of Section 2

Article 116. Article 116, permitting high seas fishing, subjects the right to fish on the high seas to the provisions of articles 63 through 67, pertaining to catadromous, anadromous, and highly migratory species; straddling stocks; and marine mammals. Article 116 therefore creates a class of interests with special protections. As a result, when that class of interests is implicated, the burden shifts to the fishery to demonstrate that its actions are not harmful to those interests.[68] This is an application of the precautionary principle. Resource management efforts must be in accord with the general obligation to protect and preserve the marine environment.

Pursuant to this article, high seas fishing states have obligations to coastal states, including the duties to conserve resources and cooperate with the coastal states. Article 116 specifies that the rights of other states to fish are subject to the sovereign rights of coastal states within their EEZs. The article may be interpreted to grant coastal state authority to prescribe conservation measures on the high seas with which the high seas fishing state must comply.[69] This article seems to anticipate deadlocks and permits unilateral action beyond a state's EEZ in certain situations.[70] This provision follows from articles 6 and 7 of the 1958 Convention on the High Seas, which permit unilateral action by a coastal state if an adjacent high seas fishery is of special interest to the coastal state. States must first negotiate; then, if negotiations fail, unilateral action may be permitted.[71] If one side refuses to negotiate in good faith or withholds relevant scientific information, the 1982 Convention's compulsory dispute settlement provisions will apply.[72]

Articles 117, 118, and 119. Article 117 emphasizes that states are obligated to adopt, or to cooperate with other states in adopting, regulations for their own nationals that are necessary to conserve the living resources of the high seas. According to article 118, states must also cooperate in the conservation and management of high seas living resources, negotiate to establish measures to conserve the resources, and cooperate to establish regional or subregional fisheries organizations. Finally, article 119 provides requirements for determining allowable catch and for establishing conservation measures for living resources of the high seas. States must take measures to maintain or restore populations that can produce the maximum sustainable yield; measures are to be based on the best scientific evidence available. Interdependence of stocks and generally recommended international minimum standards must be considered.[73]

Although scientific data play an important role in designing conservation measures, the amount of data available is not great.[74] The lack of data should not, however, lead to a failure to establish conservation measures. As discussed earlier, article 116 creates a class of interests with special protections. When that class of interests is implicated, the burden shifts to the fishery to demonstrate that its actions are not harmful.[75]

Summary

Much of part VII on the high seas codifies customary international law.[76] Since the 1982 Convention is codifying existing customary law as well as establishing new rules of international law,[77] once it comes into force it will affect third parties to the extent that it reflects customary international law.[78] In that event, rights and duties will not depend on the provisions of the 1982 Convention,[79] which has exerted a positive, considerable influence on the development of international and national law of the sea. Many states, including African, Latin American, and Pacific Island states,[80] have specifically adopted 1982 Convention provisions into new national laws.[81]

CONCLUSION

In the past, the oceans appeared endless; most uses of the high seas were permissible, if not entirely reasonable. The maritime powers had a *laissez-faire* attitude toward fishing and overfishing.[82] In the past, however, humans lacked the technology actually to occupy the sea. Now, humans can occupy the sea, extract its resources, and dominate it as they do the land.[83]

High seas fishing states have the following obligations:

1. To conserve the living resources of the high seas.
2. To cooperate with other states in conserving living resources of the high seas.
3. To negotiate with other states over conservation measures.
4. To generate and exchange scientific and other information.
5. To employ the best scientific evidence available to establish the needed conservation measures.
6. To refrain from discrimination in conservation.

The frequency with which the legal duties to conserve, cooperate, and negotiate appear gives them an "imperative significance that cannot be ignored."[84] A high seas fishing state cannot make conservation and management decisions on its own; coastal and other interested states must also be involved.[85]

The use of high seas drift nets violates both customary international law and provisions of the 1982 Convention. The use is unreasonable and constitutes an abuse of rights. The practice of drift net fishing is fundamentally inconsistent with international obligations to conserve living marine resources of the high seas.

NOTES

1. Arvid Pardo, *The Law of the Sea: Its Past and Future*, 63 Or. L. Rev. 7, 8–9 (1984).
2. United Nations Convention on the Law of the Sea, Dec. 10, 1982, U.N. Doc. A/CONF.62/122, 21 I.L.M. 1261 (1982) [hereinafter 1982 Convention]. As of April 1993, fifty-five states had ratified the 1982 Convention; sixty ratifications are required for it to enter into force. *See* Council on Ocean Law, Oceans Pol'y News, Dec. 1989, at 1, for a list of the first forty-two states. Angola was the forty-third. Council on Ocean Law, Oceans Pol'y News, Apr. 1990, at 1.
3. The principal synthetic compounds are polyamide (nylon), polyester, poly-ethylene, and polypropylene. J. Coe, Derelict Fishing Gear: Disaster or Nuisance? 6 (1986). Nylon may persist for tens to many hundreds of years. *Id.* at 12.
4. Conner & O'Dell, *The Tightening Net of Marine Plastics Pollution*, 30 Env't, Jan.–Feb. 1988, at 17–18; Coe, *supra* note 3, at 6.
5. Eisenbud, *Problems and Prospects for the Pelagic Driftnet*, 12 B.C. Envtl. Aff. L. Rev. 474, 478 (1985).
6. Squid nets have a 3- to 6-inch-diameter mesh; billfish and albacore nets have a mesh 6 to 18 inches in diameter.
7. *Curtains of Death*, 1 Buzzworm 23, 27–29 (1989).
8. Swartzman, *Present and Future Potential Models for Examining the Effect of Fisheries on*

Marine Mammal Populations in the Eastern Bering Sea, in PROCEEDINGS OF THE WORKSHOP ON BIOLOGICAL INTERACTIONS AMONG MARINE MAMMALS AND COMMERCIAL FISHERIES IN THE SOUTHEASTERN BERING SEA (ANCHORAGE, ALASKA, OCT. 18-21, 1983) 157 (Apr. 1984).

9. Taylor, *Regulation of High Seas Driftnet Fisheries in the North Pacific*, 7TH ANN. NAT'L FISHERY L. SYMP., Oct. 12-13, 1989, at 11, 12.

10. *Recent Trends in World Fish Harvests*, 50 MARINE FISHERIES REV. 57, 60; R. B. CLARKE, MARINE POLLUTION 155-156 (2d ed. 1989).

11. R. P. ANAND, ORIGIN AND DEVELOPMENT OF THE LAW OF THE SEA: HISTORY OF INTERNATIONAL LAW REVISITED 225 (1983).

12. *Id.* at 235.

13. 2 D. O'CONNELL, THE INTERNATIONAL LAW OF THE SEA 796 (1984).

14. E. Brown, *Freedom of the High Seas Versus the Common Heritage of Mankind: Fundamental Principles in Conflict*, 20 SAN DIEGO L. REV. 521, 536, 549 (1985).

15. The 1958 Convention on the High Seas, Apr. 29, 1958, 13 U.S.T. 2312, 450 U.N.T.S. 82, embodying customary international law, provides that "[t]he high seas being open to all nations, no State may validly purport to subject any part of them to its sovereignty" (art. 2(2)). The 1982 Convention, *supra* note 2, provides that "[t]he high seas are open to all States, whether coastal or land-locked" (art. 87(1)).

16. A state's perceptions of the limits on high seas freedoms and the nature of national and international rights on the high seas is influenced in part by that state's geographic relationship to the sea and how it benefits from access. E. HULL, THE INTERNATIONAL LAW OF THE SEA: A CASE FOR A CUSTOMARY APPROACH 4 (1976).

17. 2 Y.B. INT'L L. COMM'N 278 (1956).

18. R. SONI, CONTROL OF MARINE POLLUTION IN INTERNATIONAL LAW 135 (1985); 1 D. O'CONNELL, THE INTERNATIONAL LAW OF THE SEA 57 (1982).

19. Brown, *supra* note 14, at 535.

20. Convention on the Territorial Sea and the Contiguous Zone, Apr. 29, 1958, 15 U.S.T. 1606, 516 U.N.T.S. 205; 1958 Convention on the High Seas, *supra* note 15; Convention on the Continental Shelf, Apr. 29, 1958, 15 U.S.T. 471, 499 U.N.T.S. 311; Convention on Fishing and Conservation of the Living Resources of the High Seas, Apr. 29, 1958, 17 U.S.T. 138, 559 U.N.T.S. 285.

21. Article 2(2) of the 1958 Convention on the High Seas arose from concerns about nuclear testing at sea, but it was not intended to be limited to that subject. 2 O'CONNELL, *supra* note 13, at 798. The text of article 2(2) is from a United Kingdom proposal substituted for a provision explicitly discussing nuclear testing on the high seas. The continental shelf provisions were also approached from the view of reasonable use; navigation and fishing interests yielded to those of industry. The commission's commentary indicated that "the progressive development of international law, which takes place against the background of established rules, must often result in the modification of those rules by reference to new interests or needs." 1953 Y.B. INT'L L. COMM'N 102-03. A balancing was conducted, and traditional freedoms were modified.

22. 1958 Convention on the High Seas, *supra* note 15, at art. 2(2).

23. 1982 Convention, *supra* note 2, arts. 87(1), (2). The concept also forms the basis of the 1982 Convention's EEZ provisions. 1 O'CONNELL, *supra* note 18, at 58.

24. BLACK'S LAW DICTIONARY (5th rev. ed. 1979) defines the noun "abuse" as "departure from reasonable use; immoderate or improper use" and the verb "abuse" as "to make excessive or improper use of a thing." The prohibition of abuse of rights is sometimes equated with equity in international law. Schwarzenberger, *Equity in International Law*, 26 Y.B. WORLD AFF. 346, 346 (1972) [hereinafter Schwarzenberger, *Equity*]. Various states have domestic laws reflecting the abuse of rights concept. *See generally* Ward et al., *The Business Purpose Test and Abuse of Rights*, 1985 BRIT. TAX REV. 68.

25. B. SMITH, STATE RESPONSIBILITY AND THE MARINE ENVIRONMENT: THE RULES OF DECISION 71-72 (1988).

26. *Id.* at 84-85; I. BROWNLIE, PRINCIPLES OF PUBLIC INTERNATIONAL LAW 444 (3d ed. 1979) [hereinafter BROWNLIE, PRINCIPLES). Brownlie states that it is not unreasonable to regard the principle of abuse of rights as a general principle of international law, although he himself believes that it is a "useful agent in the progressive development of the law" rather than a general principle of law, in part because he doubts that the concept can be used in the absence of a treaty that defines the principle for particular circumstances. *Id.* at 444-45. The 1982 Convention's prohibition on the abuse of rights (article 300) is defined in the context of the convention's other provisions.

27. Ward et al., *supra* note 24, at 68 n.1; Bigger & Kagedan, *Notes from Other Nations*, 76 Trademark Rep. 157, 162 n.13 (1986).

28. Great Britain v. Norway, 1951 I.C.J. 116, 142.

29. SONI, *supra* note 18, at 136.

30. R. SHINN, THE INTERNATIONAL POLITICS OF MARINE POLLUTION CONTROL 48 (1974).

31. M. VILLIGER, CUSTOMARY INTERNATIONAL LAW AND TREATIES 15-16 (1985).

32. Louis Sohn, *Implications of the Law of the Sea Convention Regarding the Protection and Preservation of the Marine Environment*, in THE DEVELOPING ORDER OF THE OCEANS 109 (R. Krueger & S. Riesenfeld eds. 1985).

33. Koester, *From Stockholm to Brundtland*, 20 ENVTL. POL'Y & L. 14, 17 (1990).

34. Legault, *The Freedom of the Seas: A License to Pollute?*, 31 U. TORONTO L.J. 211, 217 (1971).

35. Bleicher, *An Overview of International Environmental Regulations*, 2 ECOLOGY L.Q. 1, 31 (1972).

36. Fisheries Jurisdiction case (United Kingdom v. Iceland), 1974 I.C.J. 3, 31.

37. *Id.* at 69; William Burke, *International Law of the Sea Concerning Coastal State Authority over Driftnets on the High Seas* 8 (manuscript prepared for the FAO, 1990) [hereinafter Burke, *Authority over Driftnets*].

38. Burke, *Authority over Driftnets*, *supra* note 37, at 8.

39. *Id.* at 12.

40. *Fisheries Jurisdiction* case, 1974 I.C.J. at 34-35.

41. Burke, *Authority over Driftnets, supra* note 37, at 14–15.

42. Oliver, *Interim Deep Seabed Mining: An International Environmental Perspective*, 8 J. LEGIS. 73, 81 (1981); Koester, *supra* note 33, at 15.

43. Harders, *In Quest of an Arctic Legal Regime: Marine Regionalism—a Concept of International Law Evaluated*, 11 MARINE POL'Y 285, 292 (1987).

44. Declaration on the Human Environment, 11 I.L.M. 1416 (1972). The declaration was adopted by UN General Assembly Resolution 2994, Dec. 15, 1972. Although convened to deal with global environmental problems as a whole, pollution was the major concern. Nevertheless, the principles and action program do relate to identified threats to all of the earth's ecosystems. Albeit general, the principles attempt to establish basic rules for international environmental law; they provide a broad consensus for action and perhaps state a common understanding or conviction among the parties to the conference.

45. 11 I.L.M. 1416, 1418 (1972).

46. *Report of the Working Group of Experts on Environmental Law on Its Second Session on Environmental Impact Assessment*, U.N. Environment Program, U.N. Doc. UNEP.WG.152/4 (1987).

47. The international community met in Nairobi in May 1982 to commemorate the tenth anniversary of the 1972 United Nations Conference on the Human Environment.

48. *News*, 21 MARINE POLLUTION BULL. 111 (1990).

49. Large-Scale Pelagic Driftnet Fishing and Its Impact on the Living Marine Resources of the World's Oceans and Seas, G.A. Res. 44/225, *adopted by consensus* Dec. 22, 1989, *reprinted in* 20 ENVTL. POL'Y & L. 36 (1990).

50. *See infra* notes 54–61 and accompanying text.

51. 101 Stat. 1477. *See* 16 U.S.C. § 1822 (Supp. 1991).

52. Nov. 1989, 29 I.L.M. 1449.

53. On November 24, 1989, the members of the Organization of Eastern Caribbean States adopted the Castries Declaration, which established a regional pelagic fisheries regime that outlaws drift nets, urges prevention of the use of indiscriminate fishing methods, and encourages global restrictions on harmful fishing practices. Council on Ocean Law, OCEANS POL'Y NEWS, Sept. 1990, at 6.

54. Lothar Gundling, *The Status in International Law of the Principle of Precautionary Action*, 5 INT'L J. ESTUARINE & COASTAL L. 23, 29 (1990). The principle of precautionary action is also a principle of law, although its scope and limits are subject to much disagreement. *Id.* at 23–24.

55. *Id.* at 26.

56. The 1984 Bremen Declaration (Conclusion A(7)); the London Declaration (VII, XV.2, XVI.1); the Third International Conference on the Protection of the North Sea; the Recommendations of the Paris Commission of June 22, 1989; the fifteenth session of the Governing Council of UNEP in 1989; the Nordic Council's Conference on Pollution of the Seas; the Barcelona Convention on the Protection of the Mediterranean Sea in 1989; and the Interparliamentary Conference on the Global Environment of 1990.

57. Gundling, *supra* note 54, at 27.

58. *Id*. at 29.

59. *Id*. at 24.

60. Brown, *supra* note 14, at 533.

61. The traditional law, in the absence of an international agreement to the contrary, is that the flag state has jurisdiction both to legislate and to enforce standards over its vessels on the high seas because each state is a sovereign. R. M. M'GONIGLE & M. W. ZACHER, POLLUTION, POLITICS AND INTERNATIONAL LAW 200 (1979). The "law of the flag" concept dates from Roman times. Gormley, *The Development and Subsequent Influence of the Roman Legal Norm of 'Freedom of the Seas,'* 40 U. DET. L.J. 561, 584 (1963). Enforcement may include rights of inspection, investigation, and prosecution.

Conversely, no state can unilaterally apply its laws to the ships of other states, with the possible exception of universal violations of international law, such as piracy. McDougal, *International Law and the Law of the Sea*, in THE LAW OF THE SEA: OFFSHORE BOUNDARIES AND ZONES 19 (L. M. Alexander ed. 1967). The 1982 Convention provides that ships are subject to the exclusive jurisdiction of the flag state on the high seas, absent "exceptional cases expressly provided for in international treaties or in this Convention." 1982 Convention, *supra* note 2, art. 92(1). Unfortunately, a flag state may not be inclined to pursue or prosecute one of its own for violations. The convention provides that there must be a genuine link between the ship and the state whose flag the ship is entitled to fly. *Id*. art. 91(1). The absence of a genuine link has caused regulatory problems in the past. On the other hand, ships flying no flag and refusing to show a flag when requested to do so may be boarded by the ships of any state. BROWNLIE, PRINCIPLES, *supra* note 26, at 253. One theory is that such a rule ensures that the high seas do not become a legal vacuum. William Kenneth Bissell, *Intervention on the High Seas: An American Approach Employing Community Standards*, 7 J. MARITIME L. & COM. 718, 725 (1976).

62. Section 1 sets forth the general provisions; section 2 sets forth provisions on conservation and management of living resources of the high seas.

63. 1982 Convention, *supra* note 2, art. 86.

64. *Id*. arts. 87(1), (2).

65. *Id*. art. 300.

66. In fact, perhaps no living marine resource exists solely in the high seas beyond 200 miles. Jacobson, *International Fisheries Law in the Year 2010*, 45 LA. L. REV. 1161, 1193 (1985).

67. Comment, *Fishery and Economic Zones as Customary International Law*, 17 SAN DIEGO L. REV. 661, 671 (1980) [hereinafter *Fishery and Economic Zones*].

68. Taylor, *supra* note 9, at 13.

69. E. Miles & W. Burke, *Pressures on the UN Convention on the Law of the Sea Arising from New Fisheries Conflicts: The Problem of Straddling Stocks*, 50 MARITIME STUD. 15 (1990).

70. W. Burke, *Fishing in the Bering Sea Donut*, 16 ECOLOGY L.Q. 285, 293 (1989).

71. *Id.* at 299 n.42.

72. *Id.*

73. Martin Belsky, *Ecosystem Model Mandate for a Comprehensive United States Ocean Policy and Law of the Sea*, 26 SAN DIEGO L. REV. 417, 467 (1989).

74. Taylor, *supra* note 9, at 12.

75. *Id.* at 13.

76. Mendelson, *Fragmentation of the Law of the Sea*, 12 MARINE POL'Y 192, 195 (1988). Article 311(1) provides that the 1982 Convention prevails over the 1958 Geneva conventions only between parties to the more recent convention. 1982 Convention, *supra* note 2; Proclamation No. 5030, 48 Fed. Reg. 10,605 (1983); *Statement by the President on United States Ocean Policy*, 19 WEEKLY COMP. PRES. DOC., Mar. 10, 1983, at 383; Boczek, *The Protection of the Antarctic Ecosystem: A Study in International Environmental Law*, 13 OCEAN DEV. & INT'L L. 347, 393 (1983).

77. Vukas, *The Impact of UNCLOS III on Customary Law*, in THE NEW LAW OF THE SEA 34 (C. L. Rozakis & C. A. Stephanou eds. 1983).

78. SONI, *supra* note 18, at 62.

79. Vukas, *supra* note 77, at 42.

80. Council on Ocean Law, OCEANS POL'Y NEWS, May 1990, at 3.

81. Tsarev, *Maritime Legislation of Coastal States and the 1982 UN Convention on the Law of the Sea*, in THE LAW OF THE SEA: WHAT LIES AHEAD? 530 (T. A. Clingan, Jr. ed. 1988).

82. Schwarzenberger, *Equity*, *supra* note 24, at 358.

83. John Knauss, *Creeping Jurisdiction and Customary International Law*, in THE DEVELOPING ORDER OF THE OCEANS 738 (R. Krueger & S. Riesenfeld eds. 1985).

84. Douglas M. Johnston, *The Driftnetting Problem in the Pacific Ocean: Legal Considerations and Diplomatic Options*, 21 OCEAN DEV. & INT'L L. 5, 22 (1990).

85. Burke, *Authority over Driftnets*, *supra* note 37, at 21.

21 Recent Developments in the Use of International Trade Restrictions as a Conservation Measure for Marine Resources*

Melinda P. Chandler

THE RECENT DRAMATIC INCREASE in the exploitation of living marine resources threatens the survival of these resources throughout the world. In response to this threat, a growing number of nations are acting unilaterally and multilaterally to take conservation and management measures that impose restrictions on trade, directly or indirectly. Such scenarios include prohibition of, or restrictions on, trade in endangered or threatened species; restrictions on imports of commercially harvested species when the harvesting has harmful incidental impacts on protected species; restrictions on commercially harvested species to ensure compliance with conservation and management measures relating to that species; and protection of ecosystems.

TRADE IN ENDANGERED SPECIES

The clearest example of the use of a trade restriction to further conservation goals is the multilateral adoption of a ban on trade in species, or products from species, that are threatened or endangered.

* The views and opinions expressed in this chapter are solely those of the author and do not necessarily reflect those of the Department of State or any other agency of the United States government. The author gratefully acknowledges the invaluable assistance of Barry Pershkow and Margaret Spring, interns at the Center for International Environmental Law–US, and Donna Darm, formerly of the Office of the Legal Adviser at the Department of State.

International Regulations

The Convention on International Trade in Endangered Species of Wild Fauna and Flora.[1] CITES is the preeminent agreement to regulate international trade as a means to protect wildlife.[2] Species and their habitats are placed in one of three appendices and are accorded various protections, based on the relative threat of their extinction. A species currently threatened with extinction is placed in appendix I,[3] which gives the highest level of protection. CITES prohibits trade in that species without the prior grant of a permit by both the importing and exporting states.[4] Appendix I also has a special application to species taken from the sea. If an appendix I species is "taken in the marine environment not under any jurisdiction of any state,"[5] a certificate authorizing "introduction from the sea"[6] is required. Such certification will be given only if a scientific authority of the state of introduction advises that such taking will not be detrimental to the survival of the species; a management authority is satisfied that other conditions for imports of appendix I species have been met; and the species is not to be used for "primarily commercial purposes."[7]

CITES also lists in appendix II those species "which although not necessarily now threatened with extinction may become so unless trade in specimens of such species is subject to strict regulation in order to avoid utilization incompatible with their survival."[8] Export controls of appendix II species are similar to those that apply to appendix I species, but the rules for imports are less strict.[9]

Finally, restrictions on appendix III species are limited to specimens originating from the jurisdiction that listed them. In other words, CITES recognizes the rights of a party to control the export of a species that it may "[identify] as being subject to regulation within its jurisdiction for the purposes of preventing or restricting exploitation."[10]

The SPAW Protocol to the Cartagena Convention. CITES-like restrictions on international trade have been incorporated in broader protection for endangered and threatened species in the wider Caribbean region by the Protocol on Specially Protected Areas and Wildlife (SPAW Protocol) to the Convention for the Protection and Development of the Marine Environment of the Wider Caribbean Region.[11] The SPAW Protocol adopts the CITES approach by prohibiting trade in species given total protection under that protocol and by regulating the sale of species given only management protection under the protocol.

Species listed in annexes I and II of the SPAW Protocol receive complete protection, and in this regard the protocol prohibits, among other things, the taking, possession, or killing of, or commercial trade in, protected fauna. With respect to species listed in annex III, each party is obligated to implement plans for their management and use. For fauna, this may include the prohibition of all nonselective means of capture, killing, hunting, and fishing of species; the institution of closed hunting and fishing seasons; and the regulation of commerce in living or dead species and their eggs, parts, or products. The annexes were signed in Kingston, Jamaica, on June 11, 1991.

United States Regulations

The Endangered Species Act.[12] In the United States, the ESA prohibits the taking of endangered species within the United States, the U.S. territorial sea, or the high seas,[13] as well as the transport or sale of such species in interstate or foreign commerce.[14] In conjunction with these prohibitions, the ESA prohibits the import into or export out of the United States of such species. Many species of marine turtles, seabirds, and cetaceans are protected under the act, but notably absent from ESA protection are corals.[15]

The Marine Mammal Protection Act.[16] Similarly, the MMPA establishes a moratorium on, among other things, the taking and importation of marine mammals and marine mammal products. There is a limited exception to this moratorium for the incidental taking of marine mammals in the course of commercial fishing operations.[17]

The Lacey Act.[18] The Lacey Act was the first major United States law aimed at controlling trade in wildlife species and products. The Lacey Act prohibits the "import, export, transport, purchase or sale of any fish or wildlife or plant taken, possessed, transported, or sold in violation of any law, treaty or regulation of the United States."[19] The Lacey Act also makes it unlawful to "import, export, transport, sell, receive, acquire, or purchase any fish or wildlife taken in violation of any foreign law."[20] The Lacey Act is a powerful weapon used to prevent and punish trade in both endangered and nonendangered wildlife. It can be invoked, however, only if a United States law, a foreign law, or a treaty prohibits the action in question.

The Pelly Amendment.[21] The ESA, the MMPA, and the Lacey Act are potent laws, but they are restricted to use in certain limited circumstances. They are less potent when the species in question is not imported into the United States. The Pelly Amendment to the Fishermen's Protective Act[22] allows the United States to respond to actions taken by a nation that diminish the effectiveness of international wildlife or fishery agreements, regardless of whether the fish or animal is imported into the United States. It provides that "[w]hen the Secretary of Commerce or the Secretary of the Interior [determines] that nationals of a foreign country . . . are engaging in trade or taking of [a] species which diminishes the effectiveness of any international program for endangered or threatened species, the Secretary [is required to certify that fact to the President]."[23] The same certification provisions apply if the secretary of commerce finds that nationals of a foreign country are conducting fishing operations in a manner that diminishes the effectiveness of an international fishery conservation program.[24] As such, the Pelly Amendment is not limited only to the protection of species that are endangered or threatened.

Other U.S. statutes have incorporated the Pelly Amendment by reference. For example, the Driftnet Impact Monitoring, Assessment, and Control Act of 1987, the MMPA, and the 1990 amendments to the Magnuson Fishery Conservation and Management Act all have provisions for certification of countries in certain circumstances, with this action being considered a certification for purposes of the Pelly Amendment.[25]

Once a certification is made, the president has the discretion to impose sanctions on the foreign country, which may include import restrictions on aquatic products (if the certification was under the fisheries provisions) or wildlife products (if the certification was made under the wildlife provisions).[26] The statute also imposes the proviso that any import prohibitions be sanctioned by the General Agreement on Tariffs and Trade.[27]

The United States has certified foreign countries ten times for diminishing the effectiveness of the International Convention for the Regulation of Whaling.[28] It has also certified Taiwan and South Korea for failing to enter into agreements for cooperative monitoring and enforcement of their North Pacific drift net fleets before June 29, 1989.

Secretary of Commerce Robert A. Mosbacher and Secretary of the Interior Manuel Lujan, Jr., "certified that Japan was contributing to the disappearance of an endangered species [the hawksbill turtle]." Under the Pelly Amendment, the president has sixty days after the certification in which to inform Congress whether trade sanctions will be imposed. On June 20, 1991, the Department of Commerce reported that negotiations between the United States and Japan had produced a pledge by Japan to reduce significantly the number of turtles it imports in 1991 and to phase out hawksbill imports completely by the end of 1992. With this commitment, the secretaries of the interior and commerce refrained from recommending specific measures to prohibit wildlife imports from Japan.

TRADE IN SPECIES WHOSE COMMERCIAL HARVEST RESULTS IN INCIDENTAL TAKING OF PROTECTED SPECIES

Many fishing operations result in the incidental taking of marine resources other than the target species. Such by-catches often include threatened and endangered species as well as other commercial species. Because the by-catch is generally discarded, prohibitions on trade in the endangered or threatened species affected do not address the problem. In this situation, there have been attempts instead to limit the trade in the target species.

International Regulations

Convention for the Prohibition of Fishing with Long Driftnets in the South Pacific.[29] In the late 1980s, drift net fleets from Japan and Taiwan began fishing heavily for albacore tuna in the high seas of the South Pacific. Alarmed over evidence that this fishery threatened the continued commercial viability of the tuna and that the fishery was damaging the marine environment of the region, a number of South Pacific states concluded the 1989 Convention for the Prohibition of Fishing with Long Driftnets in the South Pacific.

The South Pacific Driftnet Convention obligates the parties to prohibit fishing with drift nets in areas subject to their fisheries jurisdiction in the Convention Area, as well as by their nationals and vessels anywhere in the Convention Area.[30] The convention also provides that parties may take additional measures, including prohibition of imports of fish caught with drift nets in the Convention Area.[31] The United States, which has territories in the Convention Area, signed the convention but has not yet ratified it. The convention entered into force on February 28, 1992.

The U.S. Response

A variety of U.S. laws impose import restrictions on fish products caught with fishing technologies that are known to harm marine mammals or sea turtles.

The Marine Mammal Protection Act.[32] For reasons not yet understood by biologists, in the eastern tropical Pacific schools of large yellowfin tuna are known to swim beneath herds of dolphins. Fishermen deploy purse seine nets around the dolphins to catch the tuna swimming underneath. Dolphins are often killed or seriously injured in such nets, and their resulting mortality has been of concern to conservationists.

The United States has for some time imposed regulations on its fleet designed to reduce dolphin mortality in this fishery. In 1984 and again in 1988, Congress passed amendments to the MMPA banning the import of tuna caught with purse seine nets in the eastern tropical Pacific unless the harvesting nation has a comparable regulatory program for protection of dolphins and a comparable take rate.[33]

At Mexico's request, a panel of the General Agreement on Tariffs and Trade[34] was convened to hear Mexico's claims that the embargo provisions of the MMPA and the tuna-labeling provisions of the Dolphin Protection Consumer

Information Act were inconsistent with GATT. In a report that has not to date been adopted by the contracting parties, the panel found that the primary embargo against Mexico was impermissible under article XI of GATT. The panel rejected the United States's argument that the embargo was an internal measure under article III of GATT (on national treatment). It also found that GATT exceptions for measures necessary to protect animal life or health (article XX[b]) or relating to the conservation of exhaustible natural resources (article XX[g]) were not available for measures applied extrajurisdictionally. The panel further reasoned that even if articles XX(b) and XX(g) could be applied extra-jurisdictionally, they would not be available in this particular case because the unpredictability of the standards under U.S. law meant that the U.S. embargo was neither "necessary" within the meaning of article XX(b) nor "primarily aimed at conservation" within the meaning of article XX(g). The panel was also of the view that the United States had not exhausted all GATT-consistent avenues for dolphin protection, particularly through negotiation of international cooperative arrangements.

In addition, the panel found the intermediary nations' embargo to be GATT-inconsistent for the same reasons as was the primary embargo. However, it found that the Pelly Amendment aspects of the MMPA were not GATT-inconsistent since the Pelly Amendment does not mandate GATT-inconsistent measures. Finally, it found that the "dolphin-safe" labeling provisions of the Dolphin Consumer Information Act were not inconsistent with GATT.

The Magnuson Act.[35] The 1990 amendments to the Magnuson Fishery Conservation and Management Act include a provision prohibiting the importation of all fish caught with drift nets on the high seas of the South Pacific after July 1, 1991, or fish caught with drift nets anywhere in the world after July 1, 1992. The dates in the Magnuson Act correspond to the dates in United Nations General Assembly Resolution 44/225, which recommends a moratorium on high seas drift net fishing unless certain conditions are satisfied.

Public Law 101-162.[36] The shrimp trawl fishery has been of growing concern in the United States because of its impact on nontarget fish and sea turtles. The United States government recently adopted regulations requiring shrimp trawl vessels in the Atlantic and Gulf of Mexico shrimp fisheries to be equipped with turtle exclusion devices (TEDs), which have been shown to be more than 95 percent effective in excluding turtles from shrimp trawl nets.

Recognizing that sea turtles are migratory and that efforts by one country to ensure their conservation will be ineffective unless they are conserved throughout their general range, the U.S. Congress recently passed a law prohibiting imports of shrimp from countries whose commercial shrimp trawl fisheries may adversely affect these sea turtles, unless the president certifies that the country has a comparable turtle conservation program. On February 24, 1992, a lawsuit was filed challenging the U.S. government's implementation of the law.[37] It alleges that the United States erred in its implementing the law only in the Gulf of Mexico, the Caribbean, and the western Atlantic; it challenges a three-year implementation period; and it alleges that the United States failed to initiate negotiations with foreign nations for the amendment and development of international agreements.

RESTRICTIONS ON TRADE IN COMMERCIAL SPECIES WHOSE HARVEST IS REGULATED OR MANAGED

In the case of straddling stocks and highly migratory species, it is most often the case that the stocks are fished by more than one nation, complicating effective management. First, it is often difficult to achieve international agreement on effective management measures, even where management bodies are in place. In addition, even if management measures are agreed to by an international management body, they may be undercut if states fish outside of the regime, either because they do not participate in the regime or because the regime allows a mechanism for them not to adopt a recommended measure.

In United States practice, it has been recognized that measures to manage a commercial fishery may include trade restrictions. The Magnuson Act[38] asserts U.S. management authority over fish in the U.S. exclusive economic zone (EEZ) and gives authority for making management recommendations to eight regional councils. The Department of Commerce has authority to adopt regulations for the conservation and management of these fisheries and generally follows the recommendations of the councils. The councils and the Commerce Department have taken the position that their mandate to manage and conserve includes the power to impose import restrictions or to take other measures that affect imports.[39] In addition, the U.S. Congress has adopted measures for conservation and management of EEZ resources that have affected international trade.

In one case, Congress adopted a provision prohibiting trade through interstate commerce of lobsters under a certain size. This provision was adopted to enforce a size limit on lobsters harvested in the U.S. EEZ.[40] The measure had the effect of preventing the importation into the United States of lobsters from Canada under that size. Canada challenged the United States measure under the United States/Canada Free Trade Agreement (FTA), which tracks GATT in its relevant provisions. The FTA panel concluded that the U.S. measure was consistent with the FTA because it was an internal regulation, subject only to a national treatment requirement, as opposed to an import restriction.

Even where international management organizations exist, their efforts to control unsustainable fishing practices often suffer from the near impossibility of enforcing conservation and management measures on the high seas. Therefore, trade measures, or the threat of such measures, often seem the most direct manner in which to halt stock declines resulting from overharvesting. Recent efforts to improve the status of western Atlantic bluefin tuna provide an example of such an approach. Atlantic tuna and tunalike species are managed internationally by the twenty-one-member International Commission for the Conservation of Atlantic Tunas.[41]

At the 1991 annual ICCAT meeting in Madrid, Spain, the member states most concerned with the severe decline in western Atlantic bluefin tuna stocks since 1970[42]—the United States, Japan, and Canada—were under pressure to agree on a strict management regime aimed at recovery of the stock. One source of this pressure was Sweden's pending petition to list the Atlantic bluefin under appendix I of CITES.[43] Despite strong statements by the United States and Japan that they would work for a 50 percent reduction in catch, they and Canada could agree only to a 10 percent cut over two years, with a promise to meet in 1992 to consider whether a 50 percent cut would "allow for a more rapid rebuilding" of the stock. Japan introduced a resolution containing strong trade provisions for nonmember countries that was not acceptable to European members—a number of which trade extensively with nonmembers—and a weaker resolution was passed.[44]

The agreements made at the ICCAT meeting were not sufficient to dissuade Sweden from proceeding with its CITES petition. At the 1992 CITES meeting, however, Sweden withdrew its petition when Japan, the United States, and Canada pledged to continue to work within ICCAT "to restore and maintain Atlantic bluefin," with an "emphasis on quota reductions,"[45] and to monitor and restrict trade in Atlantic bluefin "taken contrary to the conservation program of ICCAT."[46]

PROTECTION OF MARINE HABITATS

Protection of marine habitats that support marine resources are, in their own way, just as critical to conservation and management as are measures to control the take of individual species. Trade restriction measures play a role where there is commercial trade in a component of the ecosystem, such as corals. Corals, for example, are listed in appendix II of CITES and do not currently receive full appendix I protection. Moreover, corals receive no protection under the ESA and are not a managed harvest as fish are, so existing international regimes are inadequate to protect or manage coral reefs or to control trade in coral. International recognition of resources as ecosystems or essential components of ecosystems is a significant advance toward protecting the biodiversity of marine resources.

The SPAW Protocol[47] is an example of a ground-breaking advancement in marine ecosystem protection and management. The Conference of Plenipotentiaries Concerning Specially Protected Areas and Wildlife in the Wider Caribbean Region lists corals, mangroves, and sea grasses for management under annex III. This would aid in, among other things, regulation of commerce in coral and coral products. The SPAW Protocol will serve as a useful prototype for the Convention for the Protection of the Natural Resources and Environment of the South Pacific Region,[48] which, like the Cartagena Convention, provides for protection of special areas and wildlife, although no protocol has yet been negotiated.

ISSUES TO CONSIDER IN UTILIZING TRADE RESTRICTIONS AS A CONSERVATION MEASURE

As demonstrated earlier, international trade restrictions play an important role in conserving and managing marine resources. It is open for debate whether trade measures should be—or even could be—further utilized as a conservation measure. As these issues are further explored in the coming years, we must remain mindful of the very real constraints that will help shape policy with respect to protection of marine resources:

United Nations Convention on the Law of the Sea.[49] Are the measures consistent with the 1982 United Nations Convention on the Law of the Sea? Apart from general dispute resolution provisions, the convention does not provide a specific enforcement mechanism for protection and management of living marine resources. Consideration of any trade measures must balance the broad obligations to protect or manage shared resources with the rights of individual nations over the resources of their EEZs.

General Agreement on Tariffs and Trade. Are the measures consistent with GATT obligations? Article XX of GATT provides a limited exception to other GATT obligations if the measure taken is necessary to protect human, animal, or plant life or health or relates to the conservation of exhaustible natural resources if such measures are made effective in conjunction with restrictions on domestic production or consumption.[50] During the Uruguay Round, however, at least one country sought to modify article XX.[51]

Effectiveness. Are the measures truly effective? There is some debate as to whether trade measures are an effective tool in protecting wildlife because in the end they are only as effective as they are enforced nationally. The recent frustration of conservationists regarding elephant ivory is but one example.

Foreign policy. Are the measures reasonably consistent with foreign policy concerns? The issue of trade measures raises troubling issues of foreign policy for *any* nation considering their use. Nations are understandably reluctant to invoke trade measures as a means to coerce international action. The line between a conservation measure and a trade sanction may not always be a clear one. In this regard, the degree of international cooperation with respect to any trade measures may be important. In any event, the practical considerations of foreign policy should not be discounted.

The role of trade as a conservation tool will undoubtedly be debated through this decade and beyond. Because of the legal and political implications of trade measures, they should be carefully considered. The successes, failures, and limitations of existing international and domestic regimes should be of guidance in this process.

NOTES

1. Convention on International Trade in Endangered Species of Wild Fauna and Flora, Mar. 3, 1973, 27 U.S.T. 1087, T.I.A.S. No. 8249 [hereinafter CITES].
2. In 1963, the International Union for Conservation of Nature and Natural Resources (IUCN) called for an international convention to regulate trade in threatened species. In 1973, ten years later, CITES became a reality, with twenty-one countries initially signing the convention. For a full discussion of CITES and its background and operation, *see* S. LYSTER, INTERNATIONAL WILDLIFE LAW 239-77 (1985). As of this writing, 108 countries are parties to CITES.
3. CITES, *supra* note 1, art. II(1).
4. *Id.* art. III. Article III(2) regulates export permits; article III(3) regulates import permits; article III(4) regulates reexport permits; and article III(5) regulates "introduction from the sea" permits.
5. *Id.* art. I(e).
6. *Id.* art. III(5).
7. *Id.* art. III(5)(a-c).
8. *Id.* art. II(2)(a).
9. In addition to the three requirements for obtaining an export permit for an appendix II species, CITES requires that appendix I permits meet the additional condition that "a Management Authority of the State of export is satisfied that an import permit has been granted for the specimen." *Id.* art. III(2)(d). Although CITES also provides for an introduction from the sea permit for an appendix II species, introduction from the sea permits for appendix I species must meet the additional condition that "a Management Authority of the State of introduction is satisfied that the specimen is not to be used for primarily commercial purposes." *Id.* art. III(5)(c).
10. *Id.* art. II(3).
11. *See* Convention for the Protection and Development of the Marine Environ-

ment of the Wider Caribbean Region, Mar. 24, 1983, 22 I.L.M. 221 (1983) [hereinafter Cartagena Convention]. The Cartagena Convention entered into force on October 11, 1986.

Seven years after adoption of the Cartagena Convention and its protocol on combating oil spills, the text of a second protocol on special protected areas and wildlife was officially adopted, on January 18, 1990. *See* Protocol on Specially Protected Areas and Wildlife, 4 CEP NEWS (newsletter of UNEP's Caribbean Environment Program) 3 (1990). The SPAW Protocol has not entered into force.

12. Endangered Species Act, 16 U.S.C. §§ 1531–43 (1988) [hereinafter ESA].
13. *See id.* § 1538(a)(1).
14. *Id.* § 1538(a)(2).
15. For a list of endangered and threatened wildlife and plants and their listed critical habitats, *see* Endangered & Threatened Wildlife and Plants, 50 C.F.R. §§ 17.11–12 (1990).
16. Marine Mammal Protection Act, 16 U.S.C. §§ 1361–1407 (1988) [hereinafter MMPA].
17. *Id.* § 1372(a)(2).
18. Lacey Act, 16 U.S.C. § 3372(a) (1988).
19. *Id.* § 3372(a)(1).
20. *Id.*
21. Enacted in 1971, the Pelly Amendment was targeted at Denmark, Norway, and Germany, all of which had refused to refrain from high seas salmon fishing in compliance with a ban on such fishing by the International Commission for the Northwest Atlantic Fisheries. After learning of the Pelly Amendment and its potential application, the three nations decided to phase out their fisheries.

The Pelly Amendment has also been used to maintain the effectiveness of the whale conservation program of the International Whaling Commission (IWC). In 1974, Japan and the USSR were certified for killing minke whales in excess of IWC quotas. The president, however, did not impose an embargo because both nations agreed to adhere to quotas in the future. In 1978, Chile, Peru, and Korea were certified for whaling in violation of IWC quotas. Again, no embargo was imposed because all three nations pledged to join the IWC and abide by its quotas. In 1985, the USSR was certified for its harvest of whales. Norway was certified in 1986 for killing minke whales, but the president did not impose sanctions because Norway promised not to resume commercial whaling. In 1988, Japan was certified for killing whales under its scientific research program. In 1990, Norway was again certified for research involving the taking of whales. Although again no embargo was imposed, allocations for fishing in the United States zone were reduced. *See* MARINE MAMMAL COMMISSION, THE MARINE MAMMAL PROTECTION ACT OF 1972 AS AMENDED OCTOBER 1990 (1990) (quoting H.R. REP. NO. 100-970 and S. REP. NO. 100-592).
22. Fishermen's Protective Act of 1947, 22 U.S.C. § 1978 (1988).
23. *Id.* § 1978(a)(2).
24. *Id.* § 1978(a)(1).
25. Driftnet Impact Monitoring, Assessment, and Control Act of 1987, *as amended by*

Act of Dec. 29, 1987, Pub. L. No. 100-220, 101 Stat. 1477 (1987), incorporates the Pelly Amendment at section 4006(b). The Marine Mammal Protection Act incorporates the Pelly Amendment at 16 U.S.C. § 1371 (1988). The 1990 Magnuson Act Amendments, *infra* note 35, incorporate the Pelly Amendment at Pub. L. No. 101-627, 104 Stat. 4436 § 107(f) (1990).

26. 22 U.S.C. § 1978(a)(4).

27. *Id.*

28. *See supra* note 21 and accompanying text.

29. South Pacific Forum: Final Act of the Meeting on a Convention to Prohibit Driftnet Fishing in the South Pacific, Including Text of Convention and Its Protocols, 29 I.L.M. 1449 (1991).

30. The Convention Area encompasses the area lying between 10 degrees north latitude and 50 degrees south latitude and 130 degrees east longitude and 120 degrees west longitude. Convention to Prohibit Driftnet Fishing in the South Pacific, *supra* note 29, art. I(a)(1).

31. *Id.* art. III(2)(c).

32. MMPA, *supra* note 16.

33. *Id.* § 1371.

34. General Agreement on Tariffs and Trade, Oct. 30, 1947, 55 U.N.T.S. 187, T.I.A.S. No. 1700 [hereinafter GATT].

35. Magnuson Act, 16 U.S.C. § 1801 (1988), *amended* by Pub. L. No. 101-627, 104 Stat. 4436 (1990).

36. 16 U.S.C. § 1537 (1988), *amended by* Pub. L. No. 101-162, 103 Stat. 1037 (1989).

37. Earth Island Institute, et al. v. James Baker, et al., Case No. C-92 0832 (N.D. Cal., Feb. 24, 1992).

38. Magnuson Act, 16 U.S.C. § 1801 (1988).

39. This approach was upheld by a federal district court in Maine in Stinson Canning Co., Inc. v. Mosbacher, 731 F. Supp. 32 (D. Me. 1990).

40. See 16 U.S.C. § 1857(1)(J) (1988).

41. International Convention for the Conservation of Atlantic Tunas, May 14, 1966, 20 U.S.T. 2887, T.I.A.S. No. 6767, 673 U.N.T.S. 63. The convention is implemented in the United States by the Atlantic Tunas Convention Act of 1975, 16 U.S.C. § 971 (1988) [hereinafter ICCAT]. Membership dropped to twenty-one in 1991, when Cuba announced that it would withdraw.

42. *See* ICCAT, 1991 REPORT OF THE STANDING COMMITTEE ON RESEARCH AND STATISTICS. Between 1970 (when the stock was already declining) and 1990, the biomass of the stock declined by at least 90 percent. The first ICCAT conservation measures for western Atlantic bluefin were implemented in 1982.

43. *See Proposal for Amendment of Appendices I and II of the CITES Convention*, in REPORT OF THE 12TH REGULAR MEETING OF THE COMMISSION (ICCAT, Madrid, Nov. 11-15, 1991) [hereinafter 1991 ICCAT Report].

44. *See Annex 11: Resolution Concerning Catches of Bluefin Tuna by Non-Member Countries*, in 1991 ICCAT REPORT, *supra* note 43.

45. In March 1992, Japan announced that it would host an ICCAT special western

Atlantic bluefin management review committee meeting with Canada and the United States in May 1992. Letter from Kazuo Shima, Japanese Commissioner to ICCAT, to Dr. Antonio Fernandez, ICCAT Executive Secretary, March 19, 1992 [hereinafter Shima Letter].

46. *See* CITES Draft Resolution (sponsored by Canada, Japan, Morocco, and the United States), Improvement in Conservation Efforts for Atlantic Bluefin Tuna (Mar. 1991). This resolution was agreed to by the sponsors but was not formally adopted at CITES because of procedural problems. Letter from John F. Turner, Head of U.S. CITES Delegation, U.S. Fish and Wildlife Service, to Dr. John A. Knauss, Undersecretary for Oceans and Atmosphere, Department of Commerce, Mar. 11, 1992. Japan was also to host a working group meeting on catches of bluefin tuna by nonmember parties in May 1992. Shima Letter, *supra* note 45.

47. *See supra* note 11 and accompanying text.

48. Convention for Protection of the Natural Resources and Environment of the South Pacific Region, Nov. 25, 1986, 26 I.L.M. 38 (1987).

49. United Nations Convention on the Law of the Sea, Dec. 10, 1982, U.N. Doc. A/CONF.62/122, 21 I.L.M. 1261 (1982).

50. *Id.* art. XX.

51. The Uruguay Round negotiations were initiated in 1986 under the auspices of the General Agreement on Tariffs and Trade. *See generally* AMERICAN BAR ASSOCIATION, SYMPOSIUM—URUGUAY ROUND TRADE NEGOTIATIONS: WHERE DO WE GO FROM HERE? (1991).

22 The Legacy and Challenge of International Aid in Marine Resource Development*

Claudia J. Carr

MODERNIZATION IN MARINE ENVIRONMENTS that does not take adequate account of the potential ecological and social ramifications of such action, particularly development directed toward industrialized production for export, is likely to result in serious problems. Much of the international aid establishment fosters this form of development and has become an active force in ecological and socioeconomic deterioration within the southern nations that it purports to be helping. Insofar as these agencies are accountable to the international community as "aid" organizations, it would seem that the burden of proof to demonstrate the allegedly positive nature of their development policies should rest with them. To date, unfortunately, data presented by these organizations regarding their goals, strategies, and intended beneficiaries in development policies have been sorely inadequate. It is possible, however, to identify some fundamental problems of aid policies for resource use and social transformation that should be of real concern to both North and South. This chapter is an attempt to identify some of these patterns for fisheries-related development, as part of the growing international effort to examine what should and can be done.

There has been a marked increase in international aid support for fisheries-

*The preparation of this article benefited enormously from the contribution of James Carr, a founder and former coordinator of a large, NGO-based international fisheries program, for his extensive information and perspective regarding fisheries and fisheries policy and his extensive comments on the manuscript. I also wish to thank Jeffrey Gritzner, director of the Public Policy Institute in Missoula, Montana, for his suggestions regarding an early manuscript and his discussions regarding coastal development and the aid process. Further, I extend my appreciation to George Leddy of the University of California, Berkeley, for his part in a long-standing and productive intellectual interchange regarding development issues, including insights from his experience with fisheries policies in Mexico, and I would like to thank David Insull of the FAO for his cooperation regarding the recent FAO fisheries surveys.

related development within southern nations since the early 1970s.[1] This aid has included the negotiation, funding, design, and implementation of projects—primarily from development banks, other multilateral organizations (e.g., the United Nations system), and bilateral programs (see figure 22.1). Fisheries-related projects funded by aid agencies vary widely. They include (1) upgrading fleets; (2) expanding supporting "services"; (3) constructing infrastructure; (4) developing postharvest operations (for example, processing and marketing); (5) establishing "fish farming," or aquaculture; and (6) increasing training, research, and institutional "capability." Much of this aid emphasizes industrialization and export trade and increases southern dependence on goods and services from the North. Rising demand and high prices for seafood and fish reduction products

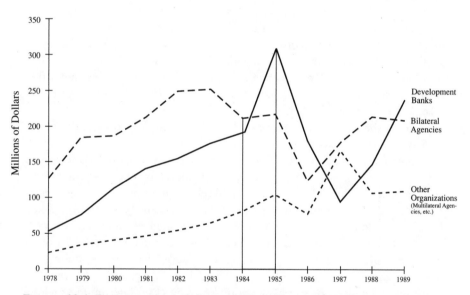

FIGURE 22.1 FISHERIES AID BY AGENCY TYPE *Data are from H. Josupeit,* A Survey of External Assistance to the Fisheries Sector in Developing Countries, *1978–1984, U.N. Food and Agriculture Organization (Fisheries Circular No. 755, Rev. 1, 1985, and Rev. 2, 1987); and D. Insull & J. Orzeszko,* A Survey of External Assistance to the Fishery Sectors of Developing Countries, *U.N. Food and Agriculture Organization (Fisheries Circular No. 744, Rev. 3, 1991). The strong discontinuity that appears from the period prior to 1985 and during 1985–1989 derives from Josupeit's calculation of project expenditure as total commitment spread over the known (or assumed) operational years; Insull and Orzeszko tabulate total project expenditure by allocating it to the project's first operational year. Because the data in Insull and Orzeszko's report, at least, exclude a number of projects reported by donors, the total values represented here are considerably reduced.*

in the developed countries—paradoxically, where fish stocks are often severely depleted—is a key element to the global economy that governs the economic strategy on the part of investment and aid interests.

The growth in export of fish[2] from southern nations is accompanied by generally declining terms of trade and by increasingly strict rules for obtaining credit. These terms commonly include demands for repayment of past-due loans and the establishment of trade and production conditions that favor northern interests. This ironic situation has culminated in *fish products being shipped from malnutrition-ridden southern nations to the dinner tables, livestock feedlots, and fertilizer factories of the affluent North.*

Major aid agencies, such as the World Bank, wield an influence over southern nations that extends far beyond merely funding development projects. This occurs through a broad range of activities by these agencies in southern nations, including assistance in the formation of new ministries or other agencies for development of marine environments, negotiation of northern fleet access to southern waters, participation in negotiations of foreign investment and cofinancing ventures, and enforcement of conditionality for loans and other finance—commonly involving privatization, deregulation, and various trade-related "liberalization" policies.

Some of the largest international aid organizations have come under severe criticism in recent years for instigating and helping to implement large, primarily export-oriented development projects. To date, this criticism has focused on aid-supported hydroelectric dam and river basin developments, tropical forestry, agro-industry, and resettlement projects. Much of this criticism has been spearheaded by northern- and southern-based nongovernmental organizations, but it has extended into major government bodies in the donor countries as well as to the boardrooms of aid organizations themselves. At issue is the constellation of destructive impacts that such projects have caused. The most common ones cited for these projects include violation of human rights, displacement of indigenous populations, overexploitation of resources, undermining of local economies and their future productivity, markedly increased debt and dependence on the North, concentration of wealth, and political conflict over dwindling resources.

The important point is that these well-documented problems are being reproduced by aid agencies in the fisheries sector and in coastal development. This is largely because the same general models and strategies for development are employed.

The few studies of international aid and its effects on fisheries-related development that have been conducted tend to emphasize the problems of purely production-oriented goals or the introduction of "inappropriate" technologies or ethics. While these are certainly important symptoms of the problems at hand, there are deeper political and economic forces from which they derive. Specifically, the most challenging questions are: Development of what? Toward what? And for whom?

This chapter examines the economic and political aspects of international aid in fisheries-related development, focusing on the problems in both coastal and oceanic regions. The analysis and discussion that follow are occasioned by the recent survey of international aid agencies conducted by the United Nations Food and Agriculture Organization (FAO), the Fisheries Project Information System (FIPIS).[3] This survey is the first of its kind for any sector of rural development.

THE ROLE OF SOUTHERN MARINE LIVING RESOURCES IN THE GLOBAL ECONOMY

Southern fisheries and related coastal resources have become threatened by global economic conditions, which are key determinants of international agency policy. Four interrelated dimensions of this crisis are particularly important: (1) an increasingly concentrated global fisheries industry, (2) rising demand for fish products in both North and South, (3) increasing export of fish from the South, and (4) increasing large-scale investment in southern fisheries and coastal resources.

The South as Producer and Consumer

A growing proportion of marine living resource production is from southern waters, which now produce slightly more than half of the world's fish landings. After a decade of strong increase in world fish catch during the 1960s, there was little, if any, increase during the 1970s. Some southern nations greatly increased their marine catch during the 1960s and 1970s.[4] A solid upturn in overall fish catch occurred during the 1980s. This was primarily from the Peruvian anchovy, the South American sardine, and other pelagic species in the west-central Pacific, the eastern Indian Ocean, and the Northwest Pacific.[5] Since the early 1980s, total landings have grown at an average of 3.8 percent per annum.[6] In some areas of highest productivity, however, growth has been accompanied by a record of serious environmental and economic problems.

At present, total tonnage of marine fish landings exceeds 85 million metric tons.[7] By the late 1980s, marine fisheries accounted for about as high a percentage of worldwide protein consumption as did beef production (about 15 percent).[8] Fish supply an even greater fraction of consumption in some southern nations (especially in coastal areas), where demand is growing for low-cost protein sources.[9] Catch for other purposes—primarily for reduction to meal and oils—accounts for less than half (about 40 percent) of the total nominal catch. This reduction product is utilized for livestock feed, fish food in aquaculture, and other industrial purposes. Its importance in international fisheries trade is likely to be sustained, if not increased.

Although both developed and developing countries have increased their imports during 1978–1987 by about the same proportion, there is a striking difference in exports and imports (see table 22.1). For Japan, western Europe, and North America, imports increased approximately twice as much as exports, with much of this increase coming from the South. Conversely, southern coastal states, while importing more fish, have significantly raised their exports of the same commodity groups (see table 22.1). Exports from the developing countries increased approximately threefold during this period, while exports from the developed countries (excluding Japan) doubled.[10]

Small-scale fishers account for a quarter of the world's output of fish and a third of the fish destined for worldwide food consumption. There are at least 12 to 15 million southern artisanal fishers, and they produce much of those nations'

TABLE 22.1 Global Trade in Seven Major Fisheries Commodity Groups

	Imports		Exports	
	1978	1987	1978	1987
Africa	650	809	415	1,270
North America	2,617	6,378	2,667	5,299
Canada / United States	2,445	6,174	1,884	3,928
Other	172	204	783	1,371
Asia				
Japan[a]	3,087	8,308	754	890
Other[b]	752	2,513	2,582	8,109
South America	154	225[c]	821	2,060
Europe	5,004	11,728	3,989	8,916
Oceania				
Australia	138	300	166	421
Other[d]	67	95	135	470
USSR	44	152	245	637

Source: Data were calculated from U.N. FOOD AND AGRICULTURE ORGANIZATION, ANNUAL STATISTICS (1987).

[a] Includes 1987 catch data, which are provisional (the Japanese tuna catch data do not correspond with data reported by ICCAT).

[b] Includes Korea, which is a very large exporter (second only to Japan) and also Thailand, India, and China.

[c] By 1987, imports were mostly by Brazil.

[d] Includes mostly Fiji, Vanuatu, and New Zealand.

total catch (about 65 percent in Asia, 85 percent in Africa, and 23 percent in Latin America). Even by conservative estimates, an additional 30 million people are involved in small-scale fishing-related activities,[11] and there are probably 100 million engaged in broader economic support activities, such as boat building and repair and harbor and port maintenance.

"Modernization" thought within international aid circles maintains that even when fish production is primarily directed toward global export markets instead of toward local consumption needs, the foreign exchange will "trickle down" to local populations. Yet the reality has been different. In many instances in which this modernization approach to development has been applied, the results have been the rapid dwindling of living marine resources, a net accumulation of debt, and a reduced supply of fish for the poorest populations.[12] Seafood once considered a staple and available to poorer local populations is now commonly exported to the North. It would certainly appear contradictory for international aid agencies to promote the development of fisheries in order to improve living conditions in the South while simultaneously pursuing new international trade relations that in fact undermine that improvement. Clearly, a management approach is needed that has the objective of improving the long-term welfare of southern local producers and consumers while maintaining sustainable ecological systems.

Overfishing and the Global Economy

A paradoxical situation has emerged whereby major aid agencies support the expansion of northern-based fisheries enterprises in the South and at the same time extend loans and grants for the buildup of southern fleets. Most major stocks of commercially targeted fish are either fully exploited or overexploited in most northern waters and are in danger of becoming so in southern ones.[13] Because the recent increase in fish landings is predominantly from highly fluctuating, shoaling pelagic species and because these stocks are more vulnerable to collapse, it should be considered doubtful that world production can continue to rise, despite possible catch increases of unconventional species. According to FAO predictions, it is more likely that production will continue to fluctuate around the past ten years' average.[14] A decrease in fish catch because of overfishing is far more probable, particularly with rising overcapacity and competition.

It is likely that the projected rise in northern demand for fish cannot be met by production in northern seas, even with better management of those fisheries.[15] In view of this, a major question emerges of whether southern seas are capable of meeting this demand and, if so, at what cost. No aid program to fisheries, even high-expenditure ones exemplified by the World Bank and the

Asian Development Bank, could provide the fourfold increase in southern productivity that would be necessary to meet the worldwide demand forecast for the year 2010.[16]

Migration of northern-owned distant-water fleets (DWFs) to southern seas, combined with increased domestic fleet activity in the South, often results in overexploitation. DWFs alone have accounted for about 50 percent of the overall catch in southern seas over the past twenty-five years.[17] In some southern ocean regions, DWFs have become a primary economic force. Generally large-scale, these fleets generate large amounts of waste and operate largely within the southern exclusive economic zones (EEZs); approximately 90 percent of commercial catches are made within these territorial waters. Domestic fleets have gained in significance vis-à-vis large foreign ones within southern waters, largely because a number of developing coastal states have effectively expanded their jurisdiction over their EEZs. In a few cases (e.g., Thailand), southern nations have established their own DWFs. Data regarding the specific activities of DWFs are, however, generally inaccessible.

In purely qualitative terms, it is clear that northern DWFs enjoy a number of advantages over southern domestic fleets. Long-distance fleets have long benefited from public subsidies and/or diplomatic support in their home states.[18] The capital base of DWF enterprises provides them with sufficient mobility to permit overfishing in one region, followed by movement to other areas. In addition, they overwhelmingly dominate access to technology, scientific data regarding available resources, processing, transport, marketing, and pricing. Furthermore, parent firms of DWFs—engaged in a multiplicity of related activities—often receive large contracts, including those from aid organizations, to provide the goods and services for development. Such contracts are well known to involve inflated prices, paid primarily through loans taken out by southern nations.

Northern fisheries and trade interests control fishing in southern EEZs through other means as well, especially since the 1982 Convention. These groups establish various contract forms, such as "over-the-side" agreements (in which national vessels transfer their catch to northern factory ships); various other contracts for a certain amount of marketable seafood by domestic fleets; and fishing access agreements—any of which may be included in comprehensive aid and investment "packages" involving loans, grants, investment, or cofinancing. The role of international aid agencies in such arrangements is sometimes straightforward.[19]

Southern coastal states are highly susceptible to these arrangements because they lack infrastructure, market access, and technical proficiency. Most of all, though, they are susceptible because of their spiraling debt and pressures for repayment. These states also lack the financial resources and the regional organizations necessary to control large northern interests or otherwise to manage their fishery resources (e.g., through collection of their own scientific data regarding fish stocks and through monitoring and enforcement of fleet compliance with regulations) or even to negotiate effectively with the large northern interests. Only a few associations, such as the South Pacific Forum Fisheries Agency, have shown any real bargaining success with foreign fleets.[20] Meanwhile, disputes may arise between foreign fishing interests and southern domestic fishers due to rapidly shrinking fish stocks and widening economic disparities. These have already occurred in Mexico, Argentina, southwestern Africa, and Southeast Asia.

Although surely unanticipated by its architects, the 1982 Convention may have adapted to the situation in which extended fisheries jurisdiction of southern coastal states and overexploitation in southern EEZs are simultaneously occurring. The convention mandates, for example, that "surplus stocks" be made available to foreign fleets if the state in question does not exploit them. It is reasonably clear that aid agencies have facilitated these so-called surplus resources being overexploited, within the context of this mandate. At the very least, aid policies have failed to help implement the convention's strengths, particularly its directive for coastal states to conserve and manage the ocean's living resources.

FISHERIES MODERNIZATION: TOWARD WHAT AND FOR WHOM?

In the face of already dwindling marine resources, primary emphasis on increased production for export and industrialization is likely to exacerbate rather than solve problems of poverty, malnutrition, and debt in the South. Yet most large aid agencies continue actively to pursue precisely this approach to fisheries development, despite the fact that most of the world's fisheries are already being exploited to the maximum they can withstand and, in many cases, well beyond it. Meanwhile, analysts within the aid establishment itself have judged the approach "unsuccessful," even by the narrow measures of volume of production and monetary return.[21]

Aid's Persistent Paradigm

Most aid agencies view fisheries development as the process of transforming relatively "backward," or "primitive," coastal-based economies of the South into more "advanced" industrial ones that are increasingly integrated into the global economy. Environments such as tropical forests and rangelands, for example, were regarded as unoccupied or empty. These environments have in common the fact that they have long been occupied by highly mobile human populations that were assumed to be of little consequence. Southern oceans and coastal areas, including mangrove communities, swamps, and coral reefs, as well as high seas, provide a prime opportunity to extend this frontier perspective.

During the past twenty-five years, the work of anthropologists, ecologists, geographers, and others has conclusively demonstrated both the sophisticated and adaptable character of traditional land use and socioeconomic systems indigenous to marine environments and the highly delicate nature of the ecosystems involved. This evidence notwithstanding, frontier-style thinking persists in public policy and investment circles. Even with its recent emphasis on the notion of scarcity and its use of concepts such as "risk minimization," "environmental protection," "appropriate technology," and "sustainable development," current development economics pays little attention to assessing available natural resources accurately, let alone to understanding the causes and dynamics of ecological and socioeconomic deterioration.

The modernization paradigm for fisheries development is rooted in a logic of brinkmanship, in which marine living resources are exploited to the maximum regardless of whether or not there are sufficient data to determine what level of extraction the stocks will bear or to predict the likely effects of target species extraction on dependent species and the ecological systems involved. Targeted fish stocks are considered "underutilized" or "suboptimally" utilized when exploited below the "maximum sustained yield," or "optimal yield." This designation is then assumed to justify agency backing for increased fishing, especially in instances of high-value species, such as tuna and shrimp.

Within aid circles, possible components of development projects for capture fisheries are evaluated primarily for their efficiency toward the ends just noted. In the economistic framework of development bank operations, for example, project efficiency is tied to "moving money" within those institutions and to generating satisfactory rates of return. A particular loan (e.g., for port construction) may be terminated if its rate of return drops below 15 or 16 percent.[22] Such figures are used to control entire projects and programs toward the political and economic objectives of their major northern constituencies.

In recent years, substantial literature in the social sciences has documented the negative effects that modernization policies have had on small-scale fishing communities.[23] Communities that are supposedly the "beneficiaries" of aid projects that direct production to greater emphasis on export instead have frequently been economically marginalized and otherwise destabilized. A good example is provided by aid projects that promote the use of larger or more "high-tech" vessels. The cash outlay (or credit) required to purchase, let alone maintain and effectively operate, the upgraded vessels or equipment is generally well beyond the means of most small-scale, or artisanal, fishers.[24] Furthermore, in the absence of regulation or management of fleet activities, the augmented fishing capability is apt to result in increased competition and overfishing. While the upgraded fleets may be assumed, or even required, by development officials to fish offshore, they in fact typically fish in *coastal* areas, where local communities are most dependent and where ecological systems are most vulnerable to disruption.[25] As a consequence, the well-being or even the survival of entire local economies may be undermined. Moreover, economic concentration is likely to occur at local and regional levels (see figure 22.2).

Industrializing fisheries operations may also have *positive* effects on local ecological and socioeconomic sustainability, depending on whether they are designed to do so. One possible example is upgrading of fishing vessel gear for reduction of waste from by-catch, or trash fish, involving altered catch methods or refrigeration. The point is that *precautionary management* of the changes introduced in any development scheme is necessary to effect positive environmental and social results and to avoid negative ones.

Small southern coastal states, including island nations, are particularly vulnerable to the problems of resource depletion, economic concentration, and dependence as a result of aid projects. Although insignificant in terms of total international aid for fisheries development, fisheries-related projects in these nations may affect a large proportion of those states' populations and some of the most fragile environments in the world.[26] These situations are unlikely to command much attention from the environmental (or social science) staffs of the major agencies, nor from their already overextended critics. Clearly, the process of selecting, designing, and implementing aid must undergo more systematic change so that these problems are not relegated to the back burner.

Finally, the problem of unplanned and unregulated industrialization in development assistance for capture-related fisheries clearly extends to ecological conditions of the EEZs and the high seas, where aid policies have failed to deal with the reality of shared stocks and highly migratory species.[27] Agency funding for increased capture and processing of highly migratory species (such as tuna), without financing of efforts to establish practical national and regional management strategies, obviously contributes to worsening overexploitation. Article 63 of the 1982 Convention enunciates the important principle of cooperation with regard to the problem of open access to shared stocks, but it is clearly inadequate as a framework for coping with it. Aid agencies have failed to help identify and address the obvious weaknesses of the convention along these lines and instead have largely reinforced them.

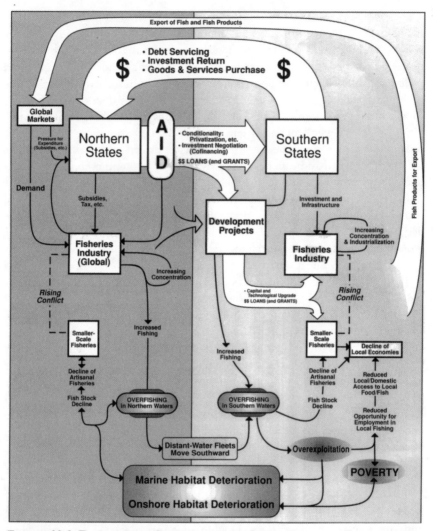

FIGURE 22.2 DYNAMICS OF INTERNATIONAL AID IN FISHERIES DEVELOPMENT IN A NORTH-SOUTH CONTEXT

The Technology Transfer Ingredient

Most modernization in marine environments involves some type of technology transfer from North to South. This transfer is a "multi-level process of communication involving a variety of senders and receivers of ideas and materials."[28] In fisheries development, this transfer consists primarily of upgrading harvest and postharvest technologies to more capital-intensive ones oriented to maximize export production. The main players in technology transfer are the aid organizations themselves, the northern and southern governments involved, and private capital. With a few notable exceptions, southern local populations have little, if any, input into aid agency deliberations regarding selection of the specific technologies to be transferred, the means of diffusion employed (whether through foreign contractors, producer incentives, or training), or the evaluation of results.

Aid is typically characterized as technical or capital aid, the latter being strongly dominant in fisheries-related development (see figure 22.3). Specific aid agencies vary widely in the transfer of specific types of technical and capital aid. A more useful classification might distinguish among physical, organizational, and conceptual technologies. This could be summarized in the following

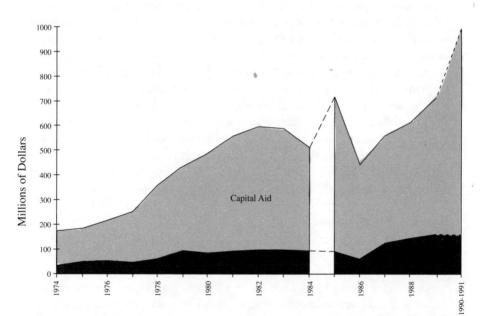

FIGURE 22.3 CAPITAL VERSUS TECHNICAL AID IN FISHERIES DEVELOPMENT
Prepared from data in H. Josupeit, A Survey of External Assistance to the Fisheries Sector in Developing Countries, 1978–1984, *U.N. Food and Agriculture Organization (Fisheries Circular No. 755, Rev. 1, 1985, and Rev. 2, 1987); and D. Insull & J. Orzeszko,* A Survey of External Assistance to the Fishery Sectors of Developing Countries, *U.N. Food and Agriculture Organization (Fisheries Circular No. 744, Rev. 3, 1991). For an explanation of the discontinuity of data representation prior to and post 1984–1985, cf. note in legend of figure 22.1.*

manner: (1) physical, including physical infrastructure (such as buildings, ports, harbors, jetties, roads, aquaculture ponds, and processing factories) and fleets (such as larger vessels, bigger engines, onboard refrigeration or processing, nets, radar, and sonar equipment); (2) organizational, including new or "modernized" ministries and agencies, credit and marketing networks, increased scientific and technical capability, training and extension; and (3) conceptual, including modeling and analytical techniques (such as cost-benefit and feasibility studies and fish stock assessment), and so-called economic reform measures.

Community-level experience has revealed that the transfer and use of technologies within modernization-oriented aid projects may create serious biological, habitat, and social stress or deterioration. Much of the documentation for this concerns capture fisheries. The adoption of new capture technologies in northern waters has already been shown to result in widespread overfishing there. In the South, where scientific assessment, regulation, and management of fisheries are far more difficult to achieve and where comprehensive resource and socioeconomic management are nearly absent, such adoption is even more likely to result in overexploitation.

Problems are generated by other dimensions of technology transfer as well. Upgraded harbor and port technologies, both for construction and for operation, generally encourage larger vessels, such as huge bottom trawlers. These ships typically produce huge amounts of waste and pollution, seriously contributing to overcapacity and overfishing.[29] Construction of jetties and dockside facilities usually involves dredging and other activities that destroy fragile coastal environments, including coral reefs, mangrove communities, and sea grass beds—environments where as much as 70 or 80 percent of commercially caught fish species may spawn or spend juvenile life cycle phases. Technologies for large-scale coastal reclamation (dredging, draining, and fill operations) to establish onshore processing, large-scale aquaculture, and marketing facilities for extensive fish farming produce a similar range of effects, outlined later in this chapter.

Recently, many agencies have increased their emphasis on "smaller-scale" technologies, as well as certain types of technical rather than capital aid (see figure 22.4). This shift has occurred within the context of political pressure from public interest groups, along with some northern officials and southern nations. In fisheries, it has often taken the form of smaller boats or engines or an increase in "indigenous capability" for absorbing new technologies—yet it remains consistent with export-directed production. The allocation of more aid to fund projects that transfer smaller-scale, or "appropriate," technologies, however, should not be viewed as having intrinsic merit. Although the transfer of such

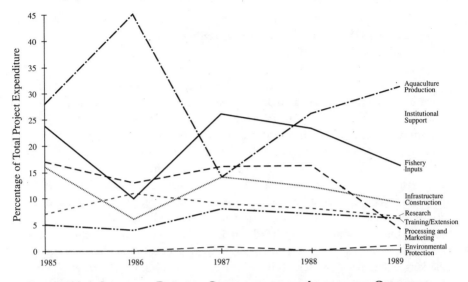

FIGURE 22.4 SELECTED PROJECT COMPONENTS OF AGGREGATE OVERSEAS EXTERNAL ASSISTANCE, 1985–1989 *Prepared from data in D. Insull & J. Orzeszko,* A Survey of External Assistance to the Fishery Sectors of Developing Countries, *U.N. Food and Agriculture Organization (Fisheries Circular No. 744, Rev. 3, 1991). The "hill and valley" appearance of expenditure values indicated in the figure partially reflects the method of tabulation used by the FAO, namely, with total project expenditure allocated to the project's first operational year. The negligible expenditures by agencies for "rural development" and for the "role of women" are omitted.*

technologies does sometimes represent a real improvement over previous, capital-intensive technologies known to produce serious problems, it may also simply present a different face for overexploitation of existing resources, economic dependence, and other socioeconomic problems—unless it is carried out within a management framework oriented to ecological and socioeconomic sustainability.

The changes associated with technology transfer in fisheries and coastal development occur as *systems*, not as isolated phenomena. There is a pressing need for in-depth examination of these systems, including examination through case studies, if the merits and harmful aspects of particular technologies are to be effectively judged and if the *process of technology transfer itself* is to be understood and possibly reformulated. In the absence of critical scientific information and a management framework to shape technology transfer policies, the basic questions persist: Technologies toward what? For whose benefit? And at what cost?

THE FISHERIES AID ESTABLISHMENT IN CAPTURE AND CULTURE

International development agencies have widely divergent approaches to the specific types of fisheries-related development, the geographic pattern of project support, and the type of financing they extend to southern nations. This variation is significant when considering possible reform measures. The primary goals of agency development projects range from generating a monetary return (primarily for northern investors), to strengthening the export component of southern economies, to serving a particular donor's (or associated recipient's) political objectives, to extending humanitarian assistance. Often these occur in combination. But the unavoidable reality of fisheries aid, as in other sectors, is its overwhelming dominance by northern interests through direct government authority, North-South government agreement, control of capital and export markets, and expertise in scientific, technical, and administrative matters. These relationships are reinforced by interagency cooperation in operations and funding within the aid establishment.

Contrasting Policies of Aid Agencies

Bilateral agencies and development banks are far more influential in southern fisheries and coastal development than all multilateral agencies combined (see figure 22.1). Together, they provided about 40 percent of total official external assistance during the latter half of the 1980s. Development bank financing in fisheries increased sharply vis-à-vis bilateral assistance in 1988 and 1989. Although 90 percent of all projects were financed through grant aid, large loans from the development banks constituted 54 percent of total project aid.[30] Loans are typically associated with large capital investment projects in infrastructure development, fleet building, marketing and processing facilities, and extensive aquaculture. By contrast, grant-based project aid is frequently correlated with a stronger technical assistance component and relatively small capital investments (with the exception of Japanese bilateral aid).

Capital aid has always definitively outweighed technical aid in fisheries development (see figure 22.3), and this situation is apparently intensifying, despite considerable rhetoric to the contrary within aid circles. Infrastructure construction (including that undertaken for processing and marketing) and upgrading of fleets constitute much of this capital aid. New physical infrastructure was particularly prominent in fisheries aid during the 1970s—primarily in Asia, although it declined sharply during the 1980s (again, with the exception of Japanese assistance). According to H. Josupeit's survey report for the FAO, about 30 percent of fisheries aid was related to upgrading of fleets between 1978 and 1984; this emphasis persisted into the mid- to late 1980s (cf. "fishery inputs" in table 22.2 and figure 22.4).

Development bank loans to southern nations are highly differentiated by region, as indicated by FIPIS data and the FAO reports. Some of the largest loan-financed projects have been instituted in Asia[31]—particularly in China, Indonesia, Malaysia, Bangladesh, the Philippines, Pakistan, Sri Lanka, and Thailand. Africa, on the other hand, has received relatively few loans. The Asian Development Bank (ADB) has always allocated the lion's share of its fisheries loans to Asian nations, especially Indonesia.[32] *The ADB and the World Bank both maintain extremely limited public access to information concerning the design, implementation, and evaluation of their development projects.*

The World Bank, the largest development institution in the world, has dominated funding of fisheries development projects for the past twenty years.[33] It classically embodies the interplay of foreign policy and private sector objectives, agency visibility to public interest constituencies, and agency experience—with the overriding operational criteria of a *bank*. A small number of the richest northern nations contribute the bulk of the World Bank's "subscription" capital (shares), and thus they dominate the voting. The United States leads, with 19.7 percent of the World Bank's voting power. The OECD countries (Japan, Canada, the United States, and European nations) account for about 60 percent of the total vote. All other nations, including some major international export fisheries powers such as Taiwan, South Korea, and China, have only a highly diffused voting power of only about 28 percent. This relative disparity results in an inherent bias in World Bank policies and a fundamental inertia constraining reform measures.

For many years, the World Bank's fisheries program largely focused on loans for construction of ports and other infrastructure—a type of loan markedly reduced in its recent policies. By the 1980s, the vast majority of its loan funds were allocated to aquaculture (see table 22.3), mostly in Asia. A much smaller fraction of loans was allocated to fleet upgrading and institutional support (see table 22.3).[34] Between 1986 and 1989, the World Bank loaned about $250 million for twenty-three different projects (our calculation from FIPIS World Bank responses suggests the much higher figure of about $350 million for forty-three new projects). According to FIPIS data, World Bank funding in recent years has included significant "institutional support," basically directed to increasing exports. Research, training, and extension are poorly represented. Finally, *funding for environmental protection was reported by the World Bank as zero.*

The other development bank of major importance in fisheries and coastal development is the ADB. The ADB is largely controlled by the United States and Japan, and it too has only insignificant input from southern nations. It has

TABLE 22.2 Selected Project Components of Official External Assistance (in Thousands of Dollars)

	1985	% of Total	1986	% of Total	1987	% of Total	1988	% of Total	1989	% of Total
Infrastructure construction	99,393	16	23,604	6	61,287	14	56,075	12	51,228	9
Aquaculture production	169,110	27	171,001	45	59,439	14	121,360	26	170,985	31
Processing and marketing	107,790	17	51,129	13	69,974	16	73,333	16	19,610	4
Fishery inputs	152,623	24	37,226	10	112,807	26	108,565	23	91,333	16
Institutional support	18,742	3	39,483	10	41,529	9	29,383	6	145,654	26
Training / extension	31,523	5	14,861	4	36,177	8	32,532	7	31,577	6
Research	42,474	7	40,359	11	40,765	9	38,106	8	30,972	6
Environmental protection	0	0	0	0	4,000	1	122	0	5,565	1
Total (including other components)	630,729	99	381,889	99	439,454	97	468,906	98	555,781	98

Source: D. Insull & J. Orzeszko, *A Survey of External Assistance to the Fishery Sectors of Developing Countries*, U.N. Food and Agriculture Organization (Fisheries Circular No. 744, Rev. 3, 1991).

Note: "Role of women" and "rural development" components are negligible in amount and are omitted, but total expenditure amounts are inclusive.

greatly increased its influence and participation in fisheries development during the past decade, such that the total of ADB loans for fisheries projects from 1986 to 1989 exceeded those of the World Bank by a sizable amount (see table 22.3). The ADB's fleet upgrading profile (about 12 percent of its fisheries aid) has generally been similar to that of the World Bank, with nearly all of it assigned to Asian nations. Its fisheries loans to Indonesia, for example, have equaled more than $300 million since 1978 and include a 1991 loan of $120 million for a complex of industrialization efforts. The ADB has a high proportion of projects with multiple components, including education and training, research, and institutional support. Its support of large-scale aquaculture is far lower than that of the World Bank, although the two banks share a primary interest in it (see table 22.3). Like the World Bank's, the ADB's funding for processing and marketing and for coastal infrastructure construction (at least that reported as part of fisheries) is low.

The other development banks were not examined in this particular study. The larger of them, the Inter-American Development Bank (IDB), has been less active than the World Bank and the ADB in fisheries-related and coastal development but is recently becoming more active. IDB funding is of major importance in some Latin American coastal states. The African Development Bank (AfDB) is not yet substantially engaged in fisheries development.

Japanese bilateral aid has now become a leading force in fisheries, with aggregate spending for 1986–1989 roughly equaling the World Bank's loans for the same period. Japan is the top bilateral aid donor to twenty-five southern nations. It is the second largest contributor to the World Bank (as well as to the International Monetary Fund and the Inter-American Development Bank) and the largest contributor to the ADB (providing more than 40 percent of total foreign aid to China). The Overseas Economic Cooperation Fund (OECF) and the Japan International Cooperation Agency (JICA) issue a wide range of loans and grants throughout the South and are overseen by a highly complicated bureaucratic system that nominally focuses on the Ministry of Foreign Affairs and the Ministry of Finance. Policies of both agencies are closely tied to Japanese foreign policy and capital interests, and Japanese assistance for southern fisheries development—almost entirely in the form of grants—is frequently linked with obligations for recipient countries to purchase Japanese products, Japanese private investment, EEZ fishing access by Japanese companies, cooperation in international organizations, and other national interests. Japanese fisheries aid is mostly for capital investment projects, with almost no funding of institutional development (see table 22.3). Its projects, though numerous, are relatively small, certainly by comparison with those of the World Bank and the ADB (see table 22.3).[35] Although parallel with the ADB and the World Bank in its primary

TABLE 22.3 Fisheries Assistance for Selected Agencies, 1986–1989

	Total Assistance (Thousands of Dollars)	% of Total Assistance	Average Project Size (Thousands of Dollars)	Number of Projects
World Bank				
Infrastructure construction	0	0.0		0
Processing/marketing	700	0.3	700	1
Aquaculture	183,000	73.1	22,875	8
Fishery inputs	31,100	12.4	3,456	9
Institutional support	29,645	11.9	14,823	2
Training/extension	490	0.2	490	1
Research	4,470	1.8	2,235	2
Environmental protection	0	0.0	0	0
Other	788	0.3	197	4
Total	250,193	100.0		
Asian Development Bank (ADB)				
Infrastructure construction	5,300	1.4	5,300	1
Processing/marketing	19,500	5.2	9,750	2
Aquaculture	78,820	21.2	13,137	6
Fishery inputs	45,500	12.2	45,500	1
Institutional support	66,954	18.0	9,565	7
Training/extension	73,300	19.7	73,300	1
Research	59,750	16.1	29,875	2
Environmental protection	0	0.0	0	0
Other	22,757	6.1	2,529	9
Total	371,881	100.0		
Japan				
Infrastructure construction	90,173	43.9	4,099	22
Processing/marketing	26,364	12.8	1,465	18
Aquaculture	3,604	1.8	1,201	3
Fishery inputs	44,231	21.5	1,923	23
Institutional support	0	0.0	0	0
Training/extension	18,357	8.9	2,622	7
Research	21,848	10.6	3,641	6
Environmental protection	0	0.0	0	0
Other	1,000	0.5	0	0
Total	205,577	100.0		
CIDA				
Infrastructure construction	5,960	10.0	2,980	2
Processing/marketing	6,215	10.4	3,108	2
Aquaculture	0	0.0	0	0
Fishery inputs	11,579	19.5	2,316	5
Institutional support	27,912	46.9	2,537	11
Training/extension	2,400	4.0	600	4
Research	0	0.0	0	0
Environmental protection	4,223	7.1	1,056	4
Other	1,220	2.1	610	2
Total	59,509	100.0		

TABLE 22.3 Fisheries Assistance for Selected Agencies, 1986–1989 (*continued*)

	Total Assistance (Thousands of Dollars)	% of Total Assistance	Average Project Size (Thousands of Dollars)	Number of Projects
NORAD				
Infrastructure construction	11,317	18.7	5,659	2
Processing/marketing	13,021	21.5	3,255	4
Aquaculture	4,704	7.8	1,176	4
Fishery inputs	100	0.2	50	2
Institutional support	4,140	6.8	518	8
Training/extension	5,100	8.4	1,020	6
Research	5,746	9.5	1,915	3
Environmental protection	0	0.0	0	0
Other	16,483	27.2	1,648	10
Total	60,611	100.0		

Source: Calculations are the author's, based on donor responses to the FAO's Fishery Project Information System (FIPIS).

Note: The FAO's summary report of the FIPIS does not treat individual agency expenditures.

involvement in Asian fisheries, Japanese aid extends throughout Latin America and Africa.[36] Recently, infrastructure construction, fleet building, processing, and marketing have dominated its fisheries program. Research and training are also relatively well represented, with a clear orientation toward increasing export production.

The Canadian International Development Agency (CIDA) departs rather strikingly from the three organizations just described, especially in its strong support for institutional development, which equaled about 47 percent of its total assistance between 1986 and 1989 (see table 22.3) and in the multidimensional nature of its projects.[37] CIDA gives substantial attention to basic human needs concerns and, in at least one known case, to women's roles. CIDA's fisheries assistance is extended almost entirely as grants, and its projects are relatively small. They are widely distributed geographically, with some special attention to West Africa. The Canadian agency's support for fleet modernization (20 percent of its total fisheries aid), although small by comparison with that of the larger agencies, is still significant (see table 22.3). Much of it goes to upgrade small-scale production. Monitoring and surveillance of fleets are frequently included in its assistance for capture fishing (such funding is not reported by the World Bank, the ADB, or Japan in FIPIS). *CIDA is the only one of the five agencies studied that reported funding for environmental protection* (see table 22.3). According to the FAO,[38] CIDA allocates 7.1 percent of its fisheries aid for environmental protection, but by our calculations (from the full set of agency responses) the figure is far higher—more than 13 percent.

The Canadian agency has been exceptional among the five agencies in another regard, namely, in its support of regional and interregional projects, including consultative committee efforts and technology transfer and training programs. In sum, although CIDA's program clearly remains in agreement with the long-standing modernization paradigm in fisheries, its more community-based, diversified, and experimental approach to development suggests some important departures from conventional aid policies.

The Norwegian bilateral agency (NORAD) funds a clear mixture of large-scale and small-scale marine-related projects. NORAD has been particularly active in freshwater fisheries and associated water development for many years. The agency's fisheries program is comparable to that of CIDA, both in its emphasis on technical aid and in its allocation of grants, not loans. By its own account, however, the Norwegian agency is more biased in favor of industrialization,[39] and its funding for institutional development is far lower. Infrastructure construction, as well as processing and marketing, are leading components of NORAD fisheries assistance, although in dollar amounts such projects are exponentially smaller than those supported by Japan (see table 22.3). Norwegian funding of training and extension, stock surveys, and feasibility studies in fisheries, moreover, are largely geared to increasing exports.[40] NORAD reported *no* support of environmental protection measures to the FAO for the FIPIS.

Multilateral organizations most important to fisheries development include parts of the United Nations system, especially the UN Development Program (UNDP), the FAO, the International Fund for Agricultural Development (IFAD), the World Food Program (WFP), the UN Environmental Program (UNEP), and the UN Industrial Development Organization (UNIDO). Outside the UN system, the European Community is the most significant multilateral entity engaged in fisheries. The European Community is also the major fish importer, accounting for 33 percent of world imports.[41]

Although far smaller than the large bilateral agencies and development banks in terms of development project funding (see figure 22.1), UN organizations play a critical technical and political role in the overall scheme of global fisheries modernization. The FAO is particularly important for maintaining the status quo in fisheries development. This organization conducts research, gathers and disseminates information, conducts training and research programs, and performs other activities associated with building scientific and technical capability. Much of the FAO's development work is done with major lending agencies such as the World Bank, and it derives much of its support from its administration of bilateral trust monies, including those managed through the FAO's Technical Cooperation Program. Other projects are funded through UNIDO, WFP, and IFAD.

Overall, the "pluralistic" image of the United Nations gives it an important role in legitimizing modernization policies, even though the most cursory examination of FAO and UNDP policies in fisheries shows them to be clearly biased in favor of a developmentalist perspective. The potential importance of these multilateral agencies to advance the adoption of fundamental changes in fisheries and coastal-based aid policies remains to be systematically explored.

Nongovernmental organizations (NGOs) have been the primary challengers of conventional aid policies and institutions. They also have exercised leadership in educating the general public about the international aid process. The political importance of NGOs is enhanced by some successful results in community-level development projects that emphasize natural resource management, sustainability of resources and local economies, and authentic participation by community members in designing, implementing, and evaluating projects. Recently, many NGOs have become informally affiliated with the major agencies in advisory capacities and cooperate with them (e.g., as community "sensitizers") in project implementation.

The "Promise" of Aquaculture

During the 1980s, international aid for aquaculture, or "fish farming," radically increased. Aggregate international aid for aquaculture between 1978 and 1984 increased more than fourfold—far more than any other fisheries assistance.[42] During this time, aid for aquaculture grew much faster than that for capture fisheries aid: 30 percent per year in the case of fish and shellfish, and over 40 percent per year when seaweeds are included.[43] Aquaculture aid remained a major emphasis during the mid to late 1980s (table 22.2), especially in the case of the World Bank where it accounted for more than seventy percent of its fisheries loans (table 22.3). Some project loans were for hundreds of millions of dollars, with high costs for large-scale infrastructural construction and import of fish feed, equipment, and chemicals such as fertilizers and antibiotics. The virtual "aquaculture movement" within fisheries aid during the 1980s was accompanied by high levels of private (and some public) investment in some southern nations.

Most of the extensive, or large-scale, aquaculture enterprises were established in Asia, where the fish-farming tradition and export potential were known to be strong.[44] Although China has received the bulk of aquaculture funding (largely from World Bank loans), countries where capture fishing is well developed, such as the Philippines, Indonesia, India, and Sri Lanka, were also targeted. Latin America and Africa, by contrast, have comparatively little fish-farming tradition and have received only limited aid support for aquaculture.[45]

The World Bank has been the top funder of aquacultural development. Its outlays escalated more than tenfold between 1978 and 1984 and made an even sharper upturn in 1985. Aquaculture constituted more than 73 percent of World Bank fisheries loans for the late 1980s. Its loans to China alone totaled $250 million, conservatively estimated by the FAO (our calculations, however, indicated the overall cost of these projects to be well over $500 million).[46] The average size of the World Bank's aquaculture projects during those years was more than $22 million (by the FAO's estimate). This far exceeds figures for the other agencies examined (see table 22.3). The ADB has been another major force in the recent impetus for aquacultural development, increasing its funding sixfold between 1978 and 1984. In the late 1980s, approximately 21 percent of ADB fisheries loans were allocated to fish farming. Some of its projects have been very ambitious in terms of scale,[47] even though on average they have been much smaller than those of the World Bank. Many ADB aquaculture projects have been multidimensional, and funding resembles that of the World Bank in that large projects are often closely in line with structural adjustment policies.

Among the bilateral agencies examined, Japanese agencies' aid for aquaculture amounted to only about 2 percent of its total fisheries assistance during the late 1980s, and this was for only a few projects of relatively small size.[48] A similar situation obtains for aquaculture project funding by CIDA, which reported only one substantial project to investigate the potential contribution of fish culture to improving nutrition of low-income groups in Sri Lanka. Finally, NORAD allocated substantial funds for a variety of large-scale and smaller, traditionally based aquaculture projects, but its average project size was only slightly more than $1 million (see table 22.3).[49] The FAO and several other UN organizations played important roles as well, particularly in offering technical assistance.

Proponents of the development of extensive (and semi-intensive) aquaculture justified their support on a number of grounds. Prominent among them were that these enterprises would contribute to (1) meeting the growing food needs and nutritional deficiencies of southern populations, (2) providing southern nations with foreign exchange for servicing existing debt and for development, (3) increasing rural income and employment, and (4) avoiding or alleviating pressures for overfishing and other overexploitation of marine living resources. Development of extensive aquaculture, including the processing and marketing associated with it, offered the global fisheries industry potentially high profits, particularly with rising international demand for fish products. To northern DWFs and their industrial parent companies, this type of investment also suggested a partial answer to the problem of reduced access to some southern EEZs.

New, export-oriented aquaculture was promoted as an essentially nondestructive means of pursuing all these ends. By and large, proponents consider these projects successful when they generate significant export revenue (usually from global marketing) and help satisfy the growing northern demand for fish products.

In some instances, one or more of these objectives has been achieved. Substantial amounts of foreign exchange have been generated, for example, in Bangladesh and Indonesia. Projects have sometimes created significant employment, although not necessarily increasing *net* local employment or real income.[50] But there is growing awareness of the fundamental inadequacy of purely profit or marketing criteria for measuring success, and some projects have been widely acknowledged as failures (e.g., in Egypt, Southeast Asia, and Central America).

Large mariculture systems are plagued by a number of common problems. These include high levels of disease and stress among culture populations; discharge of internal farm wastes (e.g., phosphates) to aquatic and terrestrial environments; water quality problems (high ammonia levels, acidic conditions); producer dependence on costly imported inputs (e.g., food for cultures, antibiotics, equipment, and larval and juvenile stock); vulnerability to global market conditions; physical and chemical destruction of mangrove communities and other coastal habitats; increased economic stress through competition with local fisheries; and political conflict resulting from significant loss of coastal resources and access to those resources; and access to jobs in culture enterprises. Traditional systems of fish farming, which supposedly were to coexist with the new systems, have been generally ignored in development design, implementation, and evaluation.

It is paradoxical that extensive mariculture operations may actually worsen the overfishing conditions they were ostensibly designed to offset. There is evidence of this occurring, largely because of the destruction of inshore fish reproductive habitats with the construction of aquaculture facilities, coupled with the problems of overproduction and decline in prices.[51] Local opposition to large-scale projects in particular has become common and has on occasion erupted into open conflict.

Toward an Extended Coastal Zone System

An "extended coastal zone system" (ECZS) may be posited as an analytical framework for understanding fisheries-related development in interlocking offshore, coastal, and interior (e.g., river basin–related or catchment-related) zones as one entity. Although the *initial* environmental and social impacts of development projects may be in one or another of these zones, these impacts frequently extend to contiguous areas. This proliferation of change occurs through a large complex of ecological and social processes involving river, delta, and nearby coastal environments; forest and agricultural systems, including vegetation, soils, and water relations; social migration patterns; resource tenure; market organization (and control); and administrative policies.

These interrelationships may affect initial social or environmental changes from development projects differently, depending on the specific conditions. Two examples may be cited here. First, negative impacts presumed to be minimal from a development project in one zone may be amplified by the effects of development projects in another. For instance, expropriation of small-scale, subsistence-oriented farmers and forest dwellers (for aid-sponsored lumbering or agricultural enterprises) may be worsened by the relocation or expropriation of comparable socioeconomic groups in nearby coastal areas for large-scale modernization projects there, by throwing them into competition for the remaining land. Another instance occurs when the destruction of coral reefs, caused by aid-funded projects such as port or other infrastructure construction, is heightened by aid-supported upgrading of the region's fleets—fleets that intensively exploit those same environments. Second, aid policies that are contradictory with regard to their "target" effects or beneficiaries of projects may also be implemented by aid agencies within an ECZS. For instance, employment gain in (aid-backed) semi-intensive or intensive aquaculture along the coast may be more than offset by job losses through mechanization (technological upgrading) within other projects in the region (e.g., fishing vessels, a processing plant, or an agricultural plantation). A number of examples of these types of problems have been encountered in investigations of river basin development, in combination with purely delta- or coastal-based aid activities. To date, aid agencies almost invariably treat such situations as separate phenomena.

An ECZS framework for analysis needs to be formulated that can be used to understand better these "socioecological" systems of change—a generalized framework that can be adapted for particular southern regions. The need for this can be further illustrated with the case of mangrove destruction in association with large-scale aquaculture. A high proportion of commercially important fish species are dependent on mangrove communities, especially for the breeding and nursery phases of their life cycles. Construction of large fish-farming facilities often involves cutting of mangrove trees on a massive scale as well as large-scale draining and dredging.[52] Large-scale deforestation (often upstream from rivers where timber operations are taking place, as in eastern Malaysia and Papua New Guinea, or caused by coastal or catchment-related agriculture) causes siltation. Chemical pollution from wood-processing plants and agricultural operations, and urban-related development, also may severely threaten mangrove communities. Aid-supported construction of roads, communication, and other infrastructure frequently produces increased access to areas, thus allowing increased cutting. These activities frequently occur together and have resulted in the loss of at least half of the mangrove communities of Central America, as well as other massive changes in deltas, along the coast and upriver. Dislocation of local populations, economic upheaval, and increased wealth polarization with the marginalization of large numbers of people have often resulted. Even with political protest, local economies are rarely compensated for the resulting stresses they experience.[53]

Any consideration of an extended coastal zone system for analyzing the likely effects of extensive aquaculture should seek to determine the maximum allowable development for a specific offshore, coastal, and contiguous interior area and the needs of local economies; the balance of different scales of aquaculture (extensive, semi-intensive, and intensive) according to carefully studied ecological and social costs; the intended beneficiaries (and the likelihood of actually serving them); and the precise nature of participation of local producers and communities in development design, implementation, and evaluation.

Destructive social and environmental outcomes could often be avoided, and genuinely positive alternatives suggested, if development projects—and the contexts in which they are being implemented—are considered in a framework like an ECZS by development officials and critics alike. Unfortunately, the incentive for such a change in outlook is largely lacking in the development agencies themselves.

Aid's Accountability Crisis

Some of the major aid agencies have undertaken certain institutional and procedural "reforms," as well as shifts in approach at the project level. Unfortunately, the majority of these reforms have been relatively insignificant, and most agencies continue to minimize environmental and social problems, if they recognize them at all.

One type of project-level shift by agencies, for example, is the increased funding of institutional and capability building (see table 22.2). This variously consists of improved credit and marketing operations, new training and extension services, scientific and technical development, and environmental protection. Depending on the character of specific policies, however, these may ameliorate—or worsen—problems of resource overexploitation and local socioeconomic decline.[54] The fundamental questions remain: Who is this capability for? And what goals is it to serve?

Another common reform—even among the development banks, which have overwhelmingly supported large-scale operations for many years—has been the allocation of more funds for small-scale fisheries. This is typically directed to the development of "nontraditional" exports, often providing larger vessels, motorization,[55] and upgrading of gear. (Some examples include recent loans by the ADB to Indonesia, Malaysia, the Philippines, Sri Lanka, and Papua New Guinea.) The FAO's FIPIS report stresses that "the major part of the support provided for capture fish production has been directed to small-scale fisheries." We would add the cautionary note, however, that "small-scale" fishing operations may actually be *large*-scale by local standards and that this action may, in fact, encourage further concentration within these economies.

In related fashion, there is a measured increase in aid expenditure for multi-component projects involving small-scale fisheries—including the fairly well publicized "rural development" and "integrated" fisheries projects. These frequently have a higher technical aid component and emphasize the linkages, for example, among production, postproduction, and service-related activities. Such efforts, at least by the largest agencies, retain the objective of promoting export production, and they typically lack a genuine management or conservation component.[56] Overall, the stated objectives and strategies for these newer-style projects are more complex than those of prior ones—some critics would charge that they are contradictory[57]—and they should receive some thorough critical analysis.

An accountability crisis may be suggested in fisheries-related development in the South, in which the major aid agencies remain fundamentally accountable to powerful northern political and economic interests but *not* to the local southern communities most affected by their actions or the (southern and northern) public interest constituencies concerned about the effects of their policies. There are important constraints on solving this accountability crisis. First, this highly skewed accountability derives not only from the structure and control of the major aid institutions but also from the modernization paradigm itself, which, through its self-perpetuating notion of northern superiority in all aspects of development, virtually prescribes continued domination by major world powers over the aid process.

Further, the sheer magnitude of development activities poses an obstacle to increased accountability in the aid process. For even the top few agencies, the number and size of projects are so large that establishing a system of adequate investigation and assessment of existing policies, let alone formulating alternative ones, is a formidable task. The situation in the World Bank is a case in point. At any one time, it is engaged in hundreds of development projects totaling hundreds of millions of dollars. The more than twenty highly complex projects the World Bank undertook in the 1980s alone[58] could easily occupy its "environmental" staff for a substantial period of time, not to mention the time and finances of its vastly overextended and underfinanced critics. In an institutional setting such as the World Bank, where the fisheries budget is extremely small compared with those for agriculture, forestry, and other types of rural development, obtaining the resources to mount an adequate response to existing and planned projects is extremely unlikely. This is particularly the case in the context of structural adjustment lending and so-called economic reform in major aid agency policies. The situation is compounded by the absence of any effective challenge by southern regional organizations or North-South organizational alliances addressing these practices.

"SUSTAINABLE" FISHERIES DEVELOPMENT: BEYOND RHETORIC

Since the concept of sustainability became popular in development policy circles during the early 1980s, a new debate regarding southern fisheries has emerged. For most members of the aid establishment, it refers to "sustainable maximum production." What is different from the earlier, productionist mentality in fisheries development is the view by these sustainability proponents that certain environmental and social variables (e.g., biological diversity, indigenous knowledge systems) *must* be considered. Thus, certain major agencies, including the World Bank, are paying some attention to the formulation of an "ecological economics" that takes at least limited account of these concerns (although not the basic ones discussed in the previous section), in an otherwise purely economic approach to development.

The viewpoint presented here supports a notion of sustainable development that might be termed "socioecological sustainability": a fisheries-related development approach designed to satisfy local socioeconomic needs and objectives. This goal would be achieved through genuine, broad participation of local communities while fostering conservation of marine and associated ecological systems and, when possible, making a contribution to genuine needs at the national level (e.g., lowering the debt, reducing regional inequities). Implementation of this plan inevitably means compromise between long-range objectives and shorter-range strategies and among different constituencies: national and local producers, large-scale and small-scale producers, and local and foreign interests.

Reform-minded organizations and individuals are increasingly focused on the systematic nature of natural resource and social problems emerging in marine and coastal development. Some progress is being made as well to establish more constructive alternative approaches to such development. These efforts will be significant for those attempting to set future research agendas, and certainly for those working to alter fundamentally the long-standing modernization paradigm. To date, concerned organizations and individuals remain in a primarily reactive position, objecting to one problematic or destructive project after another. This model must be replaced by one in which NGOs play a leadership role, fostering a fundamentally changed accountability of aid agencies—a *preemptive accountability* whereby the major aid agencies themselves will bear responsibility for demonstrating the beneficial and sustainable nature of any planned development projects—*prior* to design and implementation activities. Such a change can contribute seriously to a new, precautionary development that gives priority to the sustainability of ecological and socioeconomic systems. It is only through a coordinated and well-strategized North-South public interest effort that the rhetoric of this reform will be made a reality.

NOTES

1. Between 1974 and 1981, for example, aid agency funding for fisheries development projects increased by at least 50 percent, and it has continued to rise. A far greater increase may be assumed for the more inclusive process of coastal zone development.
2. As defined here, "fish" includes finfish, mollusks, and crustaceans.
3. D. Insull & J. Orzeszko, *A Survey of External Assistance to the Fishery Sectors of Developing Countries*, U.N. Food and Agriculture Organization (Fisheries Circular No. 755, Rev. 3, 1991) [hereinafter *FIPIS Report*]. Although the FIPIS survey has clear limitations, it nevertheless contributes to an urgently needed analysis of the basic patterns of aid policies. It is preceded by FAO surveys of a less ambitious nature, the most useful of which was summarized in 1987. H. Josupeit, *A Survey of External Assistance to the Fisheries Sector in Developing Countries, 1978–1984*, U.N. Food and Agriculture Organization (Fisheries Circular No. 755, Rev. 1, 1985, and Rev. 2, 1987).

    The computerized data base of FIPIS is far more extensive than prior FAO survey work; approximately 3,000 projects were reported, about *half* of which were utilized for the 1991 survey analysis. Development projects are characterized with regard to their objectives and components, expenditure type (loan or grant) and amount, and dates of implementation. Projects are categorized according to *lead activity*. Cofinanced projects were not recorded; nor were private sector arrangements in general, structural adjustment or conditionality terms, or informal negotiations. Nongovernmental organizational aid and government aid to intergovernmental organizations were also omitted from the FIPIS. *Fundamental differences in methodology exist between the FIPIS effort and preceding FAO surveys*, making comparison of data extremely difficult. For instance, *the earlier surveys considered aid funding averaged over the total number of project years, whereas the FIPIS recorded total project cost for initial outlay*. Moreover, there are basic typological differences between the FIPIS and earlier survey reports, for example, regarding aid organization categories (assignment of "bilateral" or "multilateral" status to particular agencies, etc.), and the identification of project "components"—some of which are noted in the discussion that follows. Development banks, while clearly multilateral in character, are treated as a separate category in the FIPIS effort. For the purposes of convenience, therefore, *the typology of "development banks," "multilateral agencies," and "bilateral agencies" is accepted for the study presented here.*

Additional data concerning five aid agencies relied on in this chapter are summarized in a memo to the author from J. Orzeszko of the FAO (July 1991) (the five aid agencies include two development banks—the World Bank and the Asian Development Bank—and the bilateral aid agencies from Japan [JICA], Canada [CIDA], and Norway [NORAD]). In addition, the author undertook a large number of independent calculations of aid expenditure patterns, based on FIPIS hard copy data. Discrepancies between the calculations of the FAO and those of the author partially result from the selection of projects for inclusion in tabulations. (Unless otherwise stated, the author is responsible for all tables and graphs.)

4. Examples are Thailand, Malaysia, Indonesia, the Philippines, Mexico, Ecuador, Argentina, Senegal, the Ivory Coast, and Sierra Leone.

5. In the Pacific Northwest, this is mostly Alaskan pollock and Japanese sardine. Altogether, these few species accounted for about half of the total catch increase during the 1980s.

6. *Review of the State of World Fishery Resources*, Marine Resources Service, U.N. Food and Agriculture Organization (Fisheries Circular No. 710, Rev. 7, Rome 1990).

7. OCEAN YEARBOOK NO. 8 (E. Borgese, N. Ginsberg, & J. Morgan eds. 1989). World landings in 1990 were about 97 million metric tons—constituting the first decrease in landings in more than a decade. Because freshwater landings are small compared with marine ones (with or without anchoveta), data available for total fisheries production have sometimes been utilized to help establish patterns for marine fisheries. In the data presented here, inland fisheries are factored out whenever possible.

8. Since fish production statistics are based on live weight, the ratio of fish to meat consumed needs to be adjusted. E. A. Huisman, *Aquaculture Research as a Tool in International Assistance*, 19 AMBIO 400–03 (1990).

9. In Papua New Guinea, for example, consumption ranges as high as 20 kilograms per person per year in coastal populations. In Africa, fish provides about a quarter of the protein consumed. The figure is higher for parts of Asia.

10. 65 Y.B. FISHERY STATISTICS—1987, U.N. Food and Agriculture Organization (Fisheries Series No. 33 and Statistics Series No. 86, 1989) [hereinafter Y.B. FISHERY STATISTICS]. In the second half of the 1980s, the value of key commodity group fish exports by developing countries more than doubled over the second half of the 1980s, and it now accounts for 47 percent of the world's total fishery exports.

11. J. Platteau, *The Dynamics of Fisheries Development in Developing Countries: A General Overview*, 20 DEV. & CHANGE 565 (1989).

12. G. KENT, FISH FOOD AND HUNGER: THE POTENTIAL OF FISHERIES FOR ALLEVIATING MALNUTRITION (1987).

13. The recent increase in marine fish landings has been primarily from a few regions with major upwelling, such as the Pacific Northwest and the Southeast Pacific.

More than half of this increase is from a few species: Peruvian anchovy, South American and Japanese sardine, and North Pacific pollock. Secondary increases in catch have occurred in the west-central Pacific and the eastern Indian Ocean.

14. Y.B. FISHERY STATISTICS, *supra* note 10.

15. Theoretically, better management, especially through reduction of waste from indiscriminate fishing practices and from low catch utilization, could *partially* satisfy the FAO's estimated 700,000-metric-ton increase in northern demand in the next few years. K. Topfer, *Assistance to Developing Countries in the Implementation of International Rules*, 14 MARINE POL'Y 259 (1990). As much as 15 million tons of fish are discarded from trawlers every year, much of which is marketable product. Ten percent of food fish caught is lost because of insufficient capability for preserving it, according to the FAO. For example, the situation in the North Atlantic has demonstrated that even when relatively abundant scientific data concerning target species and an institutional framework for comprehensive management are established— conditions that are generally missing in southern regions—progress in managing fish stocks does not necessarily result.

16. Hypothetically speaking, this would require at least $2 billion annually of World Bank expenditure. Yet the World Bank's total spending for fisheries development was $674 million over the fifteen years prior to 1984.

17. U.N. FOOD AND AGRICULTURE ORGANIZATION, THE STATE OF FOOD AND AGRICULTURE (1990). DWFs operating in the South are primarily from the main fishing nations in North America and Europe, the former Soviet Union, and Asia (especially Taiwan, Korea, and Japan).

18. Japanese fleets in southern waters, for example, have had special institutional and financial support from their government since before World War II. Recently, however, Japanese policies may be favoring international fish-trading firms over distant-water fleet interests.

19. For example, Japan has done very well in maintaining access to EEZs (even in the North). It has accomplished this in a number of ways, one of which is the threat of imposing import quotas. This threat constitutes a powerful diplomatic tool for developed countries with DWFs, particularly against southern nations, which have a high dependence on them for marketing their fisheries products in Japan. For example, Papua New Guinea markets about 70 percent of its fisheries exports in Japan; many other southern nations face similar circumstances.

20. SOUTH PACIFIC COMMISSION, REPORT OF TWENTY-FIRST REGIONAL TECHNICAL MEETING ON FISHERIES, NOUMÉA, NEW CALEDONIA, AUG. 7–11, 1989; D. J. Doulman & P. Terawasi, *The South Pacific Regional Register of Foreign Fishing Vessels*, 14 MARINE POL'Y 324–32 (1990).

21. *See* Josupeit, *supra* note 3. The World Bank's own documents indicate substantial failure in fisheries production, based on its own economic projections prior to project implementation. *Cf.* WORLD BANK, HARVESTING THE WATERS—A REVIEW OF BANK EXPERIENCE WITH FISHERIES DEVELOPMENT (Operations Evaluation Dept. Rep. No. 4984, 1984). Some updated discussions of the World Bank's experience are presented in a series of World Bank technical papers (Fisheries Series) in 1991.

22. J. C. G. de Matons, *Economic and Financial Appraisal of Port Projects at the World Bank: A Review of Policy and Practice*, 13 MARITIME POL'Y & MGMT. 4 (1988).

23. J. Platteau, *The Dynamics of Fisheries Development in Developing Countries: A General Overview*, 20 DEV. & CHANGE 565 (1989); J. Platteau, *Penetration of Capitalism and Persistence of Small-Scale Organizational Forms in Third World Fisheries*, 20 DEV. & CHANGE 621 (1989); W. Meynen, *Fisheries Development, Resource Depletion and Political Mobilization in Kerala: The Problem of Alternatives*, 20 DEV. & CHANGE 735 (1989); J. Cordell, *Carrying Capacity Analysis of Fixed Territorial Fishing*, 17 ETHNOLOGY 1 (1978); D. K. Emerson, Rethinking Artisanal Fisheries Development: Western Concepts, Asian Experiences (World Bank Staff Working Paper No. 423, 1980); C. Bailey, D. Cycon, & M. Morris, *Fisheries Development in the Third World: The Role of International Agencies*, 14 WORLD DEV. 1269 (1986); R. E. Johannes, *Traditional Marine Conservation Methods in Oceania and Their Demise*, 9 ANN. REV. ECOLOGICAL SYSTEMS 349 (1978). A related literature describes the ecological and economic effectiveness of traditional, or community-based, common property management of marine fisheries. A useful compendium of this literature may be found in NATIONAL ACADEMY OF SCIENCES, PROCEEDINGS, CONFERENCE ON COMMON PROPERTY RESOURCE MANAGEMENT (1987).

24. Compared with the $80 million cost of a 100-meter supertrawler or the $15 million cost of a 65-meter tuna seiner, a cost of $600,000 for a 15-meter gill-netter or $120,000 for a 10-meter fast potter may appear to be small, but they are out of the financial reach of nearly all local fishers.

25. Whether the new trawlers, for example, are owned by foreigners or by relatively large-scale local or domestic fishers (or as part of a joint venture) is of secondary importance to small-scale fishing communities. The point is that *they* are excluded from even short-term benefit. Fish stocks within the EEZs and the high seas may be affected as well by overexploitation caused in this way, since offshore and highly migratory species are often dependent on coastal breeding habitats and nurseries.

26. Some examples in recent fisheries aid include the loans made by the ADB to the Solomon Islands, totaling about $24 million since 1985; a $13 million loan to Somalia by the World Bank for upgrading coastal fisheries; a $5 million grant to the Marshall Islands by Japan for pier construction; and a $170,000 loan to Tonga by the ADB for its tuna export industry.

27. In Southeast Asia alone, more than thirty shared stocks—single and grouped species—are subject to aid-supported increasing fishing, yet there is no parallel management system, let alone one of international dimension, as is certainly required if these stocks are to be sustained.

28. E. Katz, M. L. Levin, & H. Hamilton, *Traditions of Research on the Diffusion of Innovations*, 18 AM. SOC. REV. 237 (1963).

29. The effects of large trawlers, which are fast (several knots per hour) and which have huge-scale netting, are exacerbated by a lowered catch *quality* due to fish compacting. Vessels equipped with gill nets, hand nets, and longlines generally produce higher-quality fish and are more selective in fish species and size.

30. Bilateral funding averaged $1.4 million per project, contrasting with $12 million per project for the banks. *FIPIS Report, supra* note 3.
31. A small number of very sizable loans have accounted for much of the development banks' fisheries lending. The ADB, for instance, has extended approximately $80 million in loans to the Philippines, $73 million to Indonesia, and $120 million proposed for 1991 for Indonesia—primarily for industrialization.
32. ADB loans to Indonesia have totaled at least $300 million since 1978, including a 1991 loan of $120 million for a complex of industrialization efforts. The ADB also has a high proportion of projects with multiple components, such as education and training, the orientation of which is nevertheless relatively homogeneous.
33. Josupeit, *supra* note 3; *FIPIS Report, supra* note 3. The World Bank's influence on fisheries is even greater when its loans to related sectors, its structural adjustment lending (in conjunction with the IMF's programs), its leadership role within the aid establishment, and its other interventionist practices are taken into account.
34. The World Bank's spending on harbor and other infrastructure construction, for example, was reportedly zero according to the *FIPIS Report; see supra* note 3. Our calculations from World Bank reporting in the FIPIS data base, however, indicated millions of dollars' expenditure. This discrepancy is presumably the result of differences in selection of projects from the data base for inclusion in calculations. A marked decrease in World Bank infrastructural spending is discernible, even taking into account the differences in data collection and tabulation among Josupeit (for 1978–1984), Insull and Orzeszko (for 1985–1989 and projected 1990 expenditures), and this author. Funding for coastal-based infrastructure is also frequently classified by agencies, with sectors and programs other than fisheries, for various reasons.
35. For 1985–1989, the FAO reports eighty Japanese, twenty-three World Bank, and twenty-nine ADB fisheries projects. The largest Japanese projects were for only $8 million to $10 million, compared with well over $100 million for the two largest development banks.
36. In Latin America, Japanese funding has been in Argentina (e.g., $7.6 million for port construction), Chile, Ecuador, and Peru (in 1983, a $20 million *loan* was issued, for port construction); in Africa, it has included Mozambique; Tanzania (troll vessels for shrimp production and for processing and marketing infrastructure); Senegal (refrigeration and processing, fleet upgrading, transport, and marketing); and Morocco (provision and promotion of engines for vessels and associated "research"). A large number of relatively small but potentially high-impact grants have been directed to South Pacific (and other) small island states, primarily for fleet building and infrastructure construction (especially ports) and for processing and marketing (e.g., in Fiji and Saint Vincent).
37. In Senegal, for example, funding has been extended for processing and marketing, fisheries cooperative development, small-scale fleet modernization, "economic planning," and surveillance. Other aid goes to Guyana, Guinea, Malaysia, Thailand (e.g., a coastal zone management project), and the Philippines (e.g., a project focusing on nutritional and other basic needs and on improved access to food and socioeconomic development).

38. *FIPIS Report, supra* note 3.
39. One of its largest projects, for example, is one for $37 million in India for small-scale fisheries industrialization. There are exceptions to this thrust of NORAD funding (e.g., a project to improve nutrition among small-scale fishermen in Mozambique and the convening of an international seminar on the experience of NORAD aid to fisheries).
40. For example, it has funded a feasibility study in Mozambique for processing techniques for anchovy production and a cost-benefit study in Tanzania for ice making and container insulation.
41. Japan ranks second, with 30 percent, and the United States ranks third, with 16 percent.
42. Josupeit, *supra* note 4.
43. Insull & Orzeszko, *supra* note 3.
44. Among the developing countries, per capita fish consumption is highest in Asia and the Pacific. Aquaculture, in fact, accounts for about 14.5 percent of the fish consumed there, compared with 2.1 percent and 0.3 percent, respectively, of the fish consumed in South America and Africa, according to the FAO. Close to 80 percent of all fish farming (and 99 percent of seaweed farming) in the world takes place in Asia, and these percentages are probably rising.
45. Mariculture aid in Latin America has primarily been focused in Chile, Peru, Ecuador, and Brazil. In Africa, projects have been relatively widespread but small in scale; the problem of inaccessibility to poor local producers has been particularly important. From 1985 to 1989, total aid for aquaculture was $669 *million* for Asia, $91 million for Africa, and $33 million for Latin America, according to FIPIS data.
46. Since 1990, World Bank aquacultural funding has decreased markedly, presumably because the initial outlays for its exceptionally large Chinese projects were completed.
47. For example, a $43 million loan to Bangladesh included funds for aquaculture and capture fisheries, and a $38 million loan to Indonesia included funds for development of private brackish shrimp farms.
48. The largest were about $10 million and $6 million, respectively, for shrimp projects in Mauritius and Ecuador.
49. This is in contrast with the average aquaculture project support levels of $22 million and $13 million, respectively, for the World Bank and the ADB from 1985 through 1989.
50. S. C. Stonich, Grassroots Movements and Environmental Management: Implications for Sustainable Development (paper presented at the 50th Annual Meeting of the Society for Applied Anthropology, Charleston, S.C., 1991).
51. The recent mission and (seven) working group reports prepared by the Study of International Fishery Research (SIFR) in the World Bank in 1989–1990— involving a series of missions and four working groups—are a good example. One particularly important report is SMALL-SCALE FISHERIES: RESEARCH NEEDS (World Bank Technical Paper No. 152, Fisheries Series, 1991).

52. In the Philippines, for example, between 60 percent and 80 percent of commercially important fish caught are dependent on mangrove environments for their reproductive and early life cycle phases. Yet mangroves there are being cut down at the rate of 4,100 hectares per year, primarily for construction of fish farms, according to a 1986 estimate by Ganapin.

53. The case study by Susan Stonich, involving the grass-roots struggle by artisanal fisher communities in southern Honduras as a result of the "boom" in shrimp mariculture, is an excellent one. Stonich describes both ecological and social effects at the local level, and some important qualifiers of assessment of such activities as "progress" at the national level.

54. The difficulty of making any generalizations about aid's capability-building efforts rests both in the fact that *specific* information about such projects is not readily available from the agencies and governments involved and in the fact that what data are available have serious limitations. In the FIPIS, for example, the FAO's typology of these types of support is combined with a number of different project components (training for processing, training for use of new fishery inputs, etc.). Therefore, these categories are only partial indicators of the activity as a whole, and they cannot be easily ascribed to the dominant "productionist" approach or to any shift toward concern for the environment and local socioeconomic systems, as distinguished from some movement toward considering the sustainment of resource and socioeconomic systems.

55. Motorization—with and without direct foreign aid—occurs at markedly different rates among southern nations, from near zero, in many contexts, to more than 80 percent (e.g., in Senegal).

56. Projects of this type include some in West Africa, the Bay of Bengal (both FAO and Trust Fund projects), and Zanzibar (a German Development Bank project).

57. For example, the objectives of a current $20 million ADB loan to Papua New Guinea include "participation" by national fishers to increase their incomes; reduction of food imports; and increase of foreign exchange earnings, with no discussion of their possible relationship.

58. These include the following: a $157 million loan to Morocco to expand its coastal fisheries; a $36 million loan to Bangladesh for brackish water shrimp exploitation; a $28 million loan to Tunisia for ports and fishing vessels; and a $22 million loan to the Philippines for jetties and ice plants.

V

The 1982 Convention on the Law of the Sea and the Nonliving Resources of the Deep Seabed

THE 1982 UNITED NATIONS CONVENTION on the Law of the Sea is widely recognized as one of the triumphs of our generation, but it has not received universal support because of its part XI, on the mining of minerals on the deep seabed. Although the discussion on this section was exhaustive during the negotiations, and although several key compromises were reached that appeared to resolve the differences, the United States, the United Kingdom, and Germany refused to sign the convention because of the provisions in this part. By March 1993, fifty-five nations had ratified the convention, and it will come into force when sixty ratifications are received. The only industrialized country among the ratifying states, however, is Iceland, and it may be unrealistic to imagine that the complicated procedure for deep seabed mining envisioned in the convention could come into place without the support of the developed world.

To address this stalemate, Artemy Saguirian has offered a creative proposal whereby part XI would be "frozen" for the time being and the rest of the convention would come into force. He argues that this approach is practical because the economics of deep seabed mining make it unrealistic that any activity will occur during the current era. He states that the basic principles agreed on at the convention—that the minerals of the seabed should be considered to be the "common heritage" of humanity and that this area should not be subject to national appropriation—should stand, but the details of the mining regime should simply be put on hold until mining becomes a more realistic possibility. Although admiring the boldness of Saguirian's proposal, Elisabeth Mann Borgese responds that it would be unacceptable in her opinion, in light of the energy that went into the design of the 1982 Convention and the ongoing exploratory activity. The chapter by Ian Townsend-Gault and Michael Smith addresses the fundamental ethical issues that should govern allocation of a "common" resource such as the deep seabed minerals.

These chapters present new and thoughtful approaches to one of the most difficult unresolved problems of ocean governance.

23 Excision of the Deep Seabed Mining Provisions from the 1982 Law of the Sea Convention: Reappraising the Principle of a Treaty's Integrity under the New Realities

Artemy A. Saguirian

A DECADE HAS PASSED since the 1982 United Nations Convention on the Law of the Sea[1] was opened for states to sign, ratify, and accede to, but its universal acceptance is still a distant prospect. Even when it does come into effect for the ratifying nations, the 1982 Convention will not be the instrument of international law that many states, particularly the former Soviet Union and the United States, intended to frame as they worked so hard to formulate its provisions. The convention cannot become a universal document in the foreseeable future unless certain elements of the regime it creates are radically altered. If such changes do not occur, the development of contemporary international law of the sea will not be orderly but will continue to be dependent on highly unpredictable actions of states.

By March 1993, the 1982 Convention had been ratified by only fifty-five nations, and it has not been ratified by any of the industrialized states, with the exception of Iceland. The meaning of this is absolutely clear. If the leading sea powers refuse to accede to the convention, not only will it be ineffective in enforcing in practice the regime proposed for the development of the seabed and ocean floor beyond national jurisdiction (the Convention Area), but also, in a broader sense, it will never become a universal international legal instrument regulating relations of states in respect to their activities in the world's oceans.

On the whole, the 1982 Convention is a major milestone in the codification and progressive development of the international law of the sea. But we have

379

been watching one single problem actually "sinking" the entire system of multilaterally established international law of the sea institutions. As David L. Larson stated after having investigated some eighty member and observer delegations to the Preparatory Commission, "The lack of progress at PrepCom has discouraged, if not prevented, many states from ratifying the Convention."[2]

The development of a single international legal instrument regulating the marine activities of states is a problem that has kept experts busy for years. I venture to say, however, that the diversity of views that we have been hearing until very recently was characterized by a lack of bold or daring proposals aimed at finding ways out of the impasse created by the convention's regime for the exploitation of deep seabed minerals. For years, we all seemed to be stunned and mesmerized by certain obsolete notions, which are clearly at variance with the present situation.

This problem is quite easy to explain. After long years of combined efforts undertaken by states that were parties to the law of the sea negotiations, it was quite difficult, if not impossible, to give up traditional patterns of thinking and immediately reevaluate the results under the political and economic realities of the postconvention period. The states were too deeply involved in backing such fundamental procedural principles as the indivisibility of the convention's text, and the "package deal" approach to its various provisions, to be ready to reappraise the very substance of this authoritative international legal instrument.

The problem involving part XI and annex III would not seem so complex, of course, were it not for the fact that most other provisions of the 1982 Convention are quite satisfactory and for the fear that it may recreate conditions for extremist maneuvers to be expanded in the ocean. We have been fighting against negative processes in national legislation caused by the lack of a single and stable international legal basis for the regulation of marine activities of states. In addition, deviations from the agreements stipulated in the convention have been snowballing. This is a historic stage in the development of the international law of the sea, and the delay of the entry into force of the 1982 Convention is producing a very negative effect on that development.

Nevertheless, as it was becoming more evident that the convention would fail to receive universal acceptance unless some substantial alterations were made in the provisions dealing with the seabed mining regime, a new thinking was gaining ground among states—both developed and developing. The statements, both formal and informal, of various states and UN representatives dealing with the subject reveal most clearly the world communities' substantial efforts to move away from a defective set of legal rules. Why do I dare insist on calling the seabed mining regulatory provisions a defective and ill-devised set of legal rules? The arguments are threefold.

First, even if we do not take into account the fundamental fact that the existing economic conditions make deep seabed mining activities scarcely probable for decades ahead (more on this subject will be said later), the proposed international regime does not reflect basic economic interests of the industrially developed countries. I do not believe I need to address here the in-depth study of the deficiencies of the deep seabed mining regime as formulated in the convention's text. The arguments against part XI are widely known, and nothing new could be added to this criticism.

In this regard, I would like to outline some basic considerations concerning the Soviet approach to the 1982 Convention on the Law of the Sea and its deep seabed mining part. This position could attract specialists' interest since it represents the approach of one of the major maritime powers of the contemporary world and because it reveals general policy options of industrially developed states.

It should be noted that the former Soviet Union retained under the convention a lot of what it might have lost. In this sense, the general political approach of the former Soviet Union to the convention was similar to the positions of other leading maritime powers.

On the basis of the long-term ocean-related interest of Russia, we could strongly state the following:

1. Beyond the deep seabed mining regime provisions, and from the viewpoint of overall Russian interests, the convention is a most acceptable international legal document regulating relations among states with regard to their activities in the world ocean.
2. Interests of Russia and the other former republics of the Soviet Union require that the convention become a universal source of international law of the sea, accepted by as many states as possible, and especially by the leading maritime nations.
3. The convention's deep seabed mining regime provisions, potentially entailing burdensome financial and economic obligations for Russia, are unacceptable as they now stand.
4. The unsettled legal issues related to the development of the Convention Area's mineral resources and the extreme difficulty of reconciling the diverse positions of the states at the Preparatory Commission have clearly delayed the entry into force of the Convention. This is highly unfavorable from the standpoint of Russia's interests and makes the future of the international law of the sea extremely uncertain.

These or similar conclusions are formulated in a survey (more than 500 pages long) prepared in the late 1980s by the Institute of the World Economy and International Relations of the Soviet Academy of Sciences. After having analyzed the entire range of financial, economic, political, and other factors, we recommended to the Soviet government that it refrain from ratifying the 1982 Convention for the time being. Our proposal was to activate diplomatic efforts aimed at finding ways of universalizing the convention and to go back to the issue of its ratification only after there emerged conditions that could lead the former Soviet Union to reappraise its general attitude toward the convention.

Although the view of the Institute of the World Economy and International Relations is not an opinion expressed on behalf of Russia or the former Soviet Union, it is hard to imagine that their official positions might be different. If, therefore, the financial obligations of the signatory states under part XI, annex III, and other relevant provisions of the convention remain intact, major doubts emerge as to the prospects for its ratification by the states that made up the former Soviet Union.

The former Soviet republics are now in a crucial phase of their economic development, and there is an acute shortage of hard currency, so I am very pessimistic about the prospect of their legislatures endorsing an irretrievable expenditure of $300 million to $600 million (even over a period of twenty-six years), which is to be invested into efforts whose results will not be desperately needed anywhere in the world for at least twenty to twenty-five years.

This general conclusion is even more true when we examine the current approach of Western countries to the 1982 Convention. Whereas the biggest problem to the former Soviet republics is the concrete financial liabilities they will incur once commercial activities begin in the Convention Area, the United States and other industrialized Western countries find part XI and annex III unacceptable primarily because of the profound conceptual controversy between the philosophy of free market enterprise on the one hand and the principles fundamental to part XI and annex III on the other.

The second reason justifying our judgment about the defectiveness of the deep seabed mining regulation stipulated by the 1982 Convention is quite new, but it is gaining support among states. Because of the continued degradation of the global environment, it becomes transparently clear that the environmental provisions of the text do not adequately deal with the risks associated with human interference in the ecologically delicate environment of the ocean floor. We simply know very little about the possible consequences to the marine environment of deep seabed mining. The global environment, and the human race as its integral part, are already bearing too many environmental burdens as a result of humanity's unwise activities. No commercial mining activities should begin until authoritative scientific conclusions are made about the possible short- and long-term impacts of deep seabed mining on the marine environment and before environmentally safe mining technology is created and proved to be effective.

Moreover, the convention's mechanism regulating deep seabed mining activities is clearly aimed at developing large-scale industrial operations in the Convention Area. The relevant provisions create an extremely detailed legal background for the development of resource activities in the area. But the convention is not so specific when it deals with issues of protecting the ocean environment while mining operations are conducted in the Convention Area. It puts forward only a few general principles that seem to be of declaratory character in the absence of relevant implementing mechanisms.

It should be noted also that full responsibility for elaborating, approving, and verifying compliance with environmental rules is vested in the International Sea-Bed Authority (to be created on entry into force of the 1982 Convention). It seems highly questionable whether the Authority would be able to assume these responsibilities properly and effectively, since under the convention it will have its own interests in the development of deep seabed resource exploitation.

Finally, the third deficiency of part XI of the 1982 Convention is that because no early commercial mining operations on the deep ocean floor can be expected, there is absolutely no reason to hurry to establish a detailed international regulatory regime for such activities.

Many of us have been wondering whether we have a right to ruin everything we have achieved just because of differences over problems that are abstract and theoretical rather than practical. In his speech at the Twenty-second Annual Conference of the Law of the Sea Institute (held in Kingston, Rhode Island, in 1988), Professor Edward L. Miles was right to remark: "There is not likely to be any commercial deep seabed mining outside national jurisdiction within the lifetime of anybody attending this Conference, even the youngest student among us."[3]

No matter what relative importance we may attach today to the regulation of deep seabed mining in the Convention Area, it cannot equal priorities in other chapters of the convention aimed at mutual coordination of national interests of states and adaptation of those interests to the needs of global development. This conclusion is all the more indisputable since deep seabed mining in the Convention Area is still a distant prospect. To put it quite frankly, the delegations of industrialized states to the Preparatory Commission hardly believed from the very beginning of its work that establishment of the International Sea-Bed Authority could proceed without essential alteration of certain provisions of the convention.

Changes in the general economic conditions regarding mining of the metals contained in polymetallic nodules (reduction of metal prices), especially so far as copper and nickel are concerned, and high production costs have long since led us to the indisputable conclusion that many provisions of part XI and annex III would substantially undermine the commercial expediency of deep seabed mining if it were to begin now. Moreover, before commercial mining actually begins, it is impossible to say with any degree of certainty how this or that legal provision of part XI and annex III of the convention meets or fails to meet conditions that may emerge in the future.

This is not the main problem, either. Earlier, industrialized countries objected to certain provisions of the convention because they imposed excessively burdensome financial and economic obligations on the mining states; but now the beginning of commercial mining is not discussed for reasons that have nothing to do with convention provisions. So the entry into force of the convention is being delayed by differences concerning regulation of activities that do not exist in practice (so far as exploitation is concerned) and that are unlikely to emerge in the foreseeable future.

We are facing an obvious paradox: The regime whose elements delegations to the Preparatory Commission are trying to formulate, with such great difficulty and apparently without much success, have neither a subject for regulation (i.e., extraction of deep seabed mineral resources) nor an object for such a regulation (i.e., relations among states in connection with deep seabed mining in the Convention Area), and it is unlikely to have them in the near future and maybe not even in the distant future.

The law cannot be above the practical relations it is supposed to regulate. This is absolutely correct and quite applicable to the situation in question. Indeed, one could not find in the history of international law any example of a situation in which relations that are far from developed have been subjected to such a detailed and specific legal regulation. But this seems to take place today in connection with the distant prospect of deep seabed mining.

All of the arguments I have just put forward are not new, and obviously I am not their sole author. What we urgently need to think about now is how to get out of this extremely difficult political and legal situation with the least adverse effect on the whole treaty system regulating ocean activities.

Let me turn now to the most important part of this discussion. The whole package of deep seabed mining provisions should be placed outside the convention, except for the relevant articles of chapters 1 and 2 of part XI, which define the international legal status of the Convention Area and the general principles regulating it. But how can this be done without the convention's integrity being affected?

The entire range of practical difficulties associated with ratification results from the principle of the document's integrity, contained in article 309, whereby the refusal to accept any single part of the convention makes it impossible for a state to be legally bound by any other part of it. In other words, the "package deal" approach, which used to be the most effective method of overcoming differences, has actually turned into the main formal legal obstacle impeding the early entry into force of the convention.

I believe, however, that we would reveal a complete lack of political foresight if we elected to discard this approach to complex international law of the sea issues. The package deal approach is absolutely indispensable, since it is fundamental to the convention. The only alternative to it would be a rupture of the integrity of treaty regulation, if not a collapse of the whole convention.

The developments involving the entry into force of the 1982 Convention have added to the fundamental problem of the correlation of two rules of law or, rather, of a rule and an exception—the principle *pacta sunt servanda* and the exception *rebus sic stantibus*. With a package deal approach to the elaboration of a multilateral international agreement, how is it possible for a state to refuse to honor certain provisions of that agreement in view of fundamental changes in one specific sector of activities regulated by it and yet retain rights and obligations arising from its other provisions? It is extremely important to answer this question, for it has to do with any internally dissoluble legal document regulating a wide range of interstate relations. In my brief presentation, I can only outline the importance of this highly interesting theoretical problem.

When a treaty of unlimited duration is concluded, it is assumed that it is binding only as long as relevant circumstances remain basically unchanged. The circumstances are legally regarded as being changed if (1) those changes are fundamental in character, that is, if they affect the objectives and the subject of the treaty; (2) the changes that have occurred were unpredictable; and (3) there have been developments affecting the very essential grounds for accord among the parties.

The *proviso rebus sic stantibus* is very controversial from the viewpoint of the effectiveness of an agreement concluded and the basic principle of international law *pacta sunt servanda*, but nonetheless we cannot disregard the influence of fundamental changes of circumstances on ratification of the international treaty by states. If we also take into account that the convention does not stipulate the right of states to any reservations or exceptions (apart from those plainly permitted by its certain articles), we might say that even before ratification, the 1982 Convention is legally much more binding to the signatories than the normal international treaty pending entry into force.

In relation to regulation of activities in the Convention Area, the delegations that negotiated the convention believed that it could and should be concluded, based on circumstances that have now ceased to exist. This means that there has been a radical change of circumstances that gives us sufficient legal arguments to withdraw carefully from the convention the mechanism regulating development of the Convention Area without destroying the convention's inner integrity.

In the light of these specific considerations, the following practical decision could be agreed on by signatories and states that are parties to the convention: A diplomatic conference is to be convened to work out and adopt a "Declaration on the Procedures of Ratification and Entry into Force of the 1982 United Nations Convention on the Law of the Sea." The declaration would thoroughly substantiate the decision that in view of basic changes in the general economic conditions of the development of deep seabed mineral resources, an indefinite postponement of the commercial phase of mining, the impossibility of forecasting with any degree of accuracy conditions for the latter's commercial expediency in the future, and the need to validate all other chapters of the convention without delay, the convention would come into effect without entry into force of the whole of part XI (with the exception of the general provisions of chapters 1 and 2 defining the international legal status of the Convention Area and its resources), annexes III and IV, and some other related provisions. The convention's seabed mining provisions would be frozen pending adequate development of financial and economic conditions needed for commercial mining to begin. At the same time, in accordance with resolution 1 of annex 1 of the Final Act of the Third UN Conference, the Preparatory Commission would pursue its efforts at studying the entire gamut of practical questions concerning activities in the Convention Area and their impact on the marine environment. At any given time, it should be ready to resume its substantial work aimed at updating the deep seabed mining provisions of the convention.

The freezing of the convention's part XI, that is, temporary excision of its deep seabed provisions, should by no means affect the basic global consensus concerning the international legal status of the Convention Area and its resources and the general principles regulating states' activities there. The accepted fundamental approach to such activities and its legal form of implementation in the convention should remain absolutely intact.

This fundamental and widely agreed-on approach of the community of states to one of humankind's global commons—the international seabed area—is reflected by the following convention principles:

1. "The Area and its resources are the common heritage of mankind" (article 136).
2. "The Area shall be open to use exclusively for peaceful purposes by all States . . . without discrimination" (article 141).
3. "No State shall claim or exercise sovereignty or sovereign rights over any part of the Area or its resources, nor shall any State or natural or juridical person appropriate any part thereof. . . . All rights in the resources of the Area are vested in mankind as a whole, on whose behalf the Authority shall act." All rights with respect to the minerals recovered from the Convention

Area should be acquired or exercised only in accordance with the convention (article 137).

4. "Activities in the Area shall . . . be carried out for the benefit of mankind as a whole . . . and taking into particular consideration the interests and needs of developing States" (article 140).

5. Any activities in the Convention Area related to exploitation of deep seabed resources shall not "affect the legal status of the waters superjacent to the Area or that of the air space above those waters" (article 135).

By retaining in the renewed convention's text these basic principles, we would be able, on the one hand, to accelerate its final and expeditious entry into force and, on the other, to address, in a clear legal environment, specific treaty provisions regulating deep seabed mining activities once the financial, economic, and other relevant conditions dictate this.

As to pioneer activities (obviously confined to the exploratory phase), they could continue to be carried out in accordance with resolution 2 of that annex and agreements concluded under the umbrella of the Preparatory Commission's work.

In conclusion, it is obvious that the development of a regime for mining activities in the Convention Area has become a major obstacle, impeding stabilization of the international legal situation in the world ocean as a whole. The solution proposed here is to freeze part XI, thereby eluding the extremely sensitive political problem of an immediate revision of the convention's text.

NOTES

1. United Nations Convention on the Law of the Sea, Dec. 10, 1982, U.N. Doc. A/CONF.62/122, 21 I.L.M. 1261 (1982).

2. David L. Larson, *When Will the U.N. Convention on the Law of the Sea Come Into Effect?* 20 OCEAN DEV. & INT'L L. 176 (1989).

3. Edward L. Miles, *Preparations for UNCLOS IV?*, 19 OCEAN DEV. & INT'L L. 423 (1988).

24 A Response to Dr. Artemy A. Saguirian

Elisabeth Mann Borgese

I HAVE STUDIED THE PREVIOUS CHAPTER, "Excision of the Deep Seabed Mining Provisions," and would like to congratulate Dr. Saguirian on this very important and profound piece of work. The problem that I have with it is that because of its first-rate quality, it may be dangerous: The danger is that it may take us backward rather than forward. As you know, I have no problem with the concept of "freezing" certain articles. As a matter of fact, I think I was the author of this concept. My problem is that this proposal goes much too far.

The application of the principle of *rebus sic stantibus* is very interesting and useful. But it requires a much more thorough analysis of what really has changed and what has not. The freezing of articles must be commensurate with the changes that have actually taken place. They cannot go beyond that. The chapter is not clear on this issue. For example, the statement that "the proposed international regime does not reflect basic economic interests of the industrially developed countries" states nothing new: nothing that has changed. In fact, Dr. Saguirian himself adds that "nothing new could be added to this criticism."

It follows that any change that would affect the basic philosophy of the provisions establishing the International Sea-Bed Authority, which indeed was intended to serve not only the "basic economic interests of the industrially developed countries" but equally those of the developing countries, is illegitimate. The fact that "the United States and other industrialized Western countries find part XI and annex III unacceptable primarily because of the profound conceptual controversy between the philosophy of free market enterprise on the one hand and the principles fundamental to part XI and annex III on the other" is nothing new. For almost two decades, however, they participated in the elaboration and negotiation of a document that transcended the philosophy of free market enterprise as it transcended the philosophy of central planning; a document that is based neither on private property nor on state property but on nonproperty—the concept of the common heritage of humankind, which integrates planning and free initiative and responsibility, production and resource conservation, development and environment.

To try to change this now, to go back to the philosophy of unbridled free enterprise of Mr. Reagan and Mrs. Thatcher (who, incidentally, brought their countries—plus Canada—to the brink of economic depression and social disruption and have already been swept away by history) has absolutely nothing to do with *rebus sic stantibus* and cannot be justified in any way, either legally or ethically.

The second argument, the environmental one, has a strong fashionable appeal today, but, again, the link to *rebus sic stantibus* is weak.

Pollution from activities in the Convention Area is explicitly dealt with in article 209, which makes it mandatory for states and for the Authority to establish international and national rules, regulations, and procedures to reduce and control pollution of the marine environment from activities in the Convention Area. And these rules, regulations, and procedures are subject to review as often as necessary. Article 145 provides a concrete framework for the rules and regulations the Authority is to adopt for prevention, reduction, and control of pollution and other hazards to the marine environment, including the coastline. Regarding interference with the ecological balance of the marine environment, particular attention is to be paid to the need for protection from harmful effects of such activities as drilling, dredging, excavation, disposal of waste, and construction and operation or maintenance of installations, pipelines, and other devices related to such activities. And, further, particular attention is to be paid to protection and conservation of the natural resources of the Convention Area and prevention of damage to the flora and fauna of the marine environment. It is also important to note that according to article 162(w), the council has the power "to issue emergency orders, which may include orders for the suspension or adjustment of operations, to prevent serious harm to the marine environment out of activities in the Area and (x) to disapprove areas for exploitation by contractors or the Enterprise in cases where substantial evidence indicates the risk of serious harm to the marine environment."

This is a firm and solid framework. The 1982 Convention could not and should not have gone further. It could not and should not have tried to lay down detailed rules when the technologies eventually to be employed for exploitation of the resources and the circumstances of exploitation are unknown. Had the Third UN Conference tried to do that, it would have fallen into the same trap into which it fell with regard to the financial regulations: It would have drawn up detailed provisions that would be obsolete by the time the 1982 Convention comes into force. Reasonably, it has left the details for the Authority itself to fill into the framework.

Dr. Saguirian's third argument, that "because no early commercial mining operations on the deep ocean floor can be expected, there is absolutely no reason to hurry to establish a detailed international regulatory regime for such activities," expresses a half-truth. True, there is no mining; hence, why should we try to regulate it at this time? However, the actual mining constitutes just one of the "activities in the Convention Area." Other activities are indeed occurring at present and will occur in the immediate future: seabed exploration and mapping, scientific research on the ocean floor, and development of technologies. All of these are ongoing activities, and the International Sea-Bed Authority has responsibilities for their coordination and regulation and is indeed empowered to carry them out itself. Under no *rebus sic stantibus* can these responsibilities be abdicated. As we have shown in a study prepared in cooperation with the Asian African Legal Consultative Committee, these activities currently involve investments of at least $100 million per year globally.

Undoubtedly, there have been some major changes—some of the more important ones could have been predicted, and have been predicted by some of us for quite some time:

1. The international community is now aware of the fact that the manganese nodules are not the only existing economic resource on the deep seabed.
2. The international community is now aware that deep-sea mineral resources exist in the international area as well as in EEZs.
3. For a variety of reasons—technological, economic, and environmental—the timetable for commercial exploitation of these resources has been drastically revised. (It has been continuously revised since the 1960s, so this delay is not really all that new.)
4. It has become evident that the private sector alone is totally unable to move ahead on what used to be called "sound commercial principles," and there has been a dramatic change in the cast of actors in seabed mining. In the 1970s, they were private companies. They failed. In the 1990s, they are states, state companies, and state-assisted companies.

Undoubtedly, plans of action and programs must adjust to these changes. What is going on today with regard to the common heritage of humankind is not commercial exploitation; its focus is on the exploration, research, and development, and the development of human resources. It is this that the international community today should facilitate and regulate. We need an interim regime, and the Pioneer Agreement provides an excellent basis.

Rebus sic stantibus gives us no justification in throwing away this entire framework. It justifies us in adjusting functions and timetables, and in using what can be used and not using what cannot be used. Obviously, when there is no production, there is no production limitation. If states do not enter into the kind of contracts described with such great detail (in the work of Mr. Leigh Ratiner of the United States), these articles will not be used. If states instead will utilize the wide-open and very general provisions for joint undertakings and develop these in accordance with ongoing and forthcoming real activities and opportunities (exploration, mapping of the seabed, research and development, development of human resources), that is what they will do.

Rebus sic stantibus certainly is applicable to the provision for the annual payment of $1 million—and the international community has already accepted the abolition of this provision. It is most certainly applicable to other financial provisions, which must all be frozen. It is applicable to article 151, on production limitation. It is applicable to all of annex III, as well as to the financial provisions in annex IV. But that is as far as you can go with *rebus sic stantibus*. I would suggest, furthermore, that "freezing" certain articles does not require a special conference. It is the Preparatory Commission itself that can—and, indeed, should—make these adjustments, which are necessary for the effective implementation of the 1982 Convention.

Last, "freezing" must go hand in hand with development of the interim regime. Its negative aspects must be balanced by positive measures in the areas in which we now can act: exploration, technology development, and development of human resources. This we can and must do jointly, for the benefit of each and all. We do not want to go back to 1967. We want to move forward into the 1990s and into the new century.

25 Environmental Ethics, International Law, and Deep Seabed Mining: The Search for a New Point of Departure

Ian Townsend-Gault and Michael D. Smith

WE LIVE IN AN AGE marked by continuing disputes over natural resources: timber, oil, and fisheries, to name a few. Most countries continue to depend on resource activities as sources of energy, employment, and wealth. The established pattern of reliance on these diminishing resources and on questionable means of exploiting them has been slow to change. Significant changes in consumptive behavior have yet to be brought about.

Few issues ignite governments and their citizens as readily as disputes concerning jurisdiction over resources. As we approach the end of the twentieth century, a major battle lies before the international community in developing an ocean regime for the deep seabed and the water superjacent to it. In a world where territorial lines of ownership define everything, the high seas as yet belong to no one.

The United Nations Convention on the Law of the Sea[1] sets forth a set of rules governing the area beyond national jurisdiction and articulates the principles on which such a regime would be based.[2] This "common heritage" principle strives toward the concept that the resources of the deep seabed should be developed for the benefit of all, regardless of a country's location, economic strength, or technological capabilities. Some developed states suggest an alternative model, basically one driven by free enterprise under national supervision in an area beyond national jurisdiction.

Yet another possibility is that of extending national jurisdiction seaward, the result being that there would be no unclaimed marine areas. Which model the international community will adopt and what will guide it in its decision making has not been resolved, but this decision will be a test of the actual commitment to provide new resource management techniques.

392

Regardless of the uncertainties, it is clear that the answer must be based on principle, as opposed to state practice, and based on an ethical foundation rather than greed. The untouched resources of the deep seabed provide humanity with the opportunity to rethink our relationship with the sea and to attempt to think beyond immediate needs. It offers humankind the chance to prove that we are capable of learning from our mistakes, our practices of overexploitation, and that we can manage the deep seabed resources in a holistic manner for the future.

Most important in this equation is that the resources of the deep seabed are beyond our reach for the moment, giving us time to plan a strategy carefully. The area beyond national jurisdiction contains living and nonliving, renewable and nonrenewable resources, with many state agendas at work in considering their future governance. These agendas, given the coastal state resource management record to date and the enormity of the decision, affecting as it does the choices and obligations of future generations, require a reassessment of approach and the identification of a firm ethic on which that governance should be based.

Any viable marine management system is required to find an effective ethical foundation for legal development. The concept of the freedom of the seas provided just such an ethical foundation in its time, and the 1982 Convention is our generation's attempt to provide a new ethical foundation for the future—the common heritage principle. If this approach is to be truly efficacious, it must be capable of engaging with and responding to national as well as international interests in the ocean. The burden laid on the principles of the freedom of the seas, enunciated centuries ago, has increased exponentially with the complexity of the international community today, and the burden on the principle of common heritage is already immense, no less so because of the opposition of the developed states, preventing the 1982 Convention and the common heritage principle from entering into force. This chapter raises the suspicion that the search for a single principle or a single ethic may be rather too simplistic, given the complex nature of the challenge. Certainly, the search must be widened considerably, going beyond the traditional sources and traditional Judeo-Christian ways of thinking about nature.

Commentators are in general agreement that the 1982 Convention marked the inauguration of the "new" law of the sea[3]—a regime evolved in little more than a decade, through a truly international process, that attempted to address the broadest possible agenda of national wishes and aspirations. In contrast, the old law of the sea evolved gradually within narrower confines and was the product of inputs from relatively few states.[4] The forces unleashed by the doctrine of the continental shelf (from 1945 on) marked the beginning of the end for the "old" law of the sea. In its early years, the doctrine was seen as marking the end of the restrictive view of coastal state jurisdiction and was enthusiastically hailed as such.[5] Of greater import was the steady stream of claims to jurisdiction (couched in various terms) that followed it. Pandora's jurisdictional box was well and truly opened. This conclusion will, of course, offend those who wish to see the guiding principles of such activities firmly situated within the confines of cooperative international responses to challenges that are, and should remain, essentially beyond the competence of any one country. To these commentators, developments like the doctrine of the continental shelf and the concept of the exclusive economic zone[6] (EEZ) are failures to be deplored. However, the resource manager sees them as merely inevitable, given the absence of effective alternatives or the difficulty in abandoning the status quo.

One of the major conditioning factors in the development of ocean policy has been the industrial dimension. Modern governments either enact measures at the behest of an industrial actor or impose measures subject to constraints that are acceptable to the industry concerned. The capital-intensive ocean industry is essentially the creature of free market economies, engaged in a special relationship with government. The Truman Proclamation[7] is a perfect example of this: The oil industry needed legal security over submarine areas beyond the limits of the territorial sea as the sine qua non for investment and hence development. Since only the adjacent coastal state can provide such guarantees, government is asked to act, and obligingly does so. Commercial pressures shaped and effectively changed U.S. policy toward the deep seabed mining provisions of the United Nations Convention on the Law of the Sea.[8] This provides a ready example of government hostility being conditioned by industrial attitudes. If commercial enterprises are to play a role in deep seabed mining, it is not difficult to see the role these considerations will play in the decision making of the international community.

The discussion to this point has focused on developments in the public international law of the sea. But paralleling this chronology, though diverging from it, are new approaches to a vast array of social-legal relationships. These ideas have had, and continue to have, profound effects on many societies, especially in the value-starved heavily industrialized states.

Those living in Western countries are aware of the widening gap between those who are content with their consumer society and those who challenge or doubt the values and beliefs on which their society is based. The latter group feels that humankind has somehow "lost its way," has not been "true to itself." Doubtless, such sentiments are a reflection of events and experiences occurring within our particular historical moment, and though skepticism and world-weariness are by no means unique features of any period in the human narrative, the depth and pervasiveness that characterize current attitudes seem to indicate that some essential changes are under way.

To put it succinctly, some societies have reached a juncture in their development that will require a fundamental reevaluation and transformation of the assumptions, values, perceptions, modes of thought, and patterns of behavior that have carried them thus far. And, the argument runs, such a reexamination is required vis-à-vis all aspects of our use of resources and our relationship with the environment: Progress is not without limits, risks, or costs. These costs have quite often been borne by elements, both natural and social, that were excluded, often deliberately and prejudicially, from consideration in the decision-making processes that govern all facets of societal and economic development.

The government-industry "partnership" is an excellent example: The leading actors see little reason to deal with the bit players. As the negative consequences of these decisions have increased in significance and begun to affect directly the interests that conventional discourses seek to protect, discontent has spread even within mainstream society. Society is beginning to perceive a threat to its quality of life, but it does not yet realize the threat to its long-term viability. Nor does society have the conceptual means to identify the source of the threat.

This emerging skeptical posture, although still somewhat exoticized and marginalized by a conservative tradition that maintains its dominance by vigorously denying the possibility of any attempt to question its legitimacy, has found expression in critical movements in numerous fields of academic endeavor. These fields address, for different purposes and in varying degrees of coherence, what Jane Flax believes "has become most problematic in our transitional state: how to understand and (re)constitute the self, gender, knowledge, social relations, and culture without resorting to linear, teleological, hierarchical, holistic, or binary ways of thinking and being."[9] For Flax and others like her, it is the overriding tendency of Western philosophical and cultural tradition to universalize, naturalize, and dichotomize in a way that inevitably nullifies the importance of specificity, context, and diversity. Theorizing revolves around the search for fixed unitary principles, necessarily defined in terms of bland generalities, and neat, mutually exclusive dichotomies, based on artificial distinctions that purport to organize and explain the world. Unfortunately, not only does this lead to cases of gross oversimplification and misrepresentation regarding the complexities and nuances of human experience in specific contexts, it also often serves as a means of establishing or reinforcing hierarchical arrangements that serve the interests and needs of dominant groups at the expense of those less powerful.

An example can be found in contemporary reaction to *Our Common Future*,[10] the report of the World Commission on Environment and Development. Through this report, the phrase "sustainable development" has entered the language. But since the phrase has become part of the vocabulary of resource exploiters as well as environmentalists, it appears that there is a danger of co-opting it, reducing it, making it serve the very traditional thought and authority pattern that it is attempting to challenge.

The need to construct a new framework for inquiry has led to the development of an increasingly influential current contemporary scholarship that can be
described as "postmodern." It should be noted that this term has been applied in
various contexts and does not refer to a single, cohesive theory but rather to a
number of loosely affiliated approaches that find commonality in their general
stance vis-à-vis traditional systems of thought. Of particular interest to this
discussion is poststructuralism, which, in its focus on the meaning of language,
on what people say and the signs they use to say it, provides a method of analysis
that can be adapted to a variety of situations.[11]

This line of thinking challenges the tendency in Western philosophy not only
to conceive of the world in terms of binary pairs of fixed opposites but also to
arrange them hierarchically so that one term is favored over, and is used to
define, the other.[12] Poststructuralism adds the further insight that this hierarchy
of concepts is very often ideologically driven.[13] By delineating rigid boundaries
between the acceptable and unacceptable, the desirable and undesirable, the
good and the bad, it is possible to create self-serving and self-perpetuating
normative categories that, although claiming to represent the world of human
nature as it is, in fact provide only a partial and privileged version of reality. The
essential interdependence of the world is obscured.

Ethics, perhaps more than other traditional branches of Western philosophy,
has relied on monolithic principles as a foundation of its prescriptions for human
conduct. Specifically, it has focused on the concept of human as rational being
and placed it in hierarchical opposition to nature. Indeed, humankind versus
nature is a familiar and powerful theme in Western culture: The triumph of
human reason, knowledge, and science and technology, as so often depicted in
popular media, is a resonant, symbolic confirmation of both our ascendancy and
our transcendence. If we scrutinize this duality more closely, we begin to
recognize inconsistencies that undermine the neatness of the distinction. More
specifically, if we deconstruct the opposition, that is, invert the hierarchy, it
becomes possible to see nature as the primary element and humans as a sub-
category of a larger biotic community. We are merely a special case of the
interconnected biological diversity that is nature. The fallacy of the person-
nature dichotomy becomes even clearer when it is considered in relation to
other traditional dualities, which, in a sense, it echoes: mind-body, reason-
emotion, and order-chaos. All of these represent conceptual distinctions that
help to ensure human control over our own destiny and allow us to locate
ourselves within some higher order.

This admittedly superficial discussion of poststructuralism and the way it
might influence the "reconstruction" (by engaging first, of course, in "deconstruction") of a viable environmental ethic is intended to serve merely as an
example of the sort of intellectual effort that may be required. Lawyers have not

been immune to these movements in other fields, or in society as a whole. The approach to theory epitomized by poststructuralism, and by postmodernism in general, has had a strong effect on legal theorists, particularly among those whose work forms the basis for what has come to be known as the critical legal studies movement.[14] Similarly, the growth of environmentalism has had an enormous effect on the development of both law and legal theory over the past twenty years. *Ways Not to Think About Plastic Trees*[15] and *Should Trees Have Standing?*[16] are obvious examples of watershed developments in legal thought with respect to the environment. People are evaluating old approaches and finding them wanting. A new ethic for ocean preservation and development must be based on an ethic that accommodates and depends on a new rationality.

Wherever the ideals of ethical development with respect to the marine environment and its resources may be heading, the route map is written in terms of regulation and control, but this control is essentially rooted in notions of territoriality. It is this tension that lies at the heart of the problem of developing a new ethic of ocean governance. The international community has yet to discover a better engine for development, monitoring, and enforcement of rules than coastal states, which have been busily extending and consolidating their jurisdiction over the ocean and its resources since 1945. Coastal states, however, act out of self-interest, defined either by their governments or by special interest groups, such as industry. How, then, can a cooperative principle based on self-denial, such as freedom of the seas and international governance, possibly survive in an international regime dominated by sovereign states?

One approach to answering this question might be to require the assessment of interests in the oceans, beyond the existing limits of national jurisdiction, with unflinching honesty. If the principle guiding the management of the high seas is to remain under pressure from coastal states, it is surely better to assess the strength of the enemy at the outset. This statement is likely to alarm those who thought that the issue had been dealt with in the 1982 Convention.[17] The problem with using the convention for this purpose lies in its uncertain standing with respect to those countries that have not signed or ratified it. Striking a balance between a new array of legitimate state interests within national jurisdiction and the areas beyond requires an accurate identification of the new realities of the situation. It therefore appears that a major ingredient of a new ethic governing the freedom of the seas is a willingness to develop requisite information for policy evolution and information sufficient to implement this policy.

One of the most formidable difficulties facing those who wish to implement a policy with an environmental ethic is the political unreality of advocating radical change within a short time frame. The same sort of difficulty afflicts resource managers and politicians, who must deal with difficult decisions affecting resource development. The first limiting factor they face is that there is no such thing as "starting over." Policies may be refined, rethought, or abandoned, but reversal is seldom attempted and in many cases is impossible.

But arguably, the deep seabed is different. Here, a new start can be made in devising regimes governing exploitation of the nonliving resources and the conduct of such activities, including protection of the marine environment. But this can be done only if the ethic that underlies the concept of the "Convention Area"[18] can be sustained. During the 1980s, many Western countries adopted policies that would allow resource activities to be conducted according to a variety of nostrums, such as the free market economy. However, examples of current difficulties in managing resource activities indicate that there is no substitute for management carried out in the public interest. In this way, managers are fully accountable to the people in whose ostensible interest the activity is being carried out.

The deep seabed is virtually unexplored and unexploited. Its development does not rely on concepts developed for the continental shelf or coastal regions. Those wishing to promote this development do not need to burden themselves with any of the debts or baggage accumulated in the areas within national jurisdiction. Of course, the deep seabed mining provisions of the 1982 Convention remain the most controversial. Dissatisfaction with them was the prime reason for the Reagan administration's refusal to accept the convention in 1982. Western attitudes toward these provisions ranged from suspicion to downright hostility, and one suspects that these countries are only waiting for the collapse of the internationalized approach to proceed with unilateral measures, which in many cases are already on the statute books.[19]

There is a sharp and irreconcilable split on grounds of principle between the two camps here: one favoring international control and the other claiming that this approach is fundamentally flawed, ineffective, and contrary to the national interests of the opposing countries. But debate is complicated by the rhetorical flourishes of the proponents of the former approach, some of whom are conscious of the fact that the deep seabed mining provisions were held out to the developing world in exchange for agreement on other parts of the convention, enormously extending national jurisdiction for all coastal states. Resource economists complicate the debate still further by claiming not only that deep seabed mining is not a commercial prospect for the foreseeable future but also in many cases that exploitation of the hard minerals of the seabed within national jurisdiction is similarly unattainable.

The commercial viability card is a strong one, and it has led to intensive development of marine resources over the past half century. It can be argued that not only are commercial considerations not necessarily a useful engine for promoting the deep seabed, but also a break with the framework of exploitation within areas of national jurisdiction is an absolute requirement, given that the foundations for activities in the Convention Area are so radically different from those in the exclusive economic zone or on the continental shelf.

But there is a third consideration: Nonliving resources in the Convention Area are one of the last repositories of their kind left in the natural world, left relatively undisturbed by humans. As the twentieth century enters its final decade, we are once again conscious of the fragile nature of our dependence on nonliving resources. In many countries, the public is becoming profoundly conscious of the misplaced liberality with which nonliving resources have been expended, often to a degree that is far advanced of their governments. These sentiments provide firm support for policies not merely of conservation but also of avoiding wasteful practices. In these countries, the public must now be convinced that certain types of development are required in the general public interest, as opposed to the needs or wishes of sectors of the community. Countries have, in short, repudiated the dangerous concept of the inevitability of development—that it has a life of its own and that this life is tied to market forces, employment opportunities, social and regional development, and so on.

Those who oppose unilateral development of the deep seabed rest their argument on firm principles. It would not be stretching the case too much to predict that some of these opponents would, nevertheless, be willing to see unnecessary development of these resources for reasons such as the distribution of the wealth to developing countries. In such an event, principles would play little part in the discussion, and necessity would take over. A decision to permit development in such cases might be totally appropriate but would be taken for questionable reasons. It cannot be argued that such an approach bodes anything but ill for the protection and conservation of deep seabed resources and the marine environment. To decide to exploit resources in order to redistribute wealth does not make the decision a principled one.

This, then, is the difficulty: The marine environment may come under threat from both international and unilateral proponents of operations involving nonliving resources of the deep seabed. The debate between these two camps is one thing; the protection of the marine environment is quite another. But it should not be assumed that once international control has been secured, marine environmental protection will inevitably follow. It will not and cannot do so unless the ethical foundation for this development differs significantly from that which dominates the exploitation of the areas within national jurisdiction.

The best approach to safeguarding the resources of the sea under international control is both to extend and to narrow the context in which they are viewed. The extension should come about by including them in a total itinerary of the world's resources and attempting to match depletion with demonstrably justified needs. The narrowing is required by stripping the discussion of the rhetoric while leaving the principles attached. However, it is only by the insistence of adherence to the principles of optimal resource management, coupled with an appreciation of the broader context of the requirements of underdeveloped countries, that a justifiable ethical position can be evolved, one that its framers can be confident of defending to their grandchildren.

The search for a principle to serve as a foundation for policies governing the high seas is driven by the identification of factors similar to those that are in themselves the ingredient of such policies and the laws that implement them. During the past half century, one of the most striking aspects of the development of both policy and law as regards the environment and resource exploitation has been the steady erosion of *laissez-faire* attitudes, or total freedom of choice, in favor of forms, regulations, and restricted choice. This does not mean that a decision has to be made between freedom and absolute regulation. The two can coexist, but as the degree of intervention increases, so does the need to formulate a philosophy of regulation that provides consensus on which areas of activity will be subject to regulation, which areas will be exempted, and why. It is the failure to evolve such a philosophy and therefore to justify intervention or nonintervention that destroys the possibility of the evolution of a steady ethical foundation for policy-making.

The central difficulty is that governments, producers, and consumers wish to postpone indefinitely long-term commitments to change. Although it is by no means desirable, or even possible, that government action should regulate or restrict choice as a means of exercising control, the inability or unwillingness to formulate a viable, liberal alternative only serves to bring the draconian even closer. In the international arena, the outcome is not so much authoritarian as immoderate in nature. Such lack of moderation may take a number of forms, but blind self-interest is surely the most common. It is difficult to explain the attitude of the European Communities to the straddling stocks question on the Grand Banks of Newfoundland, or those countries that permit drift net fishing in the Pacific, in any other way.

Although it is possible to inveigh against selfishness and self-interest such as this, it is surely also necessary to inquire more closely into the forms that it takes and the conditions that permit it (if not encourage it) to flourish. The leading characteristic is a combination of evident domestic benefits and barely discernible domestic drawbacks. The benefits are obvious: One way of replacing a domestic supply of fish extinguished by reprehensible depletion policies of previous years is to find alternative sources and start depleting them. Any questions concerning the viability of the domestic fishing industry can be conveniently shelved until an indefinite future. At the same time, there is little chance of adverse domestic repercussions. Indeed, the only source of domestic criticism with respect to the degradations of straddling stocks or the practice of drift net fishing would be those arguing against wasteful ocean exploitation practices. The chance of these arguments making any headway are comparatively remote, and the domestic political dangers posed by them are virtually nonexistent. The facts here speak for themselves.

Such views betray an evident double standard. There is one rule for exploitation of resources within the areas subject to national jurisdiction, but a very different set of rules apply to those without. The same is true of attitudes toward the marine environment; protection of shoreline and living resources seems to concentrate the regulatory mind too wonderfully, in contrast to practices either ignored or actively promoted in remote areas.

Such views are not only inconsistent but also fallacious. The fallacy lies in the assumption that the artificially determined limits to national jurisdiction have some natural function, in that it is possible to contemplate inconsistent policies on either side of the magic line. The objections to these practices take many forms—moral-ethical and conservationist. If it is true that activities contemplated with equanimity in the areas beyond national jurisdiction may adversely affect or take place within those limits, then the degree of self-delusion that acquiesces in such activities is little short of criminal. In the oceans, sovereign states have passed well beyond the point at which intervention in the uses of the sea, and the exploration of these resources, can be described as minimal. They have certainly passed the threshold below which a holistic philosophy of regulation and use is not required. Such a philosophy *is* required. If policies within the areas of national jurisdiction are in line with conservation and control, then the same philosophy must apply to the areas beyond.

NOTES

1. United Nations Convention on the Law of the Sea, Dec. 10, 1982, U.N. Doc. A/Conf.62/122, 21 I.L.M. 1261 (1982) [hereinafter 1982 Convention].
2. Part XI of the 1982 Convention lays out the regime for the deep seabed, or the "Convention Area," as it is referred to throughout the convention (article 1, paragraph 1). Section 2 is titled "Principles Governing the Area." It is this section that contains the controversial article 136, which states that "[t]he Area and its resources are the common heritage of mankind."
3. *See, e.g.*, D. M. JOHNSTON, CANADA AND THE NEW LAW OF THE SEA (1986), for a discussion of this "new" regime in the context of the existing ocean policies of Canada.
4. The first United Nations Conference on the Law of the Sea in 1958 was attended by 86 states, and roughly 150 participated in the final sessions of the Third UN Conference in 1982.

5. *See, e.g.*, Clarke & Remmer, SATURDAY EVENING POST, Sept. 30, 1945 (the view of two academics at the time in the popular media).

6. 1982 Convention, *supra* note 1, pt. V, arts. 55–75.

7. Proclamation No. 2667, 10 Fed. Reg. 12,303 (1945).

8. *See generally* CONSENSUS AND CONFRONTATION: THE UNITED STATES AND THE LAW OF THE SEA (Jon M. Van Dyke ed. 1985).

9. Flax, *Postmodernism and Gender Relations in Feminist Theory*, in FEMINISM/POSTMODERNISM (L. Nicholson ed. 1990).

10. WORLD COMMISSION ON ENVIRONMENT AND DEVELOPMENT, OUR COMMON FUTURE (1987).

11. The subsequent discussion of postmodern and poststructuralist theory owes much to the ideas developed in Baldwin, *Deconstructive Practice and Legal Theory*, 96 YALE L.J. 743 (1987); Dalton, *An Essay in the Deconstruction of Contract Doctrine*, 94 YALE L.J. 997 (1985); Scott, *Deconstructing Equality-versus-Difference: Or the Uses of Post-Structuralist Theory for Feminism*, 14 FEMINIST STUD. 33 (1988); Kennedy, *A New Stream of International Law Scholarship*, 7 WIS. INT'L L.J. 1 (1988); and Kennedy, *Theses on International Law Discourse*, 23 GERMAN Y.B. INT'L L. 353 (1980). For a general introduction, *see* NORRIS, DECONSTRUCTION: THEORY AND PRACTICE (1982); and CULLER, ON DECONSTRUCTION: THEORY AND PRACTICE AFTER STRUCTURALISM (1982). *See also* HUTCHINSON, DWELLING ON THE THRESHOLD: CRITICAL ESSAYS ON MODERN LEGAL THOUGHT (1988).

12. This predisposition becomes intuitively obvious if one considers the paired opposites that dominate everyday discourse: inside-outside, unity-diversity, person-thing, man-woman, and so on. The favored concept is often then used as a foundational principle for theory building because it embodies what Derrida referred to as "presence," the quality of being most apparent to the consciousness.

13. The "healthy" sign . . . is one which draws attention to its own arbitrariness—which does not try to palm itself off as "natural" but which, in the very moment of conveying a meaning, communicates something of its own relative, artificial status as well. The impulse behind this . . . is a political one: signs which pass themselves off as natural, which offer themselves as the only conceivable way of viewing the world, are by that token authoritarian and ideological. It is one of the functions of ideology to "naturalize" social reality, to make it seem as innocent and unchanging as nature itself . . . . Ideology, in this sense, is a king of contemporary mythology, a realm which has purged itself of ambiguity and alternative possibility.

   EAGLETON, LITERARY THEORY: AN INTRODUCTION 135 (1983).

14. *See* DAVID KAIRYS, THE POLITICS OF LAW (2d ed. 1990), for an excellent introduction to critical legal theory.

15. Laurence H. Tribe, *Ways Not to Think About Plastic Trees: New Foundations for Environmental Law*, 83 YALE L.J. 1344 (1974).

16. Christopher Stone, *Should Trees Have Standing?—Towards Legal Rights for Natural Objects*, 45 S. CAL. L. REV. 450 (1972).

17. 1982 Convention, *supra* note 1.
18. " 'Area' means the sea-bed and ocean floor and subsoil thereof, beyond the limits of national jurisdiction."
19. *See* U.S. Deep Seabed Hard Minerals Resources Act of 1980, 30 U.S.C. § 1401 *et seq.* (1982).

VI

Military Activities
and Peaceful Uses
of the High Seas

THE MAJOR MARITIME POWERS have always used the high seas for military activities, and the current era is no exception. The navies of the United States and the former Soviet Union have used the oceans for military maneuvers of all sorts and also for the testing of ballistic missiles. The United States, the United Kingdom, and France have all used atolls in the Pacific for the atmospheric testing of nuclear bombs, and the French were continuing to conduct underground nuclear tests in the volcanic atolls of French Polynesia until they "suspended" this activity in April 1992.

Although the ending of the cold war may reduce the tensions created by these naval activities, the threats to the marine environment continue, and the legal issues raised by unrestrained military activity on the high seas are challenging ones. Andrew Mack's chapter begins this part by providing an overview of the naval situation and an examination of how the problems of a "commons" become particularly difficult to unravel when national security interests are at stake. Joshua Handler focuses on the threats to the marine environment created by naval vessels, particularly nuclear-powered and nuclear weapons vessels. Joseph Morgan presents a survey of the world's navies and compares the risks to the environment they create with the risks created by other sources of pollution. Jon Van Dyke examines the claim by the major maritime powers that they can create "exclusionary" or "warning" zones on the high seas—and thereby exclude other uses of these areas—in order to conduct their ballistic missile tests or nuclear bomb tests.

These chapters provide new insights into the risks created by military activities on the high seas and offer suggestions on how to reduce these risks and thereby protect the marine environment.

26 Security Regimes for the Oceans: The Tragedy of the Commons, the Security Dilemma, and Common Security

Andrew Mack

AN UNDERLYING THEME OF THIS CHAPTER is the central human dilemma spelled out by Garrett Hardin's depiction of the "tragedy of the commons,"[1] which demonstrates how individual actors, each pursuing their "rational" self-interest, may create a situation in which all lose. In Hardin's 1968 allegory, a group of cattle-owning villagers have access to a finite area of common grazing land. The tragedy arises because each villager recognizes that if he adds an extra cow to the commons, he will gain *all* the benefit (in increased yields of milk, hides, meat, etc.) of rearing the extra animal, while the community as a whole will share the costs its pasture consumption imposes on the commons.

In Hardin's example, a "free market" situation exists—there is no authority to enforce or even create rules as to how many cattle should be allowed to exploit the commons. The "tragedy of the commons" is actually closer to the reality of an international system, in which there is no international authority and international law is relatively weak, than to intrastate relations, in which some form of authority normally exists and national and local laws are usually binding and enforceable. Moreover, in the situation described by Hardin, one may assume that the villagers share a common culture and values, which should facilitate cooperation. In the international system, this assumption cannot be made.

The classic examples of "commons" tragedies in the international system are found in the devastating consequences of uncontrolled marine pollution—especially in closed seas like the Mediterranean and Baltic—and in overexploitation of ocean fisheries.

In the absence of cooperation or effective and enforceable ocean use regulation, and with the growth of the high-technology mass-harvesting techniques of "industrial" fishing, overfishing of the world's oceans, once seen as inconceivable, has become a reality. Restraint on high seas operations by one fishing nation tends to encourage "free riding" by others—which discourages restraint in the first place. In the case of pollution, "market" imperatives can have a similarly disastrous consequence. Imagine ten competing wood-chipping plants located on a large lake. Management and workers in all the plants would prefer a nonpolluted lake, but competition is fierce, and pollution control systems for wood chip plants are expensive. Any one plant that implements pollution controls while its competitors do not will be at a competitive disadvantage—and could be forced out of business. As a result, no plant institutes pollution controls, and the lake "dies." Had cooperation been possible, or had enforceable pollution control regulations been instituted (i.e., no possibility of free riding), no plant would have been economically disadvantaged compared with the others, and the lake would not have been killed. Once again, what may be "rational" for an individual actor under market conditions may be dysfunctional for society as a whole.

THE SECURITY DILEMMA

Is there any parallel in the realm of ocean security to the dilemmas that the "tragedy of the commons" syndrome pose for ocean resource management? The so-called security dilemma is certainly one possible analogue.

The security dilemma arises when nation-states seek to maximize their security via policies of "peace through strength"—that is, by creating a military capability that will enable them to defeat (either alone or in concert with allies) any opponent bent on aggression. However, a global system in which all states seek security by increasing their military power vis-à-vis potential opponents is prone to arms races and is inherently unstable. It is logically impossible for both antagonists to feel secure at once.

First, the greater the offensive capability of the antagonists, the greater the degree of mutual suspicion and fear. By "offensive" I mean those forces that can be used to launch major assaults against the homeland of an opponent.

Second, the forces necessary for what might be called "offensive defense" are identical to the forces necessary for aggression. In a situation of increasing fear, suspicion, and hostility, worst-case defense planners on one side invariably see the offensive force structures and strategies of opponents as evidence of aggressive intentions.

Third, the nature of offensive forces is such that there is often a considerable strategic advantage to be gained by preemption—that is, by shooting first—in a crisis. This is particularly true in the modern age, when preemptive strikes against an opponent's command, communication, control, and intelligence assets can cripple the opponent's military capability. So if either side believes that war is imminent—*or if it believes that its opponent believes that war is imminent*—a war may be started that neither side originally wanted.

Fourth, arms races increase the absolute level of armaments on both sides. This in turn increases the human cost of any wars that break out, while also increasing the economic burden of peacetime military expenditure. The domestic social and political costs induced by high levels of defense expenditure may themselves become a cause of insecurity.

The net effect of the security dilemma is that the states that are subjected to it become both poorer and less secure. As with the tragedy of the commons, we can see how states acting in their "rational" self-interest can produce outcomes that are in the interest of none of them.

THE SECURITY DILEMMA AT SEA

The security dilemma may manifest itself in a number of different ways in a maritime environment—some of them particularly risk prone. The high seas are, after all, the only part of the earth's surface in which national boundaries do not inhibit the movement of military forces. Antagonistic states that are physically separated from each other on land interact closely at sea—often practicing warfare at only a few hundred meters' distance. In addition, although they are invariably defended as intended to enhance deterrence and thus reduce the threat of war, offensive naval exercises could, in a crisis, trigger the very hostilities they are supposed to prevent.

What might be done to reduce the risks imposed by the security dilemma as it affects the oceans?

SOLUTIONS

If we return to the dilemma of the commons, we note that there are a number of possible solutions to the problem. They are as follows:

1. *Community ownership*, whereby the cattle are publicly owned and the profits from keeping them are distributed among the villagers.[2]
2. *Partition*, whereby the common land is divided into small plots assigned to individual villagers. Under this system, there is no incentive for the villager to introduce more cattle than the plot can carry.

3. *Voluntary restraint*, whereby each villager agrees to restrain the number of cattle that are introduced into the commons.
4. *Regulation*, whereby a set of rules is either agreed to or imposed. Regulation then governs the number of cattle that can be introduced into the commons. For this system to work, there has to be overview ("verification" in the language of arms control) and inducements to compliance.

These possible solutions, which also have relevance for the maritime security dilemma, are considered in the sections that follow. The least practicable solutions are considered first.

Community Ownership

It is not clear what it would mean to talk about "community ownership" of the oceans—although one might wish to talk about international ownership of the ocean's *resources*. The obvious difficulty with the concept of community ownership of the oceans would be in deciding how to construct a distribution mechanism for the harvest of those resources.

Ocean resources already "belong" to the international community, at least in the sense that (outside 200-mile EEZs) they are theoretically open for all to exploit. In practice, of course, only nations able to afford modern fishing fleets, underwater mining and drilling technology, and the like can exploit ocean resources effectively.

Community ownership would also seem to presuppose some form of world government—which is hardly a realistic prospect for the foreseeable future.

Partition

Partition of the oceans—the "ocean enclosure movement," or "creeping jurisdiction"—is a process that has been under way for centuries. Indeed, "[s]ince Grotius' time, the ocean enclosure movement has claimed nearly one-third of global ocean space, with the greatest 'gains' having taken place since 1945."[3] The most obvious recent example of creeping jurisdiction is in the rapid proliferation of 200-mile EEZs over the past decade.

There is, however, a downside to the enclosure movement. Because states now have 200-mile EEZs to protect, they also have a new naval mission. That will require more ships and maritime patrol aircraft. In the Pacific, a perceived need to safeguard EEZs has led to a major increase in surface combatants. Take the case of frigates and patrol boats—the naval vessels most suited to EEZ patrol tasks. Between 1980 and 1989—the period in which 200-mile EEZs were being most rapidly introduced—the following increases took place:

	Frigates		Patrol Boats	
	1980	1989	1980	1989
China	17	37	678	915
Japan	15	57	26	14
South Korea	7	17	57	79

Similar increases took place in the Australian and most Association of Southeast Asian Nations (ASEAN) navies.

There is a further problem here—namely, surveillance. Most states, especially Third World states with large EEZs, simply do not have the numbers of ships or maritime patrol aircraft (airships may be a better bet) necessary to monitor their zones effectively.

Over-the-horizon (OTH) radar might, however, be able to provide the surveillance capability that is necessary. Indeed, one network of OTH radars could in principle provide surveillance for a number of states' EEZs. Because of the long return signal "skip" distance, OTH radars may not detect shipping at distances of *less* than 1,000 to 1,500 kilometers. This means that an OTH radar may well have to be located in country B in order to provide surveillance of the EEZ of country A, and vice versa. This suggests that EEZ surveillance (the critical ingredient for protection) may best be achieved on a cooperative, multi-lateral, and regional basis rather than on a national basis.

If navies saw their sole task as protecting their territorial and EEZ waters, the risks of naval confrontation on the high seas would disappear. But the major powers have no intention of reducing the scope and function of their naval forces. The high seas remain for them a realm from which they can project—or threaten to project—naval power against adversaries. The ability to do this is seen as enhancing deterrence and thus reducing the risk of aggression. Sea lines of communication (SLOCs), along which vital trade flows, also pass through the high seas. Blue water navies are said to be necessary for SLOC protection.

These claims are questionable. First, even at the height of the first and second cold wars, the overwhelming security imperative of the superpowers was not to enhance deterrence, which was robust and needed no enhancing, but to avoid inadvertent war. Offensive naval strategies, for the reasons suggested earlier, were antithetical to this goal.

Second, the utility of offensive naval power in Third World contingencies is difficult to establish—quite apart from any moral reservations one might have about its use. What determines the utility of such missions is less their success in achieving their immediate tactical goal than their long-term political consequences. In these terms, use of U.S. naval power in Lebanon was disastrous; in Libya, dubious. More generally, a number of academic studies have raised questions about the overall efficacy of coercive naval diplomacy.

Third, the vulnerability of seaborne trade to military interdiction has been much exaggerated. The submarine is the key weapons platform here, but the extraordinarily high price of modern submarines has meant that the number of submarines in the hands of major powers is sharply down compared with that during World War II, while the number of blue water merchant ships is sharply up. Moreover, SLOCs are almost impossible to protect along their total length. There are simply too many merchant ships to form convoys and too few escort vessels to protect them. The most sensible way to protect maritime trade from interdiction is to protect the "focal areas" of the ports of disembarkation and embarkation and have the cargo-carrying ships adopt the tactic of "evasive routing" between ports.

Evasive routing will, however, become more problematic if increasing numbers of states acquire over-the-horizon radar capabilities. Once this happens, ships on the high seas *will* be detectable and thus potentially vulnerable. There is little that a single state—even one as powerful as the United States—can do about this. Again, the solution would appear to lie in the realm of cooperative— rather than national—security.

Voluntary Restraint

The suspicion and hostility that characterize serious conflicts and give rise to arms races make voluntary restraint difficult. Verification of such restraint is almost impossible without a *negotiated* arms control verification regime (i.e., some form of regulation). In the absence of such a regime, "worst-case" thinking will tend to prevail. In the name of strategic prudence, worst-case thinking tends to overemphasize the military strength of potential or actual adversaries—the classic cases are the so-called bomber gap of the 1950s and the "missile gap" that followed it—and also tends to assume the worst of the adversary's intentions.[4]

Yet, as was the case between the United States and the Soviet Union, there may be occasions when voluntary restraint—sometimes called "unilateral" or "nonnegotiated" arms control—does appear to work. Between 1987 and 1990, the Soviets deactivated three cruisers, nine frigates, ten submarines, and three corvettes from the Soviet Pacific Fleet. The U.S. Navy also reduced its naval forces in the region. Although these reductions certainly reflected budgetary pressures, they would not have happened, at least not as rapidly, if political relationships between the superpowers had not improved dramatically.

This is also the problem with voluntary restraint: It is a hostage to the good relationship between the superpowers.

Regulation

As with other forms of ocean stewardship, regulation seems currently to offer the most realistic and hopeful approach to reducing, if not resolving, the security dilemma. Regulation in this case means the creation of security regimes incorporating naval arms control and confidence- and security-building measures (CSBMs).

There are many obstacles to progress in this area. First, the arms control paradox suggests that arms control regimes are most difficult to implement when they are most needed (i.e., in times of crisis, when suspicion and hostility are high). The arguments for negotiating CSBMs when political relationships are good are that (1) it is easier to do so when times are good than when times are bad, and (2) a negotiated CSBM regime reduces the risk of inadvertent war in times of high tension and crisis.

It should be noted here that CSBMs are *not* a panacea. They do not address the root causes of conflicts that divide nations, and they have only limited relevance in respect to unprovoked aggression. CSBMs can do little to prevent the rise of Hitlers, or Saddam Husseins, although even in these cases a CSBM regime can provide early warning of aggressive intent and reduce the risk of surprise attack.

In the Pacific, the former Soviet Union and its allies strongly supported the negotiation of naval CSBM regimes; the United States and its major ally, Japan, have rejected or ignored all of the initiatives. Australia and Canada have made cautious and qualified affirmations of support for very modest CSBM regimes.

The United States produced a series of arguments against negotiation of arms control agreements in the Pacific and against naval CSBMs in particular. For example:

1. In the Pacific, both superpowers had a series of bilateral security alliances with regional states—in contrast with the bloc system in Europe. These alliances complicate negotiations.
2. Superpower force structures in the region are asymmetric: The Soviets were weak at sea but strong on land. Knowing what Soviet weapons systems to trade off against quite different U.S. systems in force level reduction talks was thus very difficult, if not impossible.
3. Important territorial disputes remain unresolved in the region. The

Helsinki/Stockholm Conference on Security and Cooperation in Europe (CSCE) was predicated on acceptance of East-West borders. In the Pacific, territorial and sovereignty disputes may be barriers to arms control negotiations. Japan, for example, refused to negotiate CSBMs with the Soviet Union until a satisfactory solution to the so-called Northern Territories dispute was reached.[5]

4. It was claimed that it makes little sense to negotiate arms control and CSBM agreements at the regional level. Other negotiations—START, chemical weapons, and the like—should take priority over naval arms control.[6]

5. Arms control was said to be an alien and largely unwanted concept in Asia.

6. Soviet arms control proposals were blatantly one-sided. If implemented, they would put the United States at a military disadvantage.

7. Because the political climate between the superpowers improved, there was no need for naval CSBMs to be negotiated.

8. Arms control, particularly CSBMs, was said to be contrary to the doctrine of the freedom of the seas because it would involve constraints on U.S. operations on the high seas.[7]

There is some point to most of these arguments—except for the last. Arms control in Asia will, of course, be different from that in Europe. The issues *are* more complex and, sometimes, more difficult. But this does not mean that arms control is either impossible or undesirable.

Two further reasons for U.S. opposition to naval arms control have not been articulated publicly. First, the United States has military superiority in the region, it feels that security is maintained by keeping that superiority, and it fears that negotiated arms control might lead to a push for a greater degree of equality in the military balance—as happened during the CFE negotiations in Europe. Even innocuous arms control measures, like the South Pacific Nuclear Free Zone Treaty, have therefore been rejected on the grounds that they constitute the "thin end of the wedge"—leading the United States toward the "slippery slope" of full-scale naval arms control. Second, the U.S. Navy cannot yet envisage an appropriate and coherent role for itself in the post-cold war era. The navy prefers to dwell on the familiarities of yesterday's problems rather than examine the requirements of tomorrow's solutions.

Notwithstanding the ongoing opposition of the U.S. Navy, there is little doubt that naval arms control for the Pacific *is* moving up the political agenda. It has been the subject of increasing numbers of conferences, workshops, and research projects—involving officials as well as scholars. It is being studied seriously in the office of the Secretary of Defense in Washington, DC, at CINCPAC, at the RAND Corporation, and at the Center for Naval Analyses.

Moreover, the United States has already negotiated two important CSBMs affecting the navy—the 1972 Incidents at Sea Agreement and the 1989 Prevention of Dangerous Military Activities Agreement.

The sort of modest naval CSBMs that could be implemented in the short term in the Pacific include the following:

1. Agreements to exchange data on force levels, weapons platform-building programs, retirement programs, and so forth.
2. Agreements for the advance notification of agreed categories of exercises.
3. Exchange of observers on agreed categories of exercises.
4. Institutionalization of ongoing high-level dialogue on military doctrine and the particular concerns each side has about the other's strategy and force structure.

The United States has argued vigorously that the Helsinki/Stockholm CSCE process is not relevant to the Pacific, which is a very different security environment from that in which the CSCE agreement was negotiated. This argument might seem to have a point. It would indeed be inappropriate to impose a model derived from the land confrontation in Europe to maritime theaters of the Asia-Pacific region. But none of the proponents of CSBMs has proposed doing this.

The Asia-Pacific region is different for all of the reasons just noted and because maritime security issues are of far greater salience in this region than in Europe. Moreover, when thinking of security regimes for the region, there is a growing consensus that a single, overarching regime—along the lines of the CSCE—would not be appropriate. Security regimes should evolve, or be negotiated, on a subregional basis (i.e., northeastern Asia, Southeast Asia, and the Southwest Pacific).

But although the Europe-derived CSCE *model* may not be appropriate to the Pacific, the security philosophy that underpins it certainly is. This philosophy is based on very different assumptions from those that guided NATO and the Warsaw Pact for forty years and that sought to achieve security by promoting policies of "peace through strength" and deterrence—policies that exacerbated suspicion and hostility, created incentives for arms races, and undermined crisis stability.

CSCE is predicated on the principle of *common security*. It is a multilateral rather than bilateral regime that embraces the former adversaries—the NATO and Warsaw Pact countries—and the neutral and nonaligned states of Europe as well. CSCE is an *in*clusive security regime. NATO and the Warsaw Pact were based on the principles of exclusion, alliance confrontation, and national, rather than common, security.

The security goal of the rival cold war alliances—"peace through strength" and deterrence—meant that when one alliance felt secure, the other, almost by definition, felt insecure. The aim of the CSCE security regime is to increase *mutual* security by radically reducing the risks of inadvertent war—the least improbable cause of war in Europe.

The issue of possible aggression is not ignored by CSCE—indeed, the CSBMs that form a core part of the regime also provide early warning of aggressive intent and reduce the risks of surprise attack. With respect to resisting aggression, the "common security" philosophy stresses the need for "non-provocative defense" strategies and for force structures that are strong on the defensive but have only modest offensive capabilities. Such structures, it is argued, offer an adequate deterrent and a strong defense. They also eradicate completely the fear and suspicion, the incentives for arms races, and the prospects for preemptive strikes that are so characteristic of offensive strategies.

Although there is now growing discussion in the Pacific about the need for modest naval CSBM regimes to be introduced into the region, the more far-reaching concepts of common security and nonprovocative defense, which are now part of the mainstream discourse on security in Europe, are still virtually unknown.

A major task for security analysts over the next decade will be to think creatively about what common security regimes might look like in the Asia-Pacific region and how the strategies of nonprovocative defense, which were developed to deal with a land confrontation on the Central Front in Europe, may be adapted to the maritime theaters of the Pacific.

NOTES

1. *See* Garrett Hardin, *The Tragedy of the Commons*, 168 SCIENCE 1243 (1968).
2. *See* Marvin S. Soroos, *The Tragedy of the Commons in Global Perspective*, in THE GLOBAL AGENDA: ISSUES AND PERSPECTIVES (Charles W. Kegley Jr. & Eugene Witkopf eds. 1984), for a useful analysis of the commons problem in global perspective and a discussion of possible solutions.
3. Lewis M. Alexander, *The Ocean Enclosure Movement: Inventory and Prospect*, 20 SAN DIEGO L. REV., 561 (1983), at 561.
4. In the 1950s, the United States, which then lacked the sophisticated satellite surveillance techniques that it developed subsequently, came quite erroneously to believe that the Soviets had a massive lead first in strategic bombers and later in missiles. This perception provided a major impetus for expansion of U.S. strategic programs, which in turn provided the Soviets with an additional rationale for expanding *their* nuclear programs.
5. This argument is increasingly outdated. China, India, the former Soviet Union, and

South Korea have established diplomatic relations. North Korea, Japan, China, Vietnam, and Indonesia are all moving to normalize their relations.

6. There is in fact no reason for refusing to negotiate CSBMs on a regional basis—the CSCE was, after all, a regional CSBM agreement.

7. This total non sequitur is still occasionally argued by U.S. Navy spokespersons. Because arms control would not be imposed on any party but would be the consequence of an agreement voluntarily negotiated and signed, it would not be contrary to the principle of "freedom of the seas."

27 Denuclearizing and Demilitarizing the Seas

Joshua Handler

THE MILITARIZATION OF THE SEAS has a long and prosperous tradition. Competition among the great powers from antiquity to modern times inevitably spilled over into the oceans, and large and widely deployed naval fleets were the result. In the postwar period, superpower competition provided the major impetus for militarization of the oceans.

The most threatening element of these navies, both to the countries involved and to the ocean environment, has been the thousands of nuclear weapons and hundreds of nuclear reactors aboard their ships and submarines. Besides the possibility of their use, nuclear weapons pose other environmental dangers. In routine superpower naval operations over the past thirty-six years, it is estimated that some fifty nuclear weapons have been lost at sea.[1] Naval nuclear reactors do not have the obvious destructive power of nuclear weapons, but they greatly assist a nation's ability to mount offensive operations far from its shores. Also, naval reactors create severe environmental hazards. As a result of accidents, some twenty-three naval nuclear reactors are on the ocean floor, mostly from nuclear-powered submarines. Hundreds of decommissioned reactors will need to be carefully disposed of by the end of the decade.

With the end of the cold war, the superpower and allied navies have experienced a process of "spontaneous disarmament." Reduced budgets and tensions have conspired to shrink fleet sizes and diminish the number of nuclear weapons and nuclear reactors at sea. This chapter examines recent trends in naval nuclear forces, naval nuclear propulsion programs, and total fleet sizes and discusses the opportunities these developments create for denuclearizing and demilitarizing the seas.

NAVAL NUCLEAR WEAPONS

During the cold war, attention centered on the land-based intercontinental ballistic missiles (ICBMs), bombers, and nuclear weapons in Europe. The arms race at sea was out of sight and therefore out of mind. Yet almost one-third of the world's nuclear weapons were available to naval forces. By 1991, approximately 13,900 nuclear weapons were assigned to the U.S. and Soviet navies and another 600 to the British, French, and Chinese navies. The widespread deployment of naval weapons also added a new political dimension to the global effects of the superpower arms race. Considerably more mobile than their land-based counterparts, U.S. naval nuclear weapons were routinely carried into all of the world's oceans and into the ports of dozens of countries.[2] By comparison, U.S. land-based nuclear weapons were deployed in only half a dozen or so countries in Europe and in South Korea.

The majority of naval nuclear weapons, approximately 9,100, were based on long-range missiles carried by the 106 strategic ballistic missile submarines operated by all five navies. The remaining 5,400 nuclear warheads were nonstrategic weapons, encompassing nuclear torpedoes and antisubmarine rockets carried by ships and submarines; nuclear surface-to-air missiles carried by ships; nuclear depth and strike bombs for delivery by ship and land-based aircraft; nuclear antiship missiles launched by submarines, ships, and aircraft; and long-range nuclear land-attack sea-launched cruise missiles. Approximately 660 ships and submarines were able to deliver nonstrategic naval nuclear weapons (see tables 27.1 and 27.2).

During the early and mid-1980s, superpower tensions were on the rise, and this was reflected in the nuclear arms race at sea. The U.S. Navy developed several new nuclear weapons, ranging from short-range antisubmarine warfare weapons to the strategic Trident II D-5 submarine-launched ballistic missile. In addition, the United States began deploying nuclear-armed Tomahawk long-range sea-launched cruise missiles in 1984. The Soviet Union also deployed new nuclear weapon systems. Nuclear torpedoes, cruise missiles, antiship missiles, and strategic submarine-launched ballistic missiles entered the fleet.

NONSTRATEGIC NAVAL NUCLEAR WEAPONS

The rapid pace of the nuclear arms race at sea began to abate by the late 1980s. With superpower tensions diminishing, the size of naval nuclear forces, although still large, also declined. During 1988–1990, the United States retired 1,100 warheads for the nuclear Terrier surface-to-air missile and the ASROC[3] and

TABLE 27.1 Naval Nuclear Weapons

	United States	USSR	United Kingdom	France	China	Total
Strategic missile warheads	4,912	3,700	96	400	26	9,134
Subtotal	4,912	3,700	96	400	26	9,134
Nonstrategic warheads						
Cruise missiles	350	600	0	0	0	950
Aircraft bombs	625	450	25	36	0	1,136
Air-to-surface missiles	0	750	0	0	0	750
Antisubmarine weapons	900	1,300	25	0	0	2,225
Anti-air weapons	0	200	0	0	0	200
Coastal missiles	0	100	0	0	0	100
Subtotal	1,875	3,400	50	36	0	5,361
Total	**6,787**	**7,100**	**146**	**436**	**26**	**14,495**

Source: Adapted from JOSHUA HANDLER & WILLIAM M. ARKIN, NUCLEAR WARSHIPS AND NAVAL NUCLEAR WEAPONS 1990: A COMPLETE INVENTORY (Greenpeace Sept. 1990).

SUBROC[4] antisubmarine rockets. The Soviet Union also retired older nonstrategic nuclear weapons, along with the ships and submarines that carried them. In mid-October 1990, the Soviet Foreign Ministry announced that the ships and submarines of the Baltic Fleet were denuclearized.[5]

In addition to reducing nuclear forces, nations slowed or canceled the deployment or development of new weapons. None of the half-dozen ocean combat nuclear weapons that the United States had on the drawing boards from the late 1970s to the mid-1980s survived. The U.S. Congress canceled the last weapon, the U.S. Navy's B90 nuclear depth/strike bomb, in the fiscal year 1991 budget. The Soviet Union halted development of the long-range SS-NX-24 land-attack cruise missile and was deploying the SS-N-21 submarine-launched long-range cruise missile more slowly than late-1980s Western intelligence estimates had anticipated. The United Kingdom decided to delay production of a new nuclear depth bomb.

Coincident with these events, the Soviet Union made a concerted effort to eliminate tactical nuclear weapons at sea through arms control proposals. President Mikhail Gorbachev first broached this idea at the U.S.-Soviet Malta Summit in December 1989. The idea of controlling nuclear weapons at sea also gained currency in the West. In 1989, U.S. government advisor extraordinaire Paul Nitze suggested that these weapons could be removed.[6] In January 1990, the recently retired chairman of the Joint Chiefs of Staff Admiral William Crowe told the *Washington Post* that it could be worthwhile to entertain negotiations

TABLE 27.2 Nuclear-Capable Ships and Submarines (1991)

	United States	USSR	United Kingdom	France	China	Total
Submarines						
Ballistic missile	33	62	4	5	2[a]	106
Cruise missile	0	31	0	0	0	31
Attack	57	138	0	0	0	195
Total submarines	90	231	4	5	2	332
Surface ships						
Aircraft carriers	12	5	3	2	0	22
Battleships	2	0	0	0	0	2
Cruisers	19	31	0	0	0	50
Destroyers	17	32	12	0	0	61
Frigates	0	97	17	0	0	114
Patrol combatants	0	81	0	0	0	81
Total surface ships	50	246	32	2	0	330
Total	**140**	**477**	**36**	**7**	**2**	**662**

Source: Adapted from JOSHUA HANDLER & WILLIAM M. ARKIN, NUCLEAR WARSHIPS AND NAVAL NUCLEAR WEAPONS 1990: A COMPLETE INVENTORY (Greenpeace Sept. 1990).

[a] China has one Golf-class diesel-powered ballistic missile submarine used for testing and training, not included in this total, that also could be nuclear capable.

about the removal of nonstrategic nuclear weapons.[7] In the spring of 1990, Senator Edward Kennedy's Armed Services Subcommittee held hearings on the possibility of naval nuclear arms control.

Until September 1991, the administration of President George Bush steadfastly refused to address these proposals. Then, on September 27, 1991, in a major reversal, President Bush announced that the United States would do the following:

1. Withdraw all tactical nuclear weapons from its surface ships and attack submarines, as well as nuclear-depth bombs used by land-based antisubmarine warfare planes.
2. Not carry tactical nuclear weapons at sea in peacetime.
3. Dismantle or destroy many of these weapons.
4. Store remaining weapons in major U.S. depots.[8]

A week later, President Gorbachev responded in kind. He announced that besides other reductions in strategic and tactical nuclear weapons, "[a]ll tactical nuclear weapons will be removed from surface ships and multipurpose submarines. These weapons, as well as weapons from ground-based naval aviation, will be stored. Part of them will be destroyed."[9] He further proposed the destruction of all tactical naval nuclear weapons on a reciprocal basis.

The removal of nuclear weapons from U.S. ships is proceeding apace. The last weapons were estimated to be withdrawn in the late spring of 1992. As for the size of nuclear forces, the two presidents' proposals will substantially shrink their navies' tactical nuclear arsenals. Former president Bush's plan reduced the U.S. force from 1,875 to 975 nuclear weapons. Of particular note, the U.S. Navy's nuclear sea-launched missile force will be capped at 350 weapons—less than half of the 758 originally planned. The effects in Russia are not so clear, but its estimated arsenal of 3,400 weapons will likely fall to 2,900 and probably will decline even further.[10]

Despite the Russian offer to eliminate tactical naval nuclear weapons, there probably will not be a surge of pressure to do so. The Gorbachev and Bush proposals went further than several nongovernmental suggestions for controlling tactical nuclear weapons. The consensus position in the liberal Washington arms control community had been to advocate just the removal of nuclear weapons from surface ships. With little domestic pressure to accept Gorbachev's offer of total elimination, the Bush administration could afford to ignore it. The removal of all tactical nuclear weapons will mitigate problems with allied countries and with citizens concerned about U.S. nuclear weapons being secretly brought into their ports during visits.

On the other hand, the tactical nuclear arms race at sea is unlikely to resume. The factors that led to its decline will remain operative for the foreseeable future. The cost of naval nuclear weapons procurement, maintenance, and associated training and logistic support will be high compared with the two governments' constrained resources. Foreign opposition to port calls by U.S. nuclear-armed ships will persist. Tactical naval nuclear weapons will continue to be of dubious military value, even in regional wars. Illustrating this point, the U.S. battleship *Wisconsin* offloaded its three nuclear-armed Tomahawk cruise missiles in exchange for three conventional bomblet versions before departing for the Persian Gulf in August 1990. Finally, although political tensions with Russia could worsen because of unforeseen circumstances, they are unlikely to prompt a wide-scale nuclear arms race at sea.

STRATEGIC NAVAL NUCLEAR WEAPONS

Strategic naval nuclear weapons are also declining in number. The U.S. Congress terminated the Trident submarine program with the eighteenth submarine in the fiscal year 1991 budget (this is two to six submarines fewer than the navy had wanted). The implementation of the July 1991 Strategic Arms Reduction Talks

(START) agreement would have meant that by the turn of the century, the current U.S. force of thirty-three submarines armed with 5,024 warheads would have declined to eighteen submarines armed with at most 3,456 warheads.

Developments in the fall of 1991 and the beginning of 1992 will cause this number to be lower still. If President Bush's September 1991 and January 1992 State of the Union address proposals are carried out, the number of nuclear weapons on U.S. strategic submarines could fall to 2,300. If President Yeltsin's January 1992 proposed cuts are finally accepted, the number may go as low as 1,150 nuclear weapons.[11]

Similar reductions are in store for the Russian strategic submarine force. Because of the START agreement, and the additional Bush-Yeltsin proposals, the Commonwealth of Independent States (CIS) navy's 1992 force of some fifty-six ballistic missile submarines with 3,608 warheads is projected to decline to approximately eighteen to twenty-seven submarines with 1,300 to 2,300 warheads.[12]

Another trend in strategic naval nuclear weapons has been blunted. In the late 1980s, the trend was to broaden the missions of strategic naval forces to take account of their improved capabilities. The new U.S. Trident II D-5 ballistic missile could carry the larger W-88 nuclear warhead and was accurate enough to attack hardened missile silos and command bunkers, a mission that had been reserved for land-based ICBMs.

Now, however, the U.S. Navy has dropped plans to retrofit Trident II D-5 missiles on the first eight Ohio-class Trident submarines. Instead, these submarines will continue to carry less powerful Trident I missiles. Also, Bush's January 1992 State of the Union address called for cancellation of the W-88 nuclear warhead for the Trident II missile. Thus, the U.S. arsenal will contain only the existing 400 W-88s rather than the 2,000 originally envisioned.

Before September 1991, the total elimination of tactical naval nuclear weapons was feasible, but the elimination of strategic nuclear weapons remained unthinkable. Political support for use of strategic weapons as a minimum deterrent insurance policy was too widespread. All that could be anticipated—if the favorable political trends continued—was a possible willingness to ratify an agreement to reduce forces to fewer than a dozen submarines, with one or two thousand warheads on board on each side.

By the spring of 1992, the situation had reversed. The remote possibility of 1991 had become almost a reality by 1992. It is now credible to picture the elimination of strategic naval nuclear weapons in the next ten to twenty years.

Two trends point in this direction. First, all strategic nuclear forces will continue to decline in number. Under current proposals, U.S. and Russian forces would decline to approximately 4,400 to 4,700 weapons. If President Yeltsin's January 1992 proposals are accepted, the numbers could fall to 3,000 nuclear weapons on each side. This would also include extensive de-MIRVing[13] of weapons (i.e., making multiple-warhead missiles into single-warhead missiles). By virtue of this, nuclear forces will increasingly rely on tactical and strategic air-launched weapons. And with low nuclear alert rates and little political tension, there will be less need for even a few nuclear ballistic missile submarines to serve as a secure second strike force. (It is interesting that President Yeltsin offered to halt ballistic missile submarine patrols as part of his package of proposals in January 1992.) Second, the numbers of ballistic missile submarines may continue to decline. Neither the United States nor Russia is building successors to its current generation of submarines, and the high costs of nuclear power at sea will constrain attempts to expand or replace the existing fleet.

NAVAL NUCLEAR REACTORS

Naval nuclear reactors were not prominent symbols of the cold war. Nevertheless, they were intrinsic to the superpower competition. The superpowers' ballistic missile submarines carried thousands of nuclear warheads, while their nuclear reactors allowed them to stay stealthily and safely submerged for extended periods of time. Cruise missile and attack submarines were the main threat to the coasts, and to the capital ships and ballistic missile submarines of each side's navy. Nuclear power allowed these vessels easily to operate undetected while submerged in the open oceans or close to foreign shores.

The first nuclear-powered vessel, the submarine USS *Nautilus*, was commissioned in 1954. Since then, the U.S. Navy has commissioned almost 200 nuclear-powered submarines and surface vessels. The former Soviet Union had built a comparable number of submarines and surface ships since 1958, when the first Soviet nuclear-powered submarines went to sea. The United Kingdom commissioned its first nuclear-powered submarine, HMS *Dreadnought*, in 1963 and has built another twenty-one nuclear-powered submarines. France has built ten nuclear-powered submarines, with the first going to sea in 1971. China has approximately six nuclear-powered submarines, with the first thought to have become operational in 1971. Finally, India briefly entered the nuclear-powered naval club by renting a nuclear-powered Charlie-class submarine from the Soviet Union in the late 1980s. In 1991, slightly more than half of the world's nuclear reactors, approximately 510, powered the ships and submarines of the U.S., Soviet, French, British, and Chinese navies (see table 27.3).

Nuclear power is thought to give naval vessels unique capabilities. It frees submarines of one of their traditional limitations: the necessity of surfacing regularly to recharge electric batteries. It also allows submarines to ignore the constraints of limited fuel supplies and to conduct long submerged high-speed

TABLE 27.3 Nuclear Reactors on Naval Vessels (1991)

	United States	USSR	United Kingdom	France	China	Total
Submarines						
Ballistic missile	33	124	4	5	2	168
Cruise missile	0	54	0	0	0	54
Attack	85	118	14	4	3	224
Total submarines	118	296	18	9	5	446
Surface ships						
Aircraft carriers	18	0	0	0	0	18
Cruisers	18	8	0	0	0	26
Other	1	22	0	0	0	23
Total surface ships and other[a]	37	30	0	0	0	67
Total	**155**	**326**	**18**	**9**	**5**	**513**

Source: Adapted from JOSHUA HANDLER & WILLIAM M. ARKIN, NUCLEAR WARSHIPS AND NAVAL NUCLEAR WEAPONS 1990: A COMPLETE INVENTORY (Greenpeace Sept. 1990).

[a] Includes icebreakers, naval research vessels, and nuclear-powered submarines without weapons.

transits. Moreover, the nuclear power plant provides additional electrical power to run sophisticated electrical fire control, sonar, and navigation systems. In the case of surface vessels, the main advantages are greater sustained speed for high-speed transits and freedom from dependence on fossil fuel supplies.

Increasingly, however, nuclear power's economic and environmental costs are outweighing its military benefits. With surface ships in the United States, this point was reached in the early 1970s, when Admiral Hyman Rickover, the energetic head of the U.S. Navy's nuclear propulsion program, failed to get a mandate to build all major surface warships with nuclear power. As a result, the only surface ships still constructed by the U.S. Navy with nuclear power are aircraft carriers.

In the United States, nuclear-powered submarines have reached a similar point. The high cost of these vessels has resulted in a critical reexamination of the need for a large force. The extraordinary $2 billion-a-copy cost for the next-generation U.S. nuclear-powered attack submarine—the SSN-21 Seawolf—led President Bush to cancel the program in the fiscal year 1993 budget.

The cancellation of the Seawolf has caused a crisis in the nuclear-powered submarine program. In April 1992, the head of U.S. submarine programs, Vice Admiral Roger Bacon, told the U.S. Congress that "our present attack submarine building rate will not sustain a force of *any size*."[14] The two construction yards for nuclear-powered submarines—Electric Boat of Groton, Connecticut, and Newport News of Newport News, Virginia—will finish constructing the Trident-class ballistic missile submarines and Los Angeles–class attack submarines by the mid-1990s. Without immediate orders for new submarines, there will be a gap in construction that the U.S. Navy claims will severely affect or even permanently cripple the U.S. nuclear submarine industrial base. The U.S. Navy now hopes that Congress will restore funding for the Seawolf or that Los Angeles–class submarines can continue to be produced until a new underdesign Centurion-class submarine can be brought into production in the late 1990s.

In any event, the current force of U.S. nuclear-powered submarines will continue to decline. As noted, the ballistic missile submarine force will shrink to eighteen boats. The mid-1980s goal of 100 front-line attack submarines has been completely abandoned. Today's force of eighty-five Permit, Sturgeon, and Los Angeles–class attack submarines will be reduced at least to sixty-two Los Angeles–class submarines and one Seawolf submarine by the end of the decade.

The primary mission of U.S. nuclear-powered submarines was to go forward during wartime to waters near the Soviet Union and engage in antisubmarine warfare. Secondary missions included barrier antisubmarine warfare, open ocean operations in support of naval task or supply groups, and attacks against surface ships. The end of the cold war and reductions in Soviet submarine forces have eradicated the need for U.S. submarines to accomplish exacting wartime missions close to Soviet shores. Although the U.S. Navy's submarine force has tried to articulate a role for itself in regional wars, there are no equivalently demanding missions for nuclear-powered submarines in the rest of the world. Only five submarines participated in Operations Desert Shield and Desert Storm,[15] and their contribution, compared with that of all other military forces, was marginal.

The Commonwealth of Independent States (CIS) navy's submarine program faces an even more acute crisis. The number of nuclear-powered submarines is declining fast as first-generation submarines are being decommissioned. The CIS navy has halted ballistic missile submarine construction and has announced that it hopes to build 1.5 general-purpose nuclear submarines per year.[16] The current force of about 150 nuclear-powered submarines will probably be cut in half by the end of the decade.

The reductions in forces are creating environmental problems for the nuclear navies. Rear Admiral Edward Sheafer, director of naval intelligence, told the U.S. Congress in February 1992:

> The CIS does not yet have a solution for disposal of nuclear submarine reactors. As a result, the number of retired nuclear submarines scrapped per year will probably remain low, and there are already over 60 discarded nuclear submarines requiring

proper storage and disposal, posing a growing environmental problem for the Russians, in whose harbors they are lying.[17]

The CIS navy will have to dispose of some 160 submarines with 300 reactors by the year 2000. The United States faces similar problems. Approximately fifty U.S. nuclear-powered submarines have been retired, and another thirty to forty will be decommissioned by the end of the decade. In addition, the U.S. Navy will also take several nuclear-powered cruisers out of service. The United Kingdom also must decide what to do with its retired submarines. Three are already retired, and several more will be taken out of service shortly. Its problem may be complicated by difficulties suffered by reactors on some U.K. submarines.[18]

Decommissionings, increased costs associated with nuclear-powered vessels (both up-front and decommissioning costs), declining shipbuilding budgets, and better political relations mean that there will be smaller nuclear-powered fleets. Given that many missions of nuclear-powered attack or cruise missile submarines are related to superpower conflict, there has been some discussion of codifying smaller nuclear-powered submarine fleets of twenty-five to fifty submarines, perhaps even as a way station to their total elimination either through formal or informal agreements.[19]

No nuclear power has endorsed this idea yet. However, new information that the Soviet navy has dumped naval nuclear waste at sea may give more impetus to the call for limits on nuclear power at sea. In July 1991, the former head of a refueling facility for the Soviet Northern Fleet recounted that low-level solid and liquid waste had regularly been dumped during his 1975-1986 tenure. Some waste had been dumped in unsealed garbage bins.[20]

In September 1991, Andrei Zolotkov, a people's deputy for the former Soviet Union from Murmansk, reported that from 1964 to 1986 vessels of the civilian Murmansk Shipping Company dumped radioactive waste off the arctic islands of Novaya Zemlya (the company works with the fleet of Russian nuclear-powered icebreakers). The dumping included the three damaged reactors from the first Soviet nuclear-powered icebreaker, the *Lenin*, which had suffered a severe reactor accident in the mid-1960s. He also said that military waste had been dumped as well.[21]

In February 1992, Alexander F. Emelyanenkov, also a people's deputy with the former Soviet Union, reported that the Soviet navy had discarded twelve damaged reactors from nuclear submarines in shallow gulfs around Novaya Zemlya. Reportedly, six of these reactors still contained fuel. Additionally, some 11,000 containers of low-level radioactive waste were also dumped.[22]

Emelyanenkov's report has had a strong political effect in Scandinavia and the United States. The Norwegian government is currently seeking cooperation from the Russian government for investigating the area around Novaya Zemlya. If necessary, the Norwegians will pursue plans for raising the dumped materials. In the United States, a Republican senator from Alaska, Frank Murkowski, asked the Bush administration to mount a scientific investigation of these reports. These activities came on top of efforts by the Icelandic government to raise questions at the United Nations about the need for a safety regime covering nuclear reactors at sea after a Soviet Mike-class nuclear-powered submarine sank off the coast of Norway in 1989.

REDUCTION IN FLEET SIZES AND OPERATIONS

The U.S. and CIS fleets and, to some extent, the French and British navies are undergoing a process of "spontaneous disarmament." Aging vessels, shrinking budgets, and improved East-West relations are all contributing to reducing the size of the fleets.

The U.S. Navy has abandoned its 1980s goal of a 600-ship navy centered on fifteen carrier battle groups. It currently envisions a twelve-carrier, 450-ship force by 1995, down from fifteen aircraft carriers and 530 ships in 1991.[23] It is quite likely that the naval force will face further reductions by the end of the century.

The CIS fleet is going through a similar retrenchment. Older ships and submarines are being scrapped at an unprecedented rate, and there are promises of more reductions in the future. British forces are also on the decline. In the summer of 1990, the British government released an "Options for Change" study that called for a reduction of the frigate and destroyer force from some fifty ships to forty ships. The diesel- and nuclear-powered attack submarine force will shrink by almost half to approximately sixteen vessels, including the retirement of some five older nuclear-powered boats.[24] Shrinking budgets are also affecting the French navy. Plans for a second Charles de Gaulle–class nuclear-powered aircraft carrier have been forsaken. The new Le Triomphant class of ballistic missile submarines may be cut from six to five boats. Finally, the last two of the planned eight-boat Rubis nuclear-powered attack submarine class have been canceled.[25]

Smaller fleets mean lower operating tempos and forward deployments. Reduced superpower tensions have led to fewer U.S. operations in the North Atlantic and the North Pacific, traditionally areas where the navies faced each other on virtually a daily basis. The United States scaled down or canceled major exercises in these areas beginning in 1992. Serious consideration is being given to reducing the time U.S. naval forces spend forwardly deployed. With the added effects of its economic crisis, the CIS has virtually halted its out-of-area operations and exercises. In December 1991, its last combatant ships left the Mediterranean, and in late 1991 the naval patrol off northwestern Africa ceased.[26]

Smaller fleets also are leading to a less extensive overseas basing structure. The CIS navy has mostly withdrawn from Cam Ranh Bay, one of the few forward naval facilities open to its ships, submarines, and aircraft. The largest change by the U.S. Navy is its withdrawal from Subic Bay in the Philippines. Yet other significant withdrawals have taken place, such as the closing of the ballistic missile submarine base at Holy Loch, Scotland, in 1991, after thirty-one years of operation.[27]

CONCLUSION

Superpower and allied navies are facing unprecedented reductions that have already affected their global presence. They have also decreased the weapons and capabilities most suited for East-West conflict: nuclear weapons and nuclear-powered vessels. These trends create some interesting opportunities to denuclearize and demilitarize the international law of the sea.

First, regarding denuclearizing, international law must consider how to eliminate the risk posed by nuclear-powered vessels. Five nuclear-powered submarines with seven nuclear reactors and some thirty-eight nuclear warheads have sunk in the Atlantic because of accidents. Another sixteen damaged reactors (one U.S. and fifteen Soviet) have been deliberately dumped in the sea. Accident-prone nuclear-powered vessels continue to roam the high seas. Five countries—Russia, the United States, England, France, and China—derive uncertain benefits from this technology, while all the world's oceans are threatened. Given the extraordinary danger this technology poses to the global commons, international law should find a way to prohibit nuclear power at sea while still allowing nations their traditional rights to freedom of the seas.

Second, regarding demilitarizing, as the major powers' navies retrench and before regional powers engage in excessive and futile naval arms races, there is a window of opportunity to strengthen or create stronger international institutions to ensure international naval cooperation rather than naval competition.

International law has ultimately had as one of its foundations international consensus and, so, custom. Navies have traditionally served as enforcer or arbiter of some concepts of the law of the sea—freedom of the seas, innocent passage, territorial waters—in which a consensus has been lacking. However, world affairs may have reached a point of widespread dependence on or acceptance of the desirability of international trade, finance, and communication. Certainly this is the case compared with the period of Columbus's time and the succeeding century, which marked the origins of the modern-day law of the sea, when people widely believed that the world was flat, when piracy was common, and when the English Crown would give commissions to private citizens to prey on the shipping of a competing great power. If there is now an extensive agreement (or enough of one that it could be readily strengthened) on the need for and nature of international commerce, the navies' role as a substitute for a lack of international accord can be reduced.

This posited inverse relationship between international maritime consensus and the need for naval forces also suggests the directions and limits of future international agreements. Declining superpower and allied competition should receive less attention, and more attention should be paid to regional questions. As a result, multilateral rather than bilateral approaches to security will be increasingly significant. If there is enough of either a global or a regional consensus on maritime issues, blue water and coastal navies can adopt the responsibilities of coast guards or maritime police forces, such as aiding mariners in distress, enforcing regional or national environmental regulations, and dealing with piracy and terrorism.

Achieving such global or regional agreements will not be easy. Of the many issues to be addressed, at least some reform of the United Nations will be needed. The United Nations' legitimacy needs to be broadened by giving more nations veto power in the Security Council. Also, national navies should abandon a go-it-alone approach and show more willingness to operate with and depend on other navies.

The inverse relationship between international accord and military power also shows the limits of demilitarizing the seas. If a consensus cannot be reached over the use of ocean resources and the role of the great powers' blue water navies, nations will continue to acquire ever larger and more deadly naval forces to protect their resources, as well as to defend themselves against larger powers.

Ultimately, the challenge to international law and institutions in this new maritime era is to ensure that superpower competition is replaced by cooperation and not by a world of even more anarchy and military competition. Some helpful steps toward this end would be to reach a consensus on regional agendas where questions of the environment, economics, and resources are more important than security questions. Also, there will be a need to increase international confidence through unilateral and multilateral transparency measures. Some thought will have to be devoted to defining the need, nature, and role of multilateral peacekeeping forces. These forces could include the major powers' blue water navies and the forces of smaller nations. As a small start toward internationalization of the navies, their considerable scientific activities—including oceanography and meteorology—could serve regional or global institutions. Finally, as part of this effort to develop legitimate multilateral naval forces, the law of the sea will need to repudiate unilateral uses of warships for upholding the freedom of the seas.

NOTES

1. For nuclear weapons and reactor accidents, *see* WILLIAM ARKIN & JOSHUA HANDLER, NAVAL ACCIDENTS 1945-1988 (Neptune Papers No. 3, Greenpeace/Institute for Policy Studies, Washington, DC, June 1989); JOSHUA HANDLER, AMY WICKENHEISER, & WILLIAM M. ARKIN, NAVAL SAFETY 1989: THE YEAR OF THE ACCIDENT (Neptune Papers No. 4, Greenpeace, Washington, DC, Apr. 1990).

2. In 1992, for example, U.S. Navy ships made port calls in seventy-three countries. H. LAWRENCE GARRETT, SECRETARY OF THE NAVY, ET AL., DEPARTMENT OF THE NAVY 1992 POSTURE STATEMENT 110 (Feb.-Mar. 1992).

3. The antisubmarine rocket (ASROC) was a surface-launched nuclear and conventional weapon with a 6-mile range.

4. The submarine rocket (SUBROC) is a submarine-launched nuclear antisubmarine nuclear weapon with a 35-mile range.

5. William M. Arkin, Joshua Handler, & Hans Kristensen, *Soviets Disarm Mysteriously*, BULL. ATOMIC SCIENTISTS, May 1990, at 7ff; Joshua Handler, *Soviets Confirm Baltic Denuclearized*, Bull. Atomic Scientists, Dec. 1990, at 5-6.

6. Michael Gordon, *Reagan Arms Adviser Says Bush Is Wrong on Short-Range Missiles*, N.Y. Times, May 3, 1989.

7. R. Jeffrey Smith, *Crowe Suggests New Approach on Naval Nuclear Arms Cuts: Admiral's Views at Odds with White House*, Wash. Post, Jan. 8, 1990.

8. *Remarks by President Bush on U.S. and Soviet Nuclear Weapons*, Wash. Post, Sept. 28, 1991.

9. *Text of Gorbachev's Statement on Nuclear Weapons*, U.P.I. Wire Serv. (Moscow), Oct. 5, 1991.

10. HANS M. KRISTENSEN, BUSH AND GORBACHEV DISARMAMENT INITIATIVES: EFFECTS ON NON-STRATEGIC NAVAL NUCLEAR WEAPONS (Greenpeace Oct. 1991).

11. Stan Norris & William Arkin, *Nuclear Notebook*, BULL. ATOMIC SCIENTISTS, May 1992, at 48-49.

12. *Id.*

13. A multiple independently targeted reentry vehicle (MIRV) carries more than one warhead.

14. Vice Admiral Roger F. Bacon, Assistant Chief of Naval Operations (Undersea Warfare), Submarine Programs 7 (statement before the House Defense Appropriations Subcommittee, Apr. 8, 1992) (emphasis in original).

15. H. Lawrence Garrett, Secretary of the Navy, et al., Department of the Navy 1992 Posture Statement 6 (Feb.-Mar. 1992).

16. Rear Admiral Edward D. Sheafer, Director of Naval Intelligence, Intelligence Issues 14-17 (statement before the House Armed Services Committee, Feb. 5, 1992). By comparison, in peak years in the 1980s, six nuclear submarines were launched.

17. Id. at 21.

18. David Farnhall, *Polaris Reactors Fear Raised by MP*, Guardian (U.K.), Nov. 19, 1990; Christopher Bellamy, *MP Says Half of Polaris Submarines Out of Action*, Independent (U.K.), Nov. 19, 1990; Norman Friedman, *World Navies in 1992*, Proceedings, Mar. 1992, at 111.

19. See, e.g., W. Philip Ellis, *Back to the Future? Assessing Structural Limits on U.S. and Soviet Naval Forces*, in Barry M. Blechman et al., The U.S. Stake in Naval Arms Control (Henry L. Stimson Center, Washington, DC, Oct. 1990); James R. Lacy, RAND Corporation, If the Soviet Union Is Serious About Naval Arms Control (paper presented at the International Peace to the Oceans Conference, Moscow, Feb. 9, 1990); Douglas M. Johnston, *Naval Arms Control: Not in the Nation's Best Interest*, Proceedings 38; Edward Rhodes, *Naval Arms Control for the Bush Era*, 10 SAIS Rev., Summer-Fall 1990, at 211-29; U.S. Senate Armed Service Committee, National Defense Authorization Act for Fiscal Year 1991, S. Rep. No. 101-384, at 246-47 (July 20, 1990).

20. Oleg Volkov, *To Catch a Reactor Big and Small: Confessions of an Eyewitness*, Komsomolskaya Pravda, July 6, 1991, at 5.

21. Andrei Zolotkov, On the Dumping of Radioactive Waste at Sea Near Novaya Zemlya (paper presented at Greenpeace Nuclear Free Seas Campaign/Russian Information Agency Seminar "Violent Peace—Deadly Legacy," Moscow, Sept. 23-24, 1991).

22. Alexander F. Emelyanenkov, *The Secret Logbook, or The Second Discovery of the Novaya Zemlya Archipelago*, Sobesednik, No. 5 (Jan. 1992). See also Alexander F. Emelyanenkov, *From the Sobesednik Files—Split Atom: The First 50 Years*, Sobesednik, No. 12 (Mar. 1992).

23. General Colin L. Powell, The National Military Strategy 1992, at 19.

24. Ian Kemp, *UK Forces Face 18% Reduction*, Jane's Def. Weekly, Aug. 4, 1990, at 152-53.

25. Norman Friedman, *World Navies in 1992*, Proceedings, Mar. 1992, at 111-12.

26. Sheafer, *supra* note 16, at 25.

27. Bacon, *supra* note 14, at 8.

28 Navies, Ocean Resources, and the Marine Environment

Joseph R. Morgan

MARITIME DEFENSE AND SECURITY in the form of navies is one legitimate use of the world's oceans, which must be shared with activities such as fishing, shipping, exploration and exploitation of mineral resources, and waste disposal. As of 1990, 159 countries had navies;[1] even some states without sea coasts maintain naval forces.[2] In 1988, the world stock of naval seagoing forces included 52 aircraft carriers, 118 strategic nuclear submarines, 16 conventionally powered strategic submarines, 256 nuclear-powered attack submarines, 257 nuclear-powered patrol submarines, 472 conventionally powered patrol submarines, and 1,139 major surface warships. In addition to these 2,310 major naval vessels, there were 51 coastal submarines and 4,659 craft classed as "light naval forces."[3] The latter two categories are listed separately, since the small size of the vessels involved makes them much less influential in affecting the ocean environment.

A NAVAL HIERARCHY

Navies range in capability and effect on the ocean environment and resources from minuscule fleets such as those of several Pacific Island countries to the powerful, mammoth forces of the superpowers. There have been a number of attempts to classify navies and establish a ranking or hierarchy of naval forces based on numbers and types of ships; weaponry; capacity for building new, improved ship types; bases; personnel numbers; states of maintenance and training; and size of the geostrategic region over which the naval forces can exercise a reasonable degree of dominance. Considering the number of factors involved and the complexity of modern navies, such rankings are rarely considered to be definitive by so-called experts. Nevertheless, there are some quite useful analyses of naval strength that permit us to place the 159 navies of the world into a manageable number of categories.

One writer classifies the maritime powers of the world into but three categories: superpowers, medium powers, and small powers. The United States and the former Soviet Union were the two superpowers. Medium powers are harder to define but include such countries as the United Kingdom, France, India, Brazil, Japan, and (surprisingly) Israel. Israel, of course, is a small nation both in population and in area, and its armed forces, particularly its navy, are minuscule by world standards. But according to the author of this valuable study, a country's power must be judged by the strength of its alliances, particularly with the United States, which guarantees Israel's continued existence.[4]

The great majority of the maritime nations are classed as small powers. The only exception to the author's neat classification system is, by his own admission, China, about which he writes: "There will be some situations in which it is appropriate to regard China as a superpower; there will be others, perhaps more in the context of this book, in which she should be regarded as less than that. It will never be appropriate to consider her a medium power. As it has so often been through the centuries, China is sui generis."[5]

The numerous small powers, many of which can be classified as developing or Third World countries, constitute a naval hierarchy of their own. A useful categorization of the strengths of these navies is based on a combination of numbers and types of ships, weapon inventories, production capabilities, budgets, and associated national power-based criteria.[6]

Third World navies have three principal roles: a constabulary-regulatory function, a territorial (coastal) defense role, and force projection at sea.[7] How effectively they carry out these roles depends on the configuration of their naval fleets. Table 28.1 presents a Third World naval hierarchy.[8]

Table 28.1 points out some interesting characteristics of navies and national perceptions of the legitimate role of seagoing forces. Token navies are "unable even to patrol national territorial seas effectively . . . [and are] impotent in the EEZ." As the rank of navies moves upward from the token category to fleets with regional force projection capabilities, the compositions of the fleets begins to include ships of larger size. Fast attack craft (FAC) and patrol craft are prevalent in the lower ranks, and major warships begin to appear in categories 3 through 6.

Some countries in this hierarchy, Brazil and India, for instance, are classed elsewhere as medium powers.[9] Other medium powers, such as France, the United Kingdom, and Japan, would almost certainly be classed in category 6 and perhaps even higher if there were another category. The United Kingdom, for instance, demonstrated its ability to project naval power far from home during

TABLE 28.1 The Third World Naval Hierarchy

Categories of Third World Navies	Naval/Naval Aviation Structure	Naval Capabilities	States in Each Rank (Alphabetical Order)
6. Regional force projection navies	All Third World naval and naval aviation equipment categories strongly represented; more than 15 major warships and/or submarines	Impressive territorial defense capabilities and some ability to project force in the adjoining ocean basin	Argentina, Brazil, India
5. Adjacent force projection navies	Most Third World naval and naval aviation equipment categories well represented; more than 15 major warships and/or submarines	Impressive territorial defense capabilities and some ability to project force well offshore (beyond EEZ)	Chile, Iran, North Korea, Peru, South Korea
4. Offshore territorial defense navies	Quite a few Third World naval and naval aviation equipment categories well represented, including some larger units at upper levels; 6–15 major warships and/or submarines	Considerable offshore territorial defense capabilities up to EEZ limits	Colombia, Egypt, Indonesia, Libya, Mexico, Pakistan, Philippines, Taiwan, Thailand, Venezuela
3. Inshore territorial defense navies	Third World naval and naval aviation equipment categories moderately represented at lower levels and only sparsely represented at upper levels, if at all; 1–5 major warships and/or submarines	Primarily inshore territorial defense, with limited offshore defense capability	Bangladesh, Burma, Cuba, Dominican Republic, Ecuador, Ethiopia, Ghana, Malaysia, Nigeria, Syria, Uruguay, Vietnam

(continues)

TABLE 28.1 The Third World Naval Hierarchy (*continued*)

Categories of Third World Navies	Naval/Naval Aviation Structure	Naval Capabilities	States in Each Rank (Alphabetical Order)
2. Constabulary navies	Sparse representation of Third World naval equipment categories, at lower levels only. Naval aviation minimal or nonexistent. No major warships, but fast attack craft (FAC)	Some ability to prevent use of coastal waters, with concentration on constabulary functions	Algeria, Gabon, Guinea, Guinea-Bissau, Iraq, North Yemen, Oman, Saudi Arabia, Singapore, Somalia, South Yemen, Tanzania
1. Token navies	Only minimal representation at lower levels of Third World naval equipment categories. No FAC; only patrol craft and/or landing craft. Naval aviation nonexistent	Unable even to patrol national territorial seas effectively. Impotent in the EEZ	62 navies

Source: Modified slightly from MICHAEL A. MORRIS, EXPANSION OF THIRD WORLD NAVIES 25–26 (1987).

the naval conflict with Argentina over the Falkland Islands. And Japan might be able to operate effectively well outside the adjoining ocean basin were it not for constitutional prohibitions against use of armed forces for other than strictly defensive missions.

NAVAL OPERATIONS AND THE ENVIRONMENT

Ocean resources, particularly living resources, depend on a healthy environment. Naval vessels ply the waters of all oceans and, like other ships, can create undesirable environmental effects. As the following discussion will indicate, however, these effects are generally of far less consequence than those caused by merchant ship operations. First, there are fewer naval vessels of all types afloat in the world's oceans than merchant ships. Second, the largest of the superpowers' ships, the aircraft carriers and battleships, are much smaller than the most environmentally dangerous of merchant vessels, the supertankers. Finally, most naval vessels do not carry cargoes; hence, a collision or grounding generally will not result in ocean pollution.

Table 28.2 summarizes the merchant fleets of the world by both number of ships and tonnage. The world total of 75,680 merchant vessels of 100 gross tons or larger can be compared with the 7,020 naval ships, which includes light naval forces of a size comparable to the smaller of the merchant ships.

Table 28.3 focuses on tankers, including oil tankers, oil and chemical tankers, chemical tankers, and liquified gas carriers, categories of ships that have contributed most to ocean pollution in the past. The number of tankers alone exceeds the total number of naval vessels by 1,239. Of the 8,259 tankers, 442 were over 200,000 deadweight tons (dwt). The largest vessel in the world is *Hellas Fos*, an oil tanker of 555,051 deadweight tons registered in Greece.[10] Compared with these giants, naval vessels are small; aircraft carriers in the U.S. Navy displace no more than 102,000 tons,[11] and the Iowa-class battleship displaces only 57,353 tons. The largest of the Russian naval vessels are the *Tbilisi* and the *Riga*, which have displacements of 67,000 tons.[12]

Since the largest naval vessels, the aircraft carriers and battleships, do not carry cargoes, a casualty to one of them would not result in widespread environmental damage. Navy tankers (oilers) are much smaller than the ships common in merchant fleets. The largest in the U.S. Navy displaces only 41,350 tons.[13] Therefore, an accident to a naval tanker would undoubtedly result in far less

TABLE 28.2 World Merchant Fleets, 1988

Fleets (Ranked by Size)	(All Types of Vessels >100 GT[a]) GT (Millions)	dwt[b] (Millions)	Number of Vessels
1. Liberia	49.7	94.0	1,507
2. Panama	44.6	71.5	5,022
3. Japan	32.1	48.4	9,804
4. USSR	25.8	29.2	6,741
5. Greece	22.0	39.7	1,874
6. United States	20.8	29.9	6,442
7. Cyprus	18.4	32.8	1,352
8. People's Republic of China	12.9	19.4	1,841
9. Norway	9.4	15.2	2,078
10. Philippines	9.3	15.5	1,483
Top ten	245.0	395.6	38,144
	(60.7%)	(62.1%)	(50.4%)
All others	158.4	241.5	37,536
World total	**403.4**	**637.1**	**75,680**

Source: LLOYD'S REGISTER OF SHIPPING STATISTICAL TABLES 1988 (London 1988).

[a] GT = gross tons.
[b] dwt = deadweight tons.

TABLE 28.3 Tankers

Fleets (Ranked by Size), 1988	Vessels
1. Liberia	605
2. Panama	782
3. Japan	1,830
4. United States	299
5. Greece	310
6. Norway	176
7. Cyprus	129
8. Bahamas	124
9. USSR	438
10. Bermuda	41
Top Ten	4,734 (57.3%)
All others	3,525
World total	**8,259**

Source: LLOYD'S REGISTER OF SHIPPING STATISTICAL TABLES 1988 (London 1988).

Note: Includes oil tankers, oil and chemical tankers, chemical tankers, and liquified gas carriers.

environmental damage than might be caused by a similar accident to a merchant tanker.

The larger naval powers operate nuclear ships, however, and these must be considered to be of environmental concern in the event of a collision, grounding, or other serious accident. Nuclear-powered ships, as well as many conventionally propelled warships, are frequently nuclear armed. The combination of nuclear power and nuclear armament has caused a number of countries to view ships of either type as hazardous, resulting in the banning of these vessels from many ports and even territorial seas.[14] Documented evidence of ocean pollution from nuclear-powered or nuclear-armed ships is scarce, but the possibility of its happening is undeniable. The risk of nuclear pollution from the operations of naval vessels must be considered to be slight, however, since there are relatively few nuclear-powered ships in the world's navies, the accident rate of their operations has not been high, and all but the most catastrophic of accidents would have little effect on the natural ocean environment.

LAW OF THE SEA CONSIDERATIONS

The 1982 United Nations Convention on the Law of the Sea strikes a careful balance between imposing restrictions on the operations of ships in order to protect the marine environment and preserving the principle of freedom of navigation. In general, little distinction is made between warships and merchant vessels in the 1982 Convention, but state practice appears to be evolving in a way that puts greater restrictions on naval vessels. The 1982 Convention provides that coastal state sovereignty "extends, beyond its land territory and internal waters and, in the case of an archipelagic State, its archipelagic waters, to an adjacent belt of sea, described as the territorial sea."[15] However, this sovereignty is restricted: "Subject to this Convention, ships of all States, whether coastal or landlocked, enjoy the right of innocent passage through the territorial sea."[16]

The meaning of "innocent passage" is spelled out in considerable detail, with twelve specific activities considered to be prejudicial to the peace, good order, or security of the coastal state.[17] These are

1. any threat or use of force against the sovereignty, territorial integrity or political independence of the coastal State, or in any other manner in violation of the principles of international law embodied in the Charter of the United Nations;
2. any exercise or practice with weapons of any kind;
3. any act aimed at collecting information to the prejudice of the defence or security of the coastal State;
4. any act of propaganda aimed at affecting the defence or security of the coastal State;
5. the launching, landing or taking on board of any aircraft;
6. the launching, landing or taking on board of any military device;
7. the loading or unloading of any commodity, currency or person contrary to the customs, fiscal, immigration or sanitary laws and regulations of the coastal State;
8. any act of wilful and serious pollution contrary to this Convention;
9. any fishing activities;
10. the carrying out of research or survey activities;
11. any act aimed at interfering with any systems of communication or any other facilities or installations of the coastal State;
12. any other activity not having a direct bearing on passage.

Clearly, items 1, 2, 5, and 6 are activities applicable only to warships, and items 3 and 4 are more likely to be carried out by government vessels than by commercial ships. Only one of these prohibited activities, item 8, is particularly relevant to protecting the marine environment, and it is equally as likely to be violated by merchant ships as by warships. Nowhere in this list is a requirement that naval vessels must ask permission to enter a territorial sea, yet many coastal states include this obligation in their domestic legislation and routinely require it of transiting warships. Recently, Brazil prohibited a U.S. nuclear submarine from entering one of its ports and its territorial sea, which under Brazilian-claimed sovereignty extends to 200 nautical miles.[18]

Part XII of the 1982 Convention is concerned with protection and preservation of the marine environment. Although naval vessels, as well as merchant ships, can cause environmental pollution, the convention exempts warships from the applicability of the provisions regarding protection and preservation of the marine environment. Article 236, concerning sovereign immunity, states:

> The provisions of this Convention regarding the protection and preservation of the marine environment do not apply to any warship, naval auxiliary, other vessels or aircraft owned or operated by a State and used, for the time being, only on government non-commercial service. However, each State shall ensure, by the adoption of appropriate measures not impairing operations or operational capabilities of such vessels or aircraft owned or operated by it, that such vessels or aircraft act in a manner consistent, so far as is reasonable and practicable, with this Convention.

The requirement that states adopt "appropriate measures" can be construed as indicative of the fact that operations of navies may damage the marine environment, but the qualifying terminology—"not impairing operations or operational capabilities" and "so far as is reasonable and practicable"—weakens the degree of environmental protection that can be expected by application of this article. Clearly, some navies will be virtually exempt from any consideration of environmental effects of their operations, while others may be subjected to rather stringent antipollution measures (at least in peacetime), depending on the adopted policies of individual countries.

COMPREHENSIVE SECURITY: NAVAL OPERATIONS AND THE ENVIRONMENT

The concept of comprehensive security is described as deriving from

> an amalgamation of environmental security . . . and political security. The political component of comprehensive security, in turn, consists of a combination of military security, economic security, and social security (the last subsuming social justice, equity, and participatory democracy). Indeed, neither of these two major components of comprehensive security is either attainable or sustainable without the other being satisfied as well.[19]

The great majority of the world's nations apparently consider the maintenance of naval forces to be an essential component of military security and, therefore, necessary for the achievement of comprehensive security as the concept is described here. But naval operations may cause harm to the environment, and "[w]anton disruption of the environment by armed conflict is a common occurrence in many ecogeographical regions of the world."[20] Are the two statements reconcilable? Can navies be both good and bad for comprehensive security? Yes, they can. Peacetime naval operations, as discussed earlier in this chapter, have relatively little deleterious effect on the ocean environment, certainly far less than merchant ship operations, which are the backbone of world trade and an important component of economic security. Armed conflict at sea, particularly if it includes nuclear weapons, can seriously affect both terrestrial and oceanic environments. Navies, therefore, serve a useful purpose if they deter conflict; indeed, most countries, including the superpowers, much prefer to use their navies for deterrence rather than actual combat. Gunboat diplomacy as a form of deterrence is a useful mission of naval forces.

SHARING THE WATERS: A SOURCE OF CONFLICT?

The area of the world ocean is vast, but much of it is "empty space" from the standpoint of both naval operations and other equally valid uses of the seas, such as shipping, fishing, mineral resource exploitation, and scientific research. Trade routes and what naval strategists call "sea lines of communication" usually coincide, and entrances to ports, including naval bases, are frequently crowded with both naval and merchant ships. Fishers object to both merchant and naval ships sharing their waters.

The problem of insufficient sea space is exacerbated by the claims of some navies to large exclusive operating areas. These are obviously necessary to ensure the safety of nonmilitary users of the marine environment, but there is an understandable resentment on the part of fishers, for instance, if what are considered to be prime fishing grounds are alienated by naval authorities for military operations.

Many countries have established warning zones of various kinds: air defense identification zones, military warning zones that extend many miles offshore, and prohibited zones in the vicinity of sensitive operating or research areas. These zones, some of considerable size, also are not available to other legitimate users of the seas.

Seemingly in retaliation for the foregoing, there is an increasing feeling on the part of many influential environmental groups that nuclear-free zones and, in some cases, a "zone of peace" should be established in which naval operations are precluded. A number of countries now will not accept the presence of nuclear-powered or nuclear-armed navy ships in their ports, and in some cases they exclude them from their territorial seas. This prohibition is bound to provoke considerable controversy, since the 1982 Convention requires that innocent passage be permitted and there is nothing in article 19 of the convention, which describes the meaning of innocent passage, that makes nuclear power or nuclear armament prejudicial to the peace, good order, or security of the coastal state.

CONCLUSION

Few would deny that comprehensive security, in which both the environment and economic and political security are protected, is desirable. Navies, principally in their role as deterrent forces, contribute to economic and political security. Moreover, navies are, in general, relatively environmentally benign. Thus, their contributions to economic and political security are not at the expense of environmental security.

Naval forces, to be sure, must share the seas with other legitimate users of the marine environment, and military activities in the oceans are frequently incompatible with fishing and mineral resource exploitation. But the oceans are vast, and sensible accommodations can be made.

NOTES

1. JANE'S FIGHTING SHIPS, 1990-91 (Richard Sharpe ed. Coulsdon, U.K. 1990).
2. Id. (listing naval forces for Austria, Bolivia, Czechoslovakia, Malawi, Mali, and Switzerland). The navies of landlocked states are designed for patrol and constabulary functions on rivers and lakes.
3. See 8 OCEAN YEARBOOK 638-45 (E. M. Borgese, N. Ginsburg, & J. R. Morgan eds. 1989).
4. See J. R. HILL, NAVAL STRATEGY FOR MEDIUM POWERS 24, 25, & n.4 (1986).
5. Id. at 19.
6. See MICHAEL A. MORRIS, EXPANSION OF THIRD WORLD NAVIES (1987).
7. Id. at 16-17.
8. Id. at 25-26.
9. See HILL, supra note 4.
10. See LLOYD'S REGISTER OF SHIPPING STATISTICAL TABLES 1988 (London 1988).
11. See JANE'S FIGHTING SHIPS, 1990-91, supra note 1.
12. Id.
13. Id.
14. See Honolulu Advertiser, Nov. 5, 1990, at D-1, for a description of recent decisions by Brazil concerning a U.S. nuclear submarine.
15. United Nations Convention on the Law of the Sea, art. 2, Dec. 10, 1982, U.N. Doc. A/CONF.62/122, 21 I.L.M. 1261 (1982) [hereinafter 1982 Convention].
16. Id. art. 17.
17. Id. art. 19.
18. Honolulu Advertiser, supra note 14.
19. COMPREHENSIVE SECURITY FOR THE BALTIC: AN ENVIRONMENTAL APPROACH 116, 117 (Arthur H. Westing ed. 1989).
20. Id. at 114.

29 Military Exclusion and Warning Zones on the High Seas*

Jon M. Van Dyke

SINCE THE DAYS OF GROTIUS, scholars have argued that navigation and fishing on the high seas should be free and unimpeded.[1] Nonetheless, several maritime nations have recently claimed the right to declare exclusionary or warning zones on the high seas to serve their military purposes. This chapter analyzes these claims in light of the norms that now govern the high seas and concludes that claims to exclude or limit free navigation can be sustained only if they have no appreciable effect on navigation or the environment and resources of the region. Furthermore, a vessel entering such a zone cannot be forcibly removed from the area except by the nation whose flag it flies, and any vessel operating with the support of its flag government that is damaged by the military activities of another nation is entitled to receive compensation.

MILITARY WARNING ZONES ON THE HIGH SEAS

Ballistic Missile Tests

Both the former Soviet Union and the United States have used broad ocean areas for testing ballistic missiles. Before each test, the Soviet Union announced its intentions and issued a statement, usually through its news agency, Tass,

*The author would like to express his appreciation to Thomas Feeney, Dale L. Bennett, and Patti Nakaji, graduates of the University of Hawaii Law School, who assisted with research. Useful comments on an earlier draft have been received from many colleagues, including Professors Ian Brownlie, William Burke, and Bernard Oxman. The author has served in recent years as consultant to the South Pacific Regional Environment Programme, the Republic of Nauru, the Office of Hawaiian Affairs, and Greenpeace, but the views expressed herein are his own. A longer version of this chapter has been published in 15 MARINE POL'Y 147 (1991) and in MOSCOW SYMPOSIUM ON THE LAW OF THE SEA 75 (Thomas A. Clingan, Jr., & Anatoly L. Kolodkin eds. 1991).

asking other nations to alert their respective mariners and pilots to keep the area clear.[2] Afterward, the United States and other governments relay the notices to their mariners and pilots, warning those concerned to take appropriate precautions.[3]

The United States currently launches missiles from Cape Canaveral over a broad area ranging from Ascension Island to the Indian Ocean and from California into the Pacific. In addition, U.S. submarines test missiles in international waters, usually near either California or Florida, with consequent warning areas encompassing both the area around the submarines and the impact zones of the launched missiles. Like the Soviet Union, the United States precedes each test with public notices requesting mariners and pilots to avoid designated areas for a period of time.[4] Some areas, like the 200-mile zone around Kwajalein Atoll, are considered permanent warning areas twenty-four hours a day, 365 days of the year.[5]

Nuclear Bomb–testing Programs

The testing of nuclear weapons probably has had an even more significant effect on freedom of the high seas and the ocean environment than the testing of ballistic missiles. Nuclear testing in the Pacific began on July 1, 1946, when the United States exploded an atom bomb on Bikini Atoll in the Marshall Islands.[6] For that test, the United States established a "danger zone" of approximately 150,000 square nautical miles, lasting from the end of June to mid-August.[7]

U.S. nuclear testing in the Pacific continued until November 1962.[8] Testing occurred in a variety of locations, including Bikini and Enewetak atolls, Johnston Atoll (about 700 miles southwest of Honolulu), and Christmas Atoll (now part of the Republic of Kiribati). These tests led to the creation of danger zones ranging in size from 30,000 to 400,000 square nautical miles.[9]

The United States is not the only country to have tested nuclear weapons in the Pacific. British nuclear tests led to the declaration of a 6,000-square-nautical-mile danger zone around the Monte Bello Islands in 1952.[10] When Britain first tested hydrogen bombs on Malden and Christmas islands, it announced a 700,000-square-nautical-mile warning zone to be in effect from March 1 to August 1, 1957.[11] The British vacated Christmas Island and ended their Pacific Ocean testing in December 1958.[12]

The French, too, have conducted nuclear weapons tests in the Pacific. Between 1966 and 1974, France tested forty-one atmospheric nuclear devices in French Polynesia.[13] Their maritime danger zone extended 150 miles from Moruroa Atoll, with a 500-nautical-mile easterly downwind corridor. The zone for aircraft extended 200 miles beyond the island, with a corridor fanning out 1,000 miles to the east.[14] The French continue to conduct underground nuclear tests in the region, which have resulted in a permanent "prohibited area" for aircraft encompassing 1,650 square nautical miles; a surface warning zone of limited range and duration was also declared for each test.[15]

Other Military Activities

Other military uses of the high seas can also limit nonmilitary activities. In the most extreme cases, countries have claimed complete jurisdiction, based on national security concerns, over areas beyond the 12-mile limit set by the 1982 Convention on the Law of the Sea.[16] Most common are temporary "security zones" established during times of armed conflict.[17]

Although not as explicitly exclusionary as security zones, warning and identification zones can also limit the freedom of other nations. The United States first claimed an air defense identification zone (ADIZ) in September 1950, extending 300 miles off its coasts.[18] An ADIZ is an area in which the identification, location, and control of civil aircraft are required in the interest of national security.[19]

Finally, warning zones are routinely established when nations conduct naval maneuvers. These activities can interfere significantly with freedom of the seas and in some cases have resulted in serious damage to commercial vessels.[20]

CONFRONTATION OVER WARNING ZONES

Although the establishment of exclusionary zones in itself seems to contradict notions allowing freedom of movement on the high seas, it is the enforcement of such zones that has elicited controversy. The establishment and maintenance of warning zones for military purposes have resulted in several celebrated and informative confrontations on the high seas. In some cases, formal protests have been lodged; in others, issues of liability have arisen.

The United States Navy used force against the M/V *Greenpeace* in December 1989, when the navy "shouldered" the Greenpeace vessel away from a submarine missile launch site off the coast of Florida. Significantly, the Netherlands did not protest the navy's use of force to remove the Greenpeace vessel (which was flying a Dutch flag). In fact, the Netherlands reproached Greenpeace, arguing that disrupting the launch would be an "abuse of freedom."[21] A U.S. Navy officer commenting on the incident stated that the U.S. action in removing the Greenpeace vessel was justified in part because "no Dutch warship was present to assert primary jurisdiction" over it and "the United States was left with the necessity of defending its maritime rights" through "self-help."[22]

The French seem the most willing to use force to maintain the integrity of their warning zones. In June 1972, to protest atmospheric nuclear testing by the French government, the *Greenpeace III* sailed into the 100,000-square-mile danger zone that the French had declared around Moruroa Atoll. The French dispatched a fleet of vessels, including a 600-foot cruiser, minesweepers, and tugboats, to charge the *Greenpeace III* from several different angles. At one point, the ketch was almost crushed between two of the warships. After the atmospheric detonation of the bomb, a French minesweeper rammed the vessel's stern, leaving it totally paralyzed. Later, the French towed the vessel to Moruroa for minimal repairs to keep it afloat and then towed it back to open sea, where the vessel was left to limp back to Rarotonga while the series of tests was completed.[23]

One year later, the *Greenpeace III* again sailed into France's declared warning zone in international waters off Moruroa. The French navy boarded the boat and physically assaulted and arrested the crew. The New Zealand government formally protested both that action and a similar action that had occurred a month earlier. New Zealand stated that the French government had no right to interfere with navigation on the high seas and complained about the illegal arrest of its citizens. The Canadian prime minister, Pierre Trudeau, also protested to the French ambassador.[24]

Conflict arises, however, not only from deliberate attempts to observe or interfere with weapons testing. For example, on December 12, 1988, a U.S. Navy jet fighter participating in training maneuvers near the Hawaiian island of Kauai mistakenly fired a missile into the radio room of an Indian merchant vessel, the *Jag Vivek*, that was transporting wheat from Canada.[25] The missile opened a gaping hole in the vessel and killed a radio operator.[26] Although the navy stated that the vessel was in a "warning zone . . . in violation of a Notice to Mariners alerting non-exercise ships to remain outside the target area,"[27] the navy also acknowledged that it had misplaced the target vessel by 30 miles and had failed to instruct the merchant vessel properly when it sought guidance.[28] The navy paid $575,000 to the family of the deceased crew member and $405,000 to the owner of the vessel.[29]

Finally, the navies of the former Soviet Union and the United States have both entered the warning zones of the other. For the most part, the two nations have simply monitored each other from as close as possible without igniting any confrontation.[30]

THE GOVERNING LEGAL PRINCIPLES

The Freedom of the High Seas

Despite the establishment of military exclusionary zones and their consequent enforcement by the world's strongest military powers, the community of nations has accepted the concept of "freedom of the seas" for most of the past 400 years,[31] primarily because none of the maritime powers has been able to place the high seas under its sovereignty. It has thus been in their best economic interests to ensure freedom of navigation for all.[32] In 1609, the Dutch diplomat Hugo Grotius wrote that the seas should be free for navigation and fishing because natural law forbids the ownership of resources that seem "to have been created by nature for common use."[33] This category includes those resources that "can be used without loss to anyone else."[34] The use of the sea for navigation, for instance, does not diminish the potential for the same use by others.

Under customary international law and the 1982 Convention, nations appear to agree on six basic freedoms of the high seas: freedom of navigation, freedom of overflight, freedom to lay submarine cables, freedom to construct artificial islands and other installations permitted under international law, freedom of fishing, and freedom of scientific research.[35] Certain other activities are clearly prohibited, such as trading in slaves or drugs, piracy, illegal broadcasting, and breaking of submarine cables.[36] In addition, articles 88 and 141 state that the high seas and the seabed below the high seas must be used exclusively for "peaceful purposes," a concept discussed in more detail in the sections that follow.[37]

Although nations may agree on freedom of the high seas in the abstract, issues arise when these freedoms conflict with claims of "national security" that underlie the use of the high seas for weapons testing. In their 1955 article, Myres McDougal and Norbert Schlei framed the proponents' argument for unilateral, exclusionary warning areas by stating: "If 'freedom of the seas' is an absolute . . . it may therefore reasonably be asked why the seas are not as 'free' for nuclear weapons tests conducted in the interests of survival of the West, as they are for navigation and fishing."[38] The next sections look at these and other arguments in more detail.

Balancing the Interests of Nations

The nuclear bomb-testing programs generated numerous protests over the legality of exclusionary danger zones.[39] Never before had activities on the high seas laid claim to areas as large as the 400,000 square miles around Bikini Atoll established for the U.S. tests. McDougal and Schlei responded to these protests by arguing that freedom of the high seas was never absolute[40] and that "reasonable" restrictions on this freedom were permissible.[41] In their view, freedom of the high seas is "a legal conclusion invoked to justify a policy preference for certain unilateral assertions as against others."[42] As long as no nation protests the unilateral claims of a party, such claims will eventually be accepted as an established way of doing things and become customary international law.[43]

Once protests arise, as they did to the nuclear tests in the Pacific,[44] the standard for settling these disputes, according to McDougal and Schlei, is the standard of what is "reasonable as between the parties."[45] In the case of weapons testing, the factors of reasonableness that they found to be most relevant were as follows:

> [T]hat it is for a purpose much honored in world prescription [self-defense], that it asserts the least possible degree of authority necessary to the achievement of its purpose, that it is limited both in area and in duration to the minimum consistent with its purpose, that the area which it affects is of relatively slight importance to international trade and fishing, and that it is asserted in a context of crises which makes its purpose of paramount importance to all who value a free world society.[46]

Later, these authors minimized the intrusion of weapons testing on the rights of other nations by stating that the claim of the United States does not seriously interfere with "freedom of the seas" and, moreover, that "[n]o ships are seized or condemned, nor is civil or criminal jurisdiction of any kind asserted."[47] McDougal and Schlei apply these arguments not only to nuclear testing but also to any weapons testing or other military activity on the high seas.[48]

Other authors question the test of "reasonableness." D. O'Connell maintains, for instance, that the concept of reasonable use is "essentially relativistic and hence susceptible of subjective evaluation. . . . The concept is, therefore, not capable of resolving specific questions: all that it is capable of is the exclusion of their automatic resolution according to rigid rules, and the requirement that resolution be based upon appraisal as distinct from mandate."[49] The relativistic approach also means that the legal analysis of a problem may change as new facts become available. McDougal and Schlei acknowledged, for instance, that little was known about the chronic and acute effects of radioactivity on water and fish.[50] Nonetheless, they discounted the dangers of radioactivity and largely ignored long-term problems posed by the nuclear weapons tests.[51]

The Official United States Position

Following generally the rationale of McDougal and Schlei, nations that test missiles and weapons maintain that the attendant warning areas are reasonable in size, duration, and location when balanced against the rights that are infringed on.[52] The U.S. Navy's position is somewhat typical. It maintains that it can use ocean areas temporarily for military purposes, even if it interferes with other uses, as long as the use is "reasonable."[53]

A navy commentator analyzing a recent 6,000-square-mile "warning zone" off the coast of Florida argued that this "relatively large area" was justified because the Trident missile "is a sophisticated, long range missile that needs an extensive operating area."[54] Because the warning zone lasted only seven hours and was both in an area previously used for missile testing and unlikely to disrupt navigation, this commentator viewed it as reasonable.[55]

The United States's position does differ from that of McDougal and Schlei, however, because it asserts no sovereignty over the danger zones, claiming only that they are warning areas, "predicated on voluntary compliance" and subject to freedom of navigation.[56] The United States has had difficulty sustaining this "voluntary compliance" position in practice. For instance, the commander of the first nuclear tests declared: "If any ship actually tried to interfere with tests in any way . . . we would use force to see that they did not interfere. We would escort them out forcibly if necessary."[57] Whether he was referring to the area within territorial waters (over which he would have jurisdiction) or to the entire warning area is uncertain.

Recent confrontations also show that the United States will act with force, at least with respect to submarine missile launch sites. For example, in July 1989 the U.S. Navy tried unsuccessfully to board a Greenpeace ship protesting a planned submarine missile launch.[58] Later that year, as noted earlier, the U.S. Navy successfully pushed a Greenpeace vessel from a submarine launch site off the coast of Florida.[59] In its subsequent defense of this latter action, the navy stated that although the Greenpeace vessel could under international law assume the risk of entering the "warning zone," the vessel could not legally enter the "launch safety zone."[60] This distinction between warning zones for impact areas and exclusionary "launch safety zones" is a relatively new one. The navy had the authority to establish the launch safety zone, it contended, because the only safe way to launch a missile is to keep other vessels from the immediate launch site.[61] Vessels above a submarine launch could disrupt the missile's trajectory, creating risks to the submarine's crew and to third parties. Other vessels must respect this safety zone, according to the navy, in order to show "due regard" for the exercise of navigational freedoms by the navy's vessels.[62]

Abuse of Right

The flip side to the requirement to act with "due regard" of others' rights is the prohibition from acting in a manner that would constitute "an abuse of right."[63] Atmospheric testing of nuclear weapons would clearly seem to qualify as an abuse of right, given the overwhelming evidence of its harmful effects on the environment. For example, the United States's 1.4-megaton blast over Johnston Island in 1962 lit the sky from Australia to Hawaii, destroyed orbiting satellites, popped street lights in Honolulu, and altered the Van Allen radiation belts that circle the earth.[64] Even the recent French nuclear testing at Moruroa, although underground, is suspected of producing health risks to the population and the fishery there.[65]

The legality of atmospheric testing of nuclear weapons on the high seas is seriously doubtful. One authority in Britain, Earl Jowett, has stated:

> I am entirely satisfied that the United States, in conducting these [nuclear test] experiments, have taken every possible step open to them to avoid any possible danger. But the fact that the area which may be affected is so enormous at once brings this problem: that ships on their lawful occasions may be going through these waters, and you have no right under international law, I presume, to warn people off.[66]

This view is supported by the Asian-African Legal Consultative Committee, which concluded in a 1964 report that nuclear tests on "the high seas and in the airspace there above also violate the principle of the freedom of the seas and the freedom of flying above the high seas, as such test explosions interfere with the freedom of navigation and of flying above the high seas and result in pollution of the water and destruction of the living and other resources of the sea."[67]

Although literally applicable only to nuclear testing programs, this conclusion would seem to apply equally to other military activities. The testing of missiles without lethal warheads is not as inherently serious as the testing of nuclear weapons, but significant damage can certainly occur when a missile strikes a vessel.[68] The environmental impact of repeatedly launching missiles into the same fragile ecosystem—for example, into Kwajalein Atoll—should also be considered.[69] This testing would seem to violate requirements that all nations "protect and preserve rare or fragile ecosystems."[70] In addition, the extended duration of many of the warning zones—forty weeks and more per year in some cases—may constitute an "abuse of right."

The "Peaceful Purposes" Clause of the 1982 Convention

In addition to confirming that all nations have broad freedoms of the high seas,[71] the 1982 Convention also states that "the high seas shall be reserved for peaceful purposes."[72] This "peaceful purposes" clause also applies to the exclusive economic zone, the seabed, and the seabed's subsoil.[73]

The meaning of the term "peaceful purposes" is unclear. Proponents of missile tests maintain that these tests are conducted for "peaceful purposes" because the tests are intended to prepare for self-defense and thus to keep the peace and maintain national security.[74] The United States took the following position during the negotiations of the 1982 Convention:

> The United States had consistently held that the conduct of military activities for peaceful purposes was in full accord with the Charter of the United Nations and with the principles of international law. Any specific limitation on military activities would require the negotiation of a detailed arms control agreement. The [Law of the Sea] Conference was not charged with such a purpose and was not prepared for such negotiations. Any attempt to turn the Conference's attention to such a complex task would quickly bring to an end current efforts to negotiate a law of the sea convention.[75]

The text of the 1982 Convention also appears to support the conclusion that although military activities inconsistent with the UN Charter are prohibited, at least some military activities on the high seas are permissible.[76] For example, the 1982 Convention confirms the freedom of military vessels to navigate the high seas.[77] Moreover, there would be no reason for the convention to forbid explicitly "any threat or use of force against the territorial integrity or political independence of any State"[78] if the "peaceful purposes" principle were intended to ban all military activities from the ocean. Similarly, because article 19 prohibits certain military activities in the territorial sea, including, most notably, "exercise or practice with weapons of any kind," such activities are arguably permissible in the ocean beyond the territorial seas.[79]

A 1985 report of the secretary-general of the United Nations supports this position, concluding that "[m]ilitary activities which are consistent with the principles of international law embodied in the Charter of the United Nations, in particular with Article 2, paragraph 4, and Article 51, are not prohibited by the Convention of the Law of the Sea."[80]

Nonetheless, the broad and elastic term "peaceful purposes" must be seen as subject to evolution over time. This hopeful phrase encompasses the aspirations of the world community that we can begin to resolve our conflicts through nonviolent means. A recent article by the Russian scholar V. F. Tsarev states that "[t]he common requirements of using the high seas for peaceful purposes imposes an obligation to perform activities, including those of a military nature, in a way so as not to threaten the peace and security of states or create an obstacle for international merchant navigation."[81]

Tsarev then lists certain activities that are prohibited because they interfere with other interests of states and other legitimate uses of the sea, including tests of nuclear weapons, combat training areas, and missile and artillery practice, especially in areas designated for scientific research.[82]

Boleslaw Adam Boczek has taken a similar approach in referring to the "constructive ambiguity" of the "peaceful purposes" clause as a conscious compromise acceptable to both the maritime powers and those nations that wanted to restrict military activities on the oceans.[83] Boczek reasons that "the peaceful purposes clauses must be allowed some legal effect other than that meant by Article 301; otherwise they would be redundant."[84] Although the "peaceful purposes" goal of demilitarizing the oceans is only "soft" law, it may provide a "legal base" for restricting military uses of the oceans.[85] Boczek also usefully points out that there already are several specific limitations on military activities at sea, including the Seabed Treaty[86] and the nuclear-free zone treaties covering Latin America[87] and the South Pacific.[88] Finally, Boczek emphasizes the customary law restraints that exist under the duties of "reasonable regard" for the interests of other nations. Among his specific conclusions, for instance, is that "even under customary international law it would be difficult to argue that a nuclear test on the High Seas was an activity showing reasonable regard to the interests of other states."[89]

The "reasonableness balance" applicable to resolving disputes among competing users of the high seas thus must be seen as an evolving one. Certainly Boczek's contemporary view of the reasonableness of nuclear tests differs radically from McDougal and Schlei's 1955 analysis. Moreover, inasmuch as treaties signify emerging customary law, the 1982 Convention's preference "for peaceful purposes" may make it less "reasonable" to use the seas for military activities.[90]

The Seabed Treaty

The 1971 Seabed Treaty is relevant to weapons testing because it prohibits nations from placing on the seabed "any nuclear weapons or any other types of weapons of mass destruction as well as structures, launching installations or any other facilities specifically designed for storing, testing, or using such weapons."[91] Nations testing nuclear-capable missiles on the high seas monitor these missile flights with satellite transponders secured to the sea floor. These transponders are immobile, implanted weeks or even months before a test. Although the Seabed Treaty is somewhat ambiguous,[92] this practice appears to violate the treaty's prohibition against use of the sea floor for the testing of weapons of mass destruction.

APPLYING CUSTOMARY INTERNATIONAL LAW TO WARNING ZONES

The Practices of Nations

In determining which rules of customary international law govern weapons testing, it is important to recognize the difference between a "habitual practice, or usage" and a "legal obligation, or custom."[93] "A habitual practice, such as the avoidance of weapons testing zones on the oceans, constitutes a custom only if states generally recognize that they are legally obliged to adhere to that practice."[94] McDougal and Schlei pointed to the historical right-of-way given to naval vessels on military maneuvers as establishing customary law, but they also acknowledged that this right-of-way was given because "mariners preferred to avoid exercise areas altogether rather than to be delayed and endangered by unexpected encounters."[95]

Because of the serious danger that can result from entering a warning or exclusionary zone, it is unrealistic to expect that many will attempt to do so. It may also be improper to say that customary law can be established in the face of such danger, as the practice may almost approach one that is followed because of coercion. The United States seemed to recognize this problem in its reaction several years ago to the 24-nautical-mile seaward security zone claimed by Vietnam and Kampuchea (Cambodia). Based on the hazard inherent in entering the area, the United States issued special warnings to mariners to avoid these areas, but the warnings also stated that "[t]he publication of this notice . . . in no way constitutes a legal recognition by the United States of the validity of any foreign rule, regulation, or proclamation so published."[96] This reservation of rights may be one way to influence the formation of international law; the other major method is through protest.

Protests

The importance of protest in influencing customary international law is widely recognized.[97] In an important distinction, O'Connell explains that "[t]he nature of the protest may well vary, depending upon whether it aims to reserve existing rights from invasion and so prevent an historic claim from maturing, or, by denying to the claimant State the benefit of the element of general consent of nations, to inhibit custom."[98] The United States's special warning described earlier exemplified the latter as it tried on the one hand to safeguard ships while protesting Vietnam's proclaimed right to a 24-mile "security zone." All protests over designated missile ranges are similar in nature. Unfortunately, "the advantage lies with the party which acts and the disadvantage with the party which must demonstrate that the action is illegal."[99]

Several recent examples can illustrate the importance of protests. After a 1954 U.S. hydrogen bomb test in the Marshall Islands contaminated the crew of the Japanese fishing vessel *Lucky Dragon* (*Fukuryu Maru*), the Japanese government protested the action.[100] In the *Nuclear Test* cases, Australia and New Zealand challenged France's atmospheric testing of nuclear devices in the International Court of Justice. The complaint charged in part that "the interference with ships and aircraft on the high seas and in the superjacent airspace [caused by the French tests] . . . constitutes infringement of the freedom of the high seas."[101]

In more recent examples (1987 and 1989), the United States protested the presence of two Soviet missile tests alarmingly close to Hawaii.[102] These protests, which were based on safety concerns, shows that even a nation engaged in missile testing may recognize some limits to the practice. The Soviet Union responded by stating that the missiles would not fly over any land territory and the testing would involve only international waters and airspace.[103]

Liability

The ultimate issue, of course, is what should happen if a vessel is damaged as a result of a missile launched into or from an exclusionary zone, assuming that proper notice has been given. This situation presents squarely the conflict between the freedom of navigation and the claimed rights to a warning zone. If the nation whose military fired the missile created an inherently hazardous situation, perhaps it should be strictly liable for the injuries that result. It could be argued, however, that the vessel was contributorily negligent by entering into a known danger zone. Whether the vessel should have restrained its free navigation through this zone is a question that strikes at the heart of the legitimacy of establishing these zones.

Some instructions to commanding officers issued by the United States and the United Kingdom have illustrated a sense of responsibility toward other vessels in the area.[104] The payments made by the United States to the victim of the 1988 *Jag Vivek* incident[105] and the $2 million given to the crew of the *Lucky Dragon* after its contamination by the 1954 hydrogen bomb test[106] also indicate a sense of responsibility. In these incidents, however, proper warning apparently had not been given to the civilian crews.

Risk-creating nations clearly have the responsibility to warn the citizens of other nations who may be affected by the actions. "A State is under a duty to notify any other State which may be threatened by harm from the abnormally dangerous activities which the State permits to be conducted within its jurisdiction."[107] Even with a proper warning, the risk-creating nation must exercise "due diligence" to protect others from its risky activities.

If a proper warning is given, the question becomes whether the temporary appropriation of an area of the high seas by one nation is legitimate and whether the contributory negligence of the vessel entering the exclusionary zone reduces the responsibility of the risk-creating nation. The concept of contributory or comparative negligence is common to many legal systems. Under this doctrine, the liability of the party creating the risk is reduced or eliminated if the injured party also contributed to the risk or failed to take standard precautions to avoid it.[108] This doctrine suggests that a vessel entering into a warning or exclusionary zone on the high seas that was established to test missiles or to engage in other hazardous military activities may be contributorily negligent if it were injured in that zone.

Evaluating this argument under the "reasonableness" standard requires examination of the size and duration of the danger zone claimed and the extent to which it imposes a burden on the otherwise lawful maritime activities of the other vessel. It is easy (and thus "reasonable") for a vessel to avoid a small safety zone established around a temporary navigational hazard, but it may be quite burdensome for a merchant vessel to skirt hundreds of thousands of square miles or for a fishing vessel to be kept from the living resources of a large area for an extended period of time. If it is truly burdensome for a civilian vessel to adhere to the requirements of a warning or exclusionary zone, then it would be "unreasonable" to expect it to do so, and its presence in such a zone should *not* be viewed as contributory negligence.

The burden would thus appear to be on the nation creating the dangerous activity to reduce the risk to the extent possible, and this nation must pay compensation for injuries caused to the vessels, property, and citizens of other nations when they suffer injuries while conducting lawful activities in international waters.

CONCLUSION

The relativistic and flexible approach offered by McDougal and Schlei in their 1955 article may be appropriate in some circumstances for new problems that are just beginning to be examined by the community of nations, but the imprecision of the approach almost invites conflict.[109] As nations focus on a problem, they develop norms that are specific and understandable in order to promote a stable and predictable world order. Strong protests have been registered against atmospheric testing of nuclear bombs on the high seas.[110] All nations have now stopped this practice, and it now appears to be prohibited by customary international law.[111] The result reached by McDougal and Schlei in 1955 with respect to nuclear testing thus is not appropriate today; not only does it fail to consider the effect on the global commons, but it also does not (and could not) take account of recent understandings of the effects of radiation on humans and the marine environment.

Missile testing on the high seas interferes with other uses of the high seas (such as navigation and fishing) somewhat less and has somewhat less of a negative impact on the environment when compared with nuclear bomb tests. Nonetheless, the protests of nations to the high seas testing of nuclear bombs and the concerns expressed about recent missile test launches[112] clearly indicate that missile testing on the high seas will continue to be accepted as a legitimate use of the sea only insofar as it does not significantly interfere with navigation and fishing or pose serious safety concerns.

Attempts to appropriate areas of the oceans for military purposes by declaring "exclusionary zones" on the high seas are now seen as improper because they do not permit other lawful and necessary maritime activities to continue in these zones. The military powers have tried to avoid this problem by declaring "warning" zones instead, which are ostensibly less intrusive. In fact, however, warning zones are essentially identical with exclusionary zones, and the military powers have acted as if the vessels of other nations are not entitled to enter such zones. Warning zones must therefore be viewed with the same degree of suspicion and concern that nations have expressed toward exclusionary zones. The *Corfu Channel* case[113] stands as a strong precedent that maritime navigational freedoms cannot be interfered with, even to serve the security concerns of other nations, and that compensation must be paid when injuries to persons and property occur.

The current state of international law is, therefore, that missile testing on the high seas and other, similar military activities on the oceans are legitimate only if they do not impede free navigation, interfere with fishing activities, cause any significant harm to the environment, or threaten human settlements. Exclusionary and warning zones that cover large areas or are extended in duration should not be viewed as acceptable under such a standard.

If a vessel chooses to enter an exclusionary or warning zone on the high seas, the nation seeking to test its missiles or to conduct other military operations in that vicinity cannot lawfully seize or remove that vessel without the permission of the nation whose flag the vessel flies.[114] Finally, if a vessel operating with the support of its flag government is damaged as a result of the missile test or other military operation, the nation causing the damage would be liable under international law.

NOTES

1. *See* HUGO GROTIUS, MARE LIBERUM (Ralph van Daman Magoffin trans. 1916); RAM PRAKASH ANAND, ORIGIN AND DEVELOPMENT OF THE LAW OF THE SEA: HISTORY OF INTERNATIONAL LAW REVISITED (1983); United Nations Convention on the Law of the Sea, art. 87(1)(a), Dec. 10, 1982, U.N. Doc. A/CONF.62/122, 21 I.L.M. 1261 (1982) [hereinafter 1982 Convention].
2. *See* John W. Finney, *Russia's Motive Puzzles Capitol*, N.Y. Times, Jan. 8, 1960, at 2.

3. HYDROPACs announce marine notices in the Pacific Ocean, and HYDRO-PLANTs announce notices in the Atlantic and Indian oceans. U.S. DEFENSE MAPPING AGENCY, HYDROGRAPHIC/TOPOGRAPHIC CENTER, NOTICE TO MARINERS (1988) [hereinafter NOTICE TO MARINERS].

4. See Tsipis, *The Operational Characteristics of Ballistic Missiles*, 1986 SIPRI Y.B. 406 (1986); *see also* NOTICE TO MARINERS, *supra* note 3.

5. Kwajalein Atoll is surrounded by a permanent warning area with a radius of 200 nautical miles. *See* Note, *Weapons Testing Zones*, 99 Harv. L. Rev. 1040, 1048 (1986) [hereinafter *Weapons Testing Zones*]. The exact language on the nautical charts reads: "Caution—intermittent hazardous missile operations will be conducted within the area 24 hours daily, on a permanent basis." Defense Mapping Agency Hydrographic/Topographic Center, Omega, Map, Marshall Islands Northern Portion.

6. 4 MARJORIE WHITEMAN, DIGEST OF INTERNATIONAL LAW 553 (1965). *See generally* Jon M. Van Dyke, Kirk Smith, & Suliana Siwatibau, *Nuclear Activities and the Pacific Islanders*, 9 ENERGY 733 (1984).

7. The notice to mariners stated: "All vessels are warned of the hazards to ships and personnel, and are cautioned against the danger entailed in entering this area." WHITEMAN, *supra* note 6, at 558.

8. S. FIRTH, NUCLEAR PLAYGROUND 23 (1986).

9. *See generally* WHITEMAN, *supra* note 6, at 557–603.

10. *Weapons Testing Zones*, *supra* note 5, at 1050 n.70 (citing H. REIFF, THE UNITED STATES AND THE TREATY LAW OF THE SEA 364-65 (1959)).

11. WHITEMAN, *supra* note 6, at 597. The "public warning" announced that the area was "dangerous to shipping and aircraft." *Id.*

12. *Id.* at 598.

13. *See* Van Dyke, Smith, & Siwatibau, *supra* note 6, at 739.

14. See 12 I.L.M. 773 (1973).

15. GREENPEACE NEW ZEALAND, FRENCH POLYNESIA NUCLEAR TESTS: A CHRONOLOGY (1982).

16. *See* 1982 Convention, *supra* note 1, art. 3 (establishing a 12-mile limit for territorial seas). Coastal states claiming these "security zones" consider them to be like internal waters, denying to other nations even the right of innocent passage. Choon-ho Park, *The 50-Mile Boundary Zone of North Korea*, 72 AM. J. INT'L L. 866, 867 (1978) ("[C]ivilian ships and civilian planes (excluding fishing boats) are allowed to navigate or fly only with appropriate prior agreement or approval").

17. Recently, such zones have included a 25-mile naval and air security zone claimed by Nicaragua in 1983, zones claimed by the Persian Gulf nations during the Iran-Iraq conflict, and a 200-nautical-mile exclusion zone declared by the United Kingdom around the Falkland (Malvina) Islands in 1982. *See* Louis B. Sohn, *International Navigation: Interests Related to National Security*, in INTERNATIONAL NAVIGATION: ROCKS AND SHOALS AHEAD? 312-15 (Jon M. Van Dyke, Lewis M. Alexander, & Joseph R. Morgan eds. 1988). North Korea declared a 50-mile security zone off its coasts in 1977, and South Korea declared a 150-mile zone into

the Sea of Japan and a 100-mile zone into the Yellow Sea in the 1970s. *Id*. at 313 (citing Park, *supra* note 16, at 866-75). In 1973, Libya claimed the 100 miles of coastal waters in the Gulf of Sidra (Sirte) as a maritime security zone or "restricted area" but later changed this claim to one of "historic" waters. *Id*.

18. Elizabeth Cuadra, *Air Defense Identification Zones: Creeping Jurisdiction in the Airspace*, 78 VA. INT'L L.J. 485, 492 (1978). Burma, Canada, Iceland, India, Japan, Korea, Oman, the Philippines, Sweden, Taiwan, the United States, and Vietnam had claimed ADIZs as of 1977.

19. *Id*. at 493. ADIZs are designed to warn coastal nations roughly one hour before the arrival of approaching planes. Although these zones do not restrict overflight officially, any airplane ignoring the identification requirements risks being either escorted to a military air base or shot down. *Id*. at 507.

20. *See infra* notes 25-29 and accompanying text.

21. Commander Charles R. Hunt, Greenpeace and the U.S. Navy: Confrontation on the High Seas 19 (paper prepared at the Naval War College, Newport, R.I., May 14, 1990).

22. *Id*. at 19.

23. *See* ROBERT HUNTER, WARRIORS OF THE RAINBOW 116-17 (1979).

24. *See* FIRTH, *supra* note 8, at 101-02.

25. *See* Tim Ryan & Mary Adamski, *Ship Arrives with Gaping Hole*, Honolulu Star-Bull., Dec. 13, 1988, at A-8; Phil Mayer, *Navy Recommends Disciplinary Action in Freighter Accident*, Honolulu Star-Bull., Feb. 22, 1990, at A-14.

26. Ryan & Adamski, *supra* note 25; Joan Conrow & Jan TenBruggencate, *Freighter Hit by Navy Missile*, Honolulu Advertiser, Dec. 13, 1988, at A-1; Tim Ryan & Harold Morse, *Captain: "Military Was Guiding Me,"* Honolulu Star-Bull., Dec. 14, 1988, at A-1.

27. Mayer, *supra* note 25. All of the navy's announcements regarding this incident referred to the several-thousand-square-mile area in which these maneuvers occurred (an area 150 nautical miles off Kauai's western coast) as a "closed" area "off limits" to civilian vessels. *See id*.; Ryan & Adamski, supra note 25, at A-8. These official statements thus dropped the pretense that this zone was simply a warning zone, rather than an exclusionary zone.

28. Jim Borg, *Series of Errors Blamed in Kauai Missile Incident*, Honolulu Advertiser, July 18, 1990, at A-4.

29. Mayer, *supra* note 25. The U.S. Navy has nearly caused similar accidents in at least two other recent events. In May 1975, a navy attack aircraft mistook a fishing boat for a target craft and just missed dropping bombs on it. See Ryan & Adamski, *supra* note 25, at A-8; Conrow & TenBruggencate, *supra* note 26, at A-4. In 1988, a U.S. Navy destroyer narrowly missed shelling a Japanese patrol boat. Japan promptly protested the action, charging that the U.S. crew violated international law by firing its guns inside Japanese territorial waters. *See* David E. Sanger, *Japan Says U.S. Salvos Almost Hit Ship*, N.Y. Times, Nov. 12, 1988, at L-3.

30. Edward H. Kalcum, *Soviet Intelligence Ship Intrudes on Trident Test*, AVIATION WK. & SPACE TECH., Jan. 25, 1982, at 21; *see also U.S. Is Accused of Spying on Soviet Rocket*

Tests, N.Y. Times, Nov. 1, 1961, at 16 (describing the activities of United States observers of a Soviet missile test); Hunt, *supra* note 21, at 5 (noting that a Soviet ship monitored a U.S. submarine launch from outside the 5,000-yard "launch safety zone").

31. *See, e.g.*, GROTIUS, *supra* note 1; 2 D. O'CONNELL, THE INTERNATIONAL LAW OF THE SEA 793 (1984); ANAND, *supra* note 1, at 129.

32. *See* 2 O'CONNELL, *supra* note 31, at 792-93.

33. GROTIUS, *supra* note 1, at 28.

34. *Id.* at 27.

35. 1982 Convention, *supra* note 1, art. 87. The 1982 Convention had been ratified by fifty-five nations as of March 1993. It will take effect when sixty nations have ratified it. A few nations, such as the United States, have refused ratification because of a disagreement over deep sea mining, and others have expressed concern about the financial obligations that might accompany ratification. *See generally* CONSENSUS AND CONFRONTATION: THE UNITED STATES AND THE LAW OF THE SEA CONVENTION (Jon M. Van Dyke ed. 1985); Jon M. Van Dyke & Christopher Yuen, *"Common Heritage" v. "Freedom of the High Seas": Which Governs the Seabed?*, 19 San Diego L. Rev. 493 (1982). The non-seabed provisions of the convention appear, however, to be widely accepted as codifying the norms that govern maritime activities, and all nations have viewed this convention as a primary source of customary international law. *See also* Convention on the High Seas, art. 2, Apr. 29, 1958, 13 U.S.T. 2312, T.I.A.S. No. 5200, 450 U.N.T.S. 82 (listing four freedoms—navigation, fishing, laying of submarine cables and pipelines, and flight over the high seas).

36. 1982 Convention, *supra* note 1, arts. 99-111.

37. *See infra* notes 71-90 and accompanying text.

38. Myres S. McDougal & Norbert A. Schlei, *The Hydrogen Bomb Tests in Perspective: Lawful Measures for Security*, 64 YALE L.J. 648, 685 n.203 (1955).

39. *See infra* note 101 and accompanying text.

40. McDougal & Schlei, *supra* note 38, at 663.

41. *Id.* at 655-61, 682-95.

42. *Id.* at 663.

43. *See id.* at 659 n.62 ("It is not the unilateral claims to use but rather the tolerances of external decision makers, including the specific decisions of international officials, which create the expectations of pattern and uniformity in decision, of practice in accord with rule, commonly called law").

44. *See* Emmanuel Margolis, *The Hydrogen Bomb Experiments and International Law*, 64 YALE L.J. 629 (1955); *see also infra* note 101 and accompanying text.

45. McDougal & Schlei, *supra* note 38, at 660.

46. *Id.* at 686. The warning zones were, in fact, quite broad. By 1954, approximately thirty-five surface, or atmospheric, nuclear bombs had been detonated, with warning areas that covered an area of approximately 400,000 square nautical miles. This warning area was in effect for at first months, and then years, at a time. WHITEMAN, *supra* note 6, at 557-60.

47. McDougal & Schlei, *supra* note 38, at 684 (citations omitted).
48. *Id.* at 652-53, 690-95.
49. 1 O'CONNELL, *supra* note 31, at 58.
50. McDougal & Schlei, *supra* note 38, at 692.
51. *Id.* at 652-53, 692-94.
52. McDougal & Schlei, *supra* note 38, at 691.
53. Hunt, *supra* note 21, at 11.
54. *Id.* at 12.
55. *Id.* at 13.
56. This position was spelled out in a paper prepared for the use of the U.S. delegation at the Conference on the Law of the Sea held in Geneva in 1958:

   > The Delegation should bear in mind, however, it does not necessarily follow as seemingly suggested by McDougal and Schlei, that a nation may unilaterally appropriate for its exclusive use a portion of the high seas for this purpose. In particular, the United States has been careful not to claim the right to establish a prohibited or restricted area which is tantamount to closing off a portion of the seas as a matter of enforceable right, action customarily taken only within the limits of territorial waters.
   >
   > In contrast, Danger or Warning areas on the high seas are predicated on the principle of voluntary compliance. As a matter of comity these areas are generally observed. . . . The nuclear testing areas in question have been established as danger areas, warning all vessels and aircraft to stay clear, but not prohibiting them from entering the hazard area.

   WHITEMAN, *supra* note 6, at 550 (quoting, in part, INTERNATIONAL LAW SITUATION AND DOCUMENTS 627 (1956)). The United States also relied on the "reasonableness of the area or zone from the standpoint of size, duration, and location" and on the "fact that there ha[d] been no protest [as of 1958] of the United States conduct." *Id.*

57. *Ships, Planes Warned to Avoid Bikini Area,* N.Y. Times, June 26, 1946, at 7.
58. *Greenpeace Succeeds in Halting Test Launch,* Honolulu Advertiser, July 29, 1989, at B-1; Hunt, *supra* note 21, at 2.
59. Jeffrey Schmalz, *After Skirmish with Protesters, Navy Tests Missile,* N.Y. Times, Dec. 5, 1989, at A-1 (nat'l ed.); *see also supra* text accompanying note 75.
60. *Id.* (quoting Vice Admiral Roger F. Bacon); Hunt, *supra* note 21, at 2.
61. This position is arguably supported by the 1982 Convention, which authorizes safety zones around artificial installations. 1982 Convention, *supra* note 1, art. 60(4)-(8). Those zones are, however, limited to a radius of 500 meters, *id.* art. 60(5), and thus are not as burdensome as the 5,000-yard zone claimed around this launch site. The navy based the size of the safety zone on the lessons learned from earlier launch failures, which scattered debris as far as 4,000 yards from the launch point. *See* Hunt, *supra* note 21, at 15.
62. *See* Hunt, *supra* note 21, at 6, 12. The "due regard" phrase comes from articles 58(3) and 87(2) of the 1982 Convention, *supra* note 1.

63. 1982 Convention, *supra* note 1, art. 300.

64. Firth, *supra* note 8, at 25.

65. *See* H. R. ATKINSON, P. J. DAVIES, D. R. DAVY, L. HILL, & A. C. MCEWAN, REPORT OF A NEW ZEALAND, AUSTRALIAN AND PAPUA NEW GUINEA SCIENTIFIC MISSION TO MURUROA ATOLL (1984); Jon M. Van Dyke, *Protected Marine Areas and Low-lying Atolls* (chapter 16 in this book).

66. S. Azadon Tiewul, *International Law and Nuclear Test Explosions on the High Seas*, 8 CORNELL INT'L L.J. 45, 48 n.9 (1974); *see also* Margolis, *supra* note 44, at 635.

67. THE WORK OF THE ASIAN-AFRICAN LEGAL CONSULTATIVE COMMITTEE 1956-1974, at 96 (1974). The delegations supporting the final report were Burma, Ceylon (Sri Lanka), India, Indonesia, Japan, Pakistan, Thailand, the United Arab Republic (Egypt), Ghana, the Philippines, Laos, and the League of Arab States. *Id.* at 24.

68. *See supra* notes 25-29 and accompanying text (discussing the U.S. Navy's accidental striking of a commercial vessel with a missile).

69. *See* Van Dyke, *supra* note 65.

70. 1982 Convention, *supra* note 1, art. 194(5).

71. *Id.* art. 87.

72. *Id.* art. 88.

73. *Id.* arts. 58, 141.

74. McDougal and Schlei, for instance, stressed the "overriding utility of the [hydrogen bomb atmospheric] tests to the free world." McDougal & Schlei, *supra* note 38, at 691. George Schultz, President Reagan's secretary of state, argued that it was the strength of the United States that brought its adversaries to the bargaining table. Speech given by George Schultz at Kennedy Theater, University of Hawaii at Manoa, July 21, 1988.

75. 5 UNCLOS III O.R. (67th plenary mtg.), para. 81, *quoted in* Francesco Francioni, *Peacetime Use of Force, Military Activities, and the New Law of the Sea*, 18 CORNELL INT'L L.J. 203, 222 (1985).

76. *See generally* David L. Larson, *Naval Weaponry and the Law of the Sea, in The UN Convention on the Law of the Sea: Impact and Implementation*, 19 L. SEA INST. PROC. 41, 56-57 (E. D. Brown & R. R. Churchill eds. 1987); Bernard H. Oxman, *The Regime of Warships Under the United Nations Convention on the Law of the Sea*, 24 VA. J. INT'L L. 809, 829-32 (1984).

77. 1982 Convention, *supra* note 1, art. 95; *see also* Rex J. Zedalis, *Foreign State Military Use of Another State's Continental Shelf and International Law of the Sea*, 16 RUTGERS L.J. 1, 95 n.393 (1984).

78. 1982 Convention, *supra* note 1, art. 301.

79. Francioni, *supra* note 75, at 223.

80. *Report of the Secretary-General, General and Complete Disarmament Study on the Naval Arms Race*, U.N. Doc. A/40/535, para. 188 (1985).

81. V. F. Tsarev, *Peaceful Uses of the Sea: Principles and Complexities*, 10 MARINE POL'Y 153, 156 (1988).

82. *Id.* at 156-57.

83. Boleslaw Adam Boczek, *The Peaceful Purposes Reservation of the Convention on the Law of the Sea*, in OCEAN YEARBOOK 8 at 329, 336 (E. M. Borgese, N. Ginsburg, & J. R. Morgan eds. 1989).

84. *Id.* at 358.

85. *Id.* at 360.

86. The Treaty on the Prohibition of the Emplacement of Nuclear Weapons and Other Weapons of Mass Destruction on the Seabed and the Ocean Floor and in the Subsoil Thereof, Feb. 11, 1971, 23 U.S.T. 701, T.I.A.S. No. 7337 [hereinafter The Seabed Treaty]. *See infra* notes 91–92 and accompanying text.

87. Treaty for the Prohibition of Nuclear Weapons in Latin America, Feb. 14, 1967, 634 U.N.T.S. 281, 6 I.L.M. 521 (1967).

88. South Pacific Nuclear-Free Zone Treaty, Aug. 6, 1985, 24 I.L.M. 1440 (1985).

89. Boczek, *supra* note 83, at 343.

90. *See* Anthony D'Amato, *Law Generating Mechanisms of the Law of the Sea Conference and Convention*, in CONSENSUS AND CONFRONTATION, *supra* note 35, at 125.

91. The Seabed Treaty, *supra* note 86, art. I(1). *See generally* INTERNATIONAL NAVIGATION: ROCKS AND SHOALS AHEAD? 352-64 (J. Van Dyke, L. Alexander, & J. Morgan eds. 1988). The original Soviet draft of the treaty would have gone even further and "banned all military uses of the seabed including submarine surveillance systems." Larson, *supra* note 76, at 56.

92. *See* 2 O'Connell, *supra* note 31, at 827; Boczek, *supra* note 83, at 338.

93. *See Weapons Testing Zones*, *supra* note 5, at 1049.

94. *Id.*

95. McDougal & Schlei, *supra* note 38, at 678.

96. Special Warning No. 45, NOTICE TO MARINERS, *supra* note 3, Jan. 2, 1987.

97. *See, e.g.*, 1 O'CONNELL, *supra* note 31, at 42. The United States has also recognized the importance of protests in influencing international law. *See* WHITEMAN, *supra* note 6, at 550 (quoting the United States delegation at the 1958 Geneva Conference in recognizing the importance of protest).

98. 1 O'CONNELL, *supra* note 31, at 39.

99. *Id.* at 40.

100. *See* Van Dyke, Smith, & Siwatibau, *supra* note 6, at 736; WHITEMAN, *supra* note 6, at 585-86 ("the United States Government has the responsibility of compensating for economic losses that may be caused by the establishment of a danger zone and for all losses and damage that may be inflicted on Japan and Japanese people as a result of the nuclear tests").

101. *Nuclear Test* cases (*Australia v. France, New Zealand v. France*), 1973 I.C.J. 253, 457, 12 I.L.M. 749, 768 (1973). For the texts of the protests lodged by Australia and New Zealand against France, see IAN BROWNLIE, SYSTEM OF THE LAW OF NATIONS: STATE RESPONSIBILITY, pt. I at 91-94, 110-11 (1983).

102. Andy Yamaguchi, *That Russian ICBM Test Was a Soviet Hit, American Miss*, Honolulu Star-Bull. & Advertiser, Nov. 1, 1987, at A-1.

103. *See generally* Stu Glauberman, *Waihee Protests Soviet Tests*, Honolulu Advertiser, Aug. 11, 1989, at A-1; Rod Ohira & Helen Altonn, *Hawaii Protests Soviet Test*,

Honolulu Star-Bull., Aug. 11, 1989, at A-1, A-8; Stu Glauberman, *Soviet Missile Tests Provoke U.S. Protest*, Honolulu Advertiser, Aug. 12, 1989, at A-1, A-4; *Soviets Halt Isle Missile Tests*, Honolulu Advertiser, Aug. 16, 1989, at A-3 (quoting State Department spokesperson Richard Boucher).

104. *See, e.g.*, NOTICE TO MARINERS, *supra* note 3, Jan. 2, 1987, at I20.

105. *See supra* notes 25-29 and accompanying text.

106. *See* Van Dyke, Smith, & Siwatibau, *supra* note 6, at 736; Personal and Property Damage Claims, T.I.A.S. No. 3160, Jan. 4, 1955.

107. John M. Kelson, *State Responsibility and the Abnormally Dangerous Activity*, 13 HARV. INT'L L.J. 197, 243 (1972) (citing the *Corfu Channel* case (*United Kingdom v. Albania*), 1949 I.C.J. 4).

108. For the way this doctrine applies in admiralty situations, *see generally* SAMIR MANKABADY, COLLISION AT SEA: A GUIDE TO LEGAL CONSEQUENCES 25-31 (Amsterdam, North Holland 1978). At common law, damages were equally divided if both vessels were negligent, regardless of the comparative degrees of fault. This rule was altered pursuant to the 1910 International Convention for the Unification of Certain Rules of Law with Respect to Collision Between Vessels, the 1910 Convention for the Unification of Certain Rules of Law Relating to Assistance and Salvage at Sea, and the United Kingdom's Maritime Conventions Act of 1911. The current rule requires a court to apportion liability "in proportion to the degree in which each vessel was at fault." 1911 Maritime Conventions Act, sec. 1; The Lucile Bloomfield, [1967] 1 Lloyd's Rep. 341, 351.

109. The standard of reasonableness "is an imprecise measure of validity whose very imprecision tends to encourage conflict." Joseph W. Dellapenna, *Canadian Claims in Arctic Waters*, 7 Land & Water L. Rev. 383, 407 (1972).

110. *See, e.g.*, *supra* note 101 and accompanying text (discussing the protests by Australia and New Zealand of French nuclear weapons testing).

111. *See* Tiewul, *supra* note 66, at 68-70; Boczek, *supra* note 83, at 343; *see also supra* notes 63-70 and accompanying text (discussing the legality of nuclear weapons testing).

112. *See supra* notes 97-103 and accompanying text.

113. The *Corfu Channel* case (*United Kingdom v. Albania*), 1949 I.C.J. 4. The United Kingdom sought compensation for the deaths of forty-five British seamen and injuries to forty-two others, as well as for the serious damage suffered by two destroyers when they struck mines while passing through the North Corfu Strait between Albania and the Greek island of Corfu. The International Court of Justice ruled that Albania was liable for the damage even though it had not laid the mine fields because Albania was in a position to know what was happening in its waters and had a duty to notify other states that might be endangered by the activity. The court stated that international law obliges every state "not to allow knowingly its territory to be used for acts contrary to the rights of other States." *Id.* at 22.

This case is particularly significant for the present discussion because the operators of the British vessels knew that dangers lurked in the Corfu Channel when they sailed through it and may, according to some views, have been acting illegally when they entered these waters. *See* BROWNLIE, *supra* note 101, at 48, citing the dissenting opinions of Judges Azevedo and Krylov. The court did not feel that the responsibility of Albania was in any way reduced because the British may have been contributorily negligent in sailing through these waters. The waters were regarded as safe because they had been swept for mines during the two previous years.

114. *Weapons Testing Zones, supra* note 5, at 1057–58.

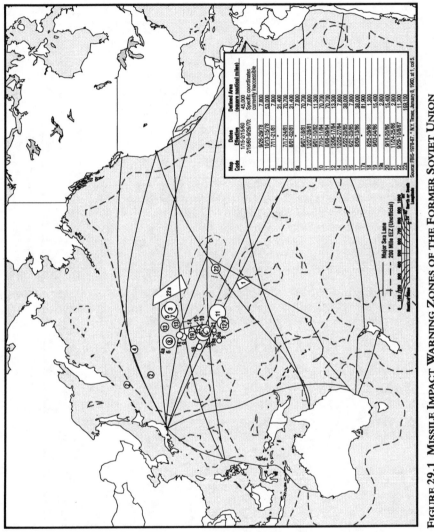

FIGURE 29.1 MISSILE IMPACT WARNING ZONES OF THE FORMER SOVIET UNION

Sources: From reports in the U.S. Foreign Broadcast Information Service (FBIS), 1978–1987, and *New York Times*, Jan. 8, 1960. (Figure prepared by Thomas Feeney.)

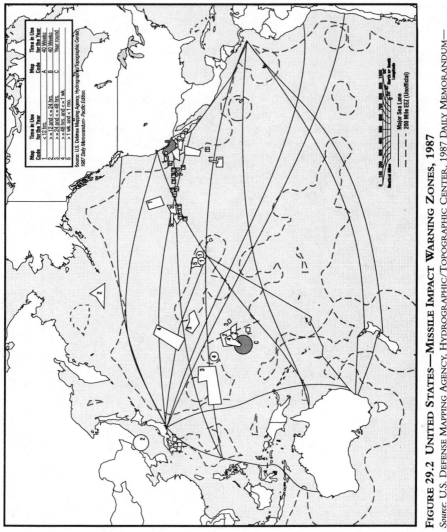

FIGURE 29.2 UNITED STATES—MISSILE IMPACT WARNING ZONES, 1987

Source: U.S. DEFENSE MAPPING AGENCY, HYDROGRAPHIC/TOPOGRAPHIC CENTER, 1987 DAILY MEMORANDUM— PACIFIC EDITION. (Figure prepared by Thomas Feeney.)

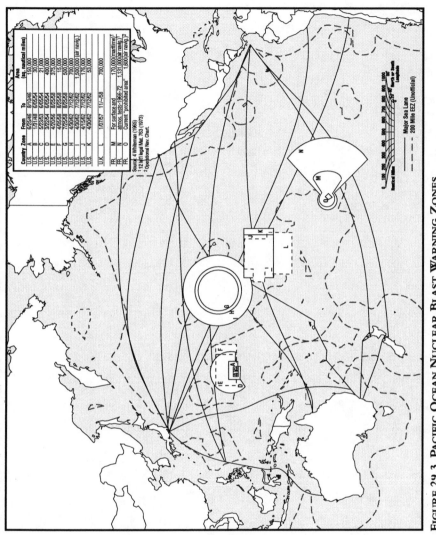

Country	Zone	From	To	Area (sq. nautical miles)
U.S.	A	6/25/46	8/10/46	150,000
U.S.	B	7/31/48	6/05/54	30,000
U.S.	C	3/22/53	6/05/54	21,000
U.S.	D	3/22/54	6/05/54	400,000
U.S.	E	4/20/56	8/25/58	375,000
U.S.	F	4/06/58	8/25/58	15,000
U.S.	G	7/28/58	8/25/58	500,000
U.S.	H	4/30/62	7/12/62	700,000
U.S.	I	4/30/62	7/12/62	1,500,000 (air navy) [1]
U.S.	J	4/15/62	7/12/62	176,000
U.S.	K	4/30/62	7/12/62	53,000
U.K.	L	1/07/57	11/—58	700,000
FR.	M	For surface and		170,000 (maritime) [1]
FR.	N	atmos. tests–1966–72		1131,000 (air navy) [1]
FR.	O	Current "prohibited area" [2]		1,600 (air navy) [2]

Source: 4 Whiteman (1965)
[1] 12 Int'l legal Mat. 763 (1973)
[2] Operational Nav. Chart.

0 100 200 300 400 500 600 700 800 900 1000
Nautical Miles

0 to 25° North or South Longitude

——— Major Sea Lane
– – – 200 Mile EEZ (Unofficial)

FIGURE 29.3 PACIFIC OCEAN NUCLEAR BLAST WARNING ZONES

Source: 4 MARJORIE WHITEMAN, DIGEST OF INTERNATIONAL LAW (553–58, 560, 572, 579, 593–94, 603, 662 (1965): 12 I.L.M. 763 (1973). (Figure prepared by Thomas Feeney).

Summary and Conclusions—
Ocean Governance: Converging
Modes of Idealism

Douglas M. Johnston

THE WORKS IN THIS VOLUME ARE SPECIAL, standing outside the mainstream literature on the law of the sea, because they are the work of idealists. Most of the scholars and observers following events in this field since the late 1960s have been stimulated by the conviction that they have been privileged witnesses to important changes in political, economic, technological, and other forms of "reality." However, most of the writers in this book may believe, as I do, that the world runs essentially on the fuel of idealism—high-octane stuff. There are, of course, lower-level forms of human energy that also contribute to social change, but over the longer haul it seems to be idealism above all that makes the difference.

Even the concept of ocean governance is an expression of idealism, if it conveys the hope that all ocean uses and users can be made subject to reasonable considerations of equity and rational concepts of efficiency or effectiveness. It may be that the ocean dilemma lies not so much in the challenges to idealism as in the coexistence of differing, and sometimes competing, brands of idealism, which in turn can be seen to project different visions of the future. In this book, at least eight such visions can be identified and distinguished.

One vision of the future, clearly presented here, is that ocean law development must proceed within the framework of the 1982 United Nations Convention on the Law of the Sea (1982 Convention) and on the basis of supplementary arrangements being negotiated under the auspices of the Preparatory Commission. It might be argued that this is a realistic vision, because the framework consists of actual negotiated outcomes. And yet it is idealistic because it assumes that there will be universal acceptance of such a framework despite all the evidences of default, deviation, and defiance throughout the history of general multilateral treaty making.

The second vision of the ocean's future is revisionist. This is a view that purports to accept economic reality and on that basis justifies the need for immediate, unscheduled review, and even radical revision, of those controversial areas of the still "unperfected" 1982 Convention and Preparatory Commission arrangements that seem incapable of securing widespread acceptance among the most active ocean users. However, this too is a kind of idealism because it assumes a high level of tolerance among proponents of the convention for radical changes after all these strenuous years of compromise diplomacy. It assumes that the package will not unravel, that further restructuring of the new law of the sea can be effected in the 1990s without fundamental demoralization in the legal community of nations.

The third (Eunomian) vision is bolder still, envisaging new manifestations of legal development for international society that would effectively repudiate the traditional statist infrastructure of international law. It embraces realism through rejection of the artificial constraints of legal formalism but projects idealism by placing faith in the possibility of finding informal as well as formal, social as well as statist, ways of dealing with the complex problems and fundamental issues that characterize the human condition.

The fourth vision is youthful, born of the conviction that the 1982 Convention has failed because of a lack of environmental virtue. It is said that we are caught up in a tangle of political contradictions and ethical inadequacies and might as well start all over again, planning for the Fourth United Nations Conference on the Law of the Sea. This is idealism of the most optimistic brand, accepting the possibility of a world community ready to reassemble and reinvest prodigious energies devoted to the replacement of the 1982 Convention with a new treaty designed holistically to capture the seamlessness of the ocean's web.

The fifth vision is that of functionalism, offering perceptions of a world order better equipped to depoliticize issues and manage problems through the sophisticated application of specialized knowledge. Solution is seen to require the breaking down of massive global problems into separate issue contexts and problem situations through regime building and other modes of cooperative management and behavior, so that functionalist logic can prevail. This perspective is idealistic because it assumes the merit of specificity and projects a benevolent future characterized by the rational application of human knowledge.

The sixth vision of ocean governance expresses faith in the continuity of legal tradition. From this perspective, ocean users need, above all, the assurance of stability offered by legal doctrine, and all other forms of idealism are unrealistic to the extent that they fail to provide that assurance. But doctrinalism, in the eyes of nondoctrinarians, is itself a form of utopianism that promises what cannot be delivered, especially in nonlitigious cultures; and arguably it is also a form of elitism bound to rigidify ocean governance by placing it firmly in the grip of inflexible, culturally insensitive, technically obsessed legal specialists, mostly of Western origin.

The seventh vision foresees a future of global divergence and prescribes that we should be preparing for an increasingly asymmetrical world order. For prophets of this kind, globally conceived and negotiated norms must be expected to be subject to diverse interpretations and applications at regional, national, and other subglobal levels, even in the name of "treaty implementation." A cooperative ethic is more likely to be expressed and realized in the "ideology" of regionalism than in that of globalism and thus may be vulnerable to the objection that it undermines the systemic ideal of universality. This perception is truly idealistic because it assumes the willingness of universalists to tolerate a high degree of relativism for the sake of cultural and regional variation and national autonomy.

The eighth vision of the ocean's future is indigenous and naturalist, pivoting on the idea of a simpler, purer, and more honest world in which power will be shared differently and more equitably. This may be the noblest, but also the most utopian, brand of antistatist idealism, offering no strategy for inducing the power centers of the world to resort to power sharing without any prospect of gain, much less of commensurate gain, but keeping alive the sense of guilt or shame in the affairs of states and providing a reproof to those who maintain the legitimacy of international law on the basis of national status.

What can we do with a list like this? Perhaps it may be interesting to juxtapose these visions with the most familiar transnational brands of idealism that challenge traditional theory and infrastructure in the law of the sea. They seem to be five in number.

ENVIRONMENTALISM

Perhaps more than any other, this type of idealism seems to be shared by all contributors to this volume. The difficulty with environmental idealism is that it means different things to different people. For some, it assumes a heroic dimension, requiring a holistic frame of reference that virtually denies the possibility of simple, moderate, and inexpensive solutions to problems perceived to be environmental in nature. For idealists of this kind, a global approach to almost any widespread environmental concern might be favored over a regional approach.

The concept of the ecosystem as the only logical "unit" of ocean management can complicate the tasks of dealing with issues of the marine environment and discourage, rather than facilitate, interstate cooperation if the ecosystem in question straddles the existing ocean boundary between neighboring states. Logically, environmentalism dictates a radical rethinking of boundary making, not least at sea. For some environmentalists, reverence for the natural order can result in the elevation of favored species, whether endangered or not, on questionable ethical grounds and in the adoption of postures that are intolerant of reasonable efforts to conserve nature through the maintenance of population levels in light of the best available scientific evidence. For still others, environmental idealism prescribes functional and contextual specificity in the treatment of environmental problems, not least in the prevention and control of marine pollution.

The tension between "theological" and "scientific" mind-sets is one of the most conspicuous features of contemporary environmentalism, reminding us that the existence of idealism among those who share concerns about the threats to the ocean environment may also be a cause for division.

STRUCTURALISM

Some of the authors of this book represent structural idealism, the view that the best future for the world at large lies with improved structures. Like environmentalism, this kind of idealism is internationalist when addressed to the largest of human welfare concerns and is variously expressed in such distinguishable modes as world federalism, functionalism, and neo-functionalism.

At the negotiations for the 1982 Convention, structuralism ostensibly carried the day with agreement to set up a number of new global institutions to carry out designated tasks of rule making (the Assembly), supervision (the Council), administration (the Secretariat), management (the Enterprise), and adjudication (Sea-Bed Disputes Chamber) in the new context of deep ocean mining under the auspices of the International Sea-Bed Authority; to perform quasi-adjudicative functions for the determination of the extent of the continental shelf beyond 200-nautical-mile exclusive economic zone (EEZ) limits (Commission on the Limits of the Continental Shelf); and to discharge litigational, arbitral, conciliatory, and other responsibilities in the face of many other kinds of disputes (the International Tribunal for the Law of the Sea and bodies referred to in annexes V, VII, and VIII).

Many of those who advocate most strongly for the 1982 Convention, in the present or a revised form, believe that one of the chief benefits of moving in that direction will be the establishment of these new institutions and a strengthening of existing ones. But, as we learned from the history of the negotiations for the 1982 Convention, the difficulty with structural idealism lies in designing, for a future that may be receding as far as ocean mining is concerned, a structure that is likely to satisfy everyone when the day arrives for making it operational.

HUMANISM

Surely all of us would concede that human welfare is the ultimate value, our most intimate concern. The authors here remind us of the different forms that humanism may take: world food production, transfer of knowledge and skills, special indigenous rights, and so on. The difficulty is in finding the best path for achieving these humanist goals and in determining a bold yet realistic time frame.

Humanism, it may be objected, is fixed so intently on the distant, dreamed-of future that it fails to help us locate the golden road. Perhaps its chief function is corrective, serving to remind legalists and institutionalists that idealism of their sort may be illusory. Formulas and structures, however enlightened, cannot feed or clothe, or battle disease and poverty. As David Hume perceived, no real or imaginable structure in human affairs saves us from our dependence on benevolence. The redistributive ideology of the new law of the sea may be nothing more than empty rhetoric if there is no political will to guarantee at least a trickle down of benefits to disadvantaged citizens and coastal committees, and not merely an empowerment of elites in the disadvantaged states.

NONSTATISM

Most of these authors seem to share the conviction that it is no longer possible, in the 1990s, to put much faith in the nation-state system. Special reference is made here to the needs and vulnerabilities of indigenous islanders and coastal traditions, but the special case underlines the general sense of inadequacy generated by the premises of interstate law, diplomacy, and organization. Reference is made to the dramatic, and potentially transforming, role of nongovernmental organizations (NGOs) in many countries today, reflecting the high standards of knowledge, skill, and awareness brought by highly educated elites outside government to their self-appointed tasks of public education and policy modification.

Given the emergence of increasingly effective NGOs in virtually all regions of the world since the late 1970s, nonstatist idealism has assumed a practical significance that was impossible even twenty years ago. Perhaps it is no longer unrealistic to envisage the creation of an influential, multipurpose forum of

NGOs in the early years after the turn of the century. An oceanic orientation for such a forum—or at least for part of its agenda—might help to expose the inadequacies of statism and nationalism as engines for securing and distributing human benefits from the resources of the earth.

TRANSNATIONAL BUREAUCRATISM

Less emphasized in this work, but perhaps equally deserving of a place among the modes of idealism relevant to ocean development, is what used to be called "transnationalism": that is, the alignment of counterpart bureaucratic specialists across national lines with a view to the development and implementation of international policies designed to promote human welfare and the general interest of states.

Progress in the environmental law of the sea that has occurred in the sector of prevention and control of marine pollution—especially in the development of oil pollution preventive measures—seems due in large part to the effectiveness of this kind of idealism *within* the nation-state system. The fact that it operates largely unseen and unreported may seem to be a disadvantage in a media-driven era, and yet its very invisibility may be the source of its success.

If depoliticization is the key to dealing pragmatically and technically with serious problems of ocean development and management, then transnational bureaucratism may be the most important idealism of all in a world that is still run by national governments and national politicians.

The themes of idealism in ocean affairs could be developed much further, but within the limits of these comments it might be appropriate to end on a note of optimism. Pessimism is easy, and always in fashion, at least among those ambitious to appear sophisticated and up-to-date. However, perhaps only those genetically determined to be pessimistic can fail to find any encouragement in some of the social, political, and diplomatic developments of the late 1980s and early 1990s. All of the five forms of idealism reflected in this book and all of the eight visions of the future discerned here are critical reactions to the status quo in ocean affairs. Taken together, in convergence, they pose a challenge to self-styled realists who would paint a different picture of the years ahead.

Perspectives on Environmental Harmony

THE MATERIALS IN THIS VOLUME provide a rich source of ideas on how to reorient ourselves toward an approach of kinship and environmental harmony in dealing with the marine environment. Along with many of the authors, the editors agree that the existing international legal system is fundamentally flawed and must be reconceived to embrace the interest of all society, not just of states. It is no longer possible to pretend that this postfeudal system is sufficient to meet the challenges of today's complex and interconnected world. Full participation by indigenous peoples and others interested in and affected by the sea is essential. We also specifically advocate a greatly expanded role for nongovernmental organizations.

In addition, the editors would like to offer their views, based on these materials, on the principles that should be adopted to guide future ocean governance.

1. The precautionary principle must govern decision making. This principle requires that when scientific information is in doubt, the party that wishes to develop a new project or change the existing system has the burden of demonstrating that the proposed changes will not produce unacceptable adverse impacts on existing resources and species. This principle is central to ensuring that decision makers are guided by an environmental protection policy designed to improve ocean resource management over time.

2. Government agencies and private parties must analyze alternatives to proposed actions that are likely to have substantial effects on the marine environment and consider alternative proposals that avoid the adverse effects. The process of preparing these environmental impact statements should include active public participation and should draw on interdisciplinary perspectives so that decision makers can understand fully the implications of each development. The results must be made available to the public.

3. Special protection must be provided to rare and fragile ecosystems and endangered and threatened species in order to ensure that the biodiversity of the ecosystem is not reduced.

4. When conflicts arise, protection of living resources should in general be given priority over exploitation of nonliving resources; nonexclusive uses should

be preferred over exclusive uses; and reversible exclusive uses should be preferred over nonreversible exclusive uses. Potential conflicts should be identified early and in an orderly fashion, and equitable solutions should be developed by processes that protect and enhance public order.

5. The public trust doctrine should govern decisions in order to protect the interests of the whole community and the interests of intergenerational equity. This doctrine requires that conflicts be resolved in favor of keeping the oceans whole and protecting the interests of the public today and in the future. Managing resources as a commons should be preferred over privatizing such resources. If private developments are allowed, financial benefits from such developments should be shared with the public. The costs and benefits of each ocean development should be understood before a project is undertaken, and the benefits should be distributed fairly. All costs arising from a development should be internalized, under the "polluter pays" principle. The public must be able to protect public trust interests in the courts and administrative agencies, either through broad public interest standing or through an adequately funded ombudsman or guardian designated to protect the oceans, its natural objects, and its living creatures.

6. The resources of the oceans should be utilized in a manner that promotes sustainable living, and resource exploitation should not be the dominating factor in decision making regarding ocean resource management. Effects of resource exploitation, including cumulative effects, should be examined holistically and should be fully understood before new developments are undertaken.

7. The historically based claims of indigenous peoples to ocean space and ocean resources should be recognized, and their traditional practices of dealing with ocean resources from a perspective of kinship and harmony should be followed whenever possible.

8. Developed countries should assist developing countries with funding and technology to enable them to undertake the responsibilities outlined in these principles.

Biographies

■■

Philip Allott teaches international law, European community law, and constitutional law at the University of Cambridge. Before returning to teach at Cambridge, where he took his first degree, he served as legal adviser in the British Foreign and Commonwealth Office. He was an alternate representative in the United Kingdom's delegation to the Third United Nations Conference on the Law of the Sea. He is the author of *Eunomia: New Order for a New World* (1990).

R. P. Anand is professor of international law at the School of International Studies at the Jawaharlal Nehru University in New Delhi, India. Among his numerous publications on the law of the sea and international law are *Legal Regimes of the Sea-Bed and Developing Countries* (1975); *Origin and Development of the Law of the Sea: Caracas and Beyond* (editor, 1978); and *International Law and Developing Countries: Confrontation or Cooperation?* (1984).

Professor Anand earned B.A. (1951), LL.B. (1953), and LL.M. (1957) degrees from Delhi University and LL.M. (1962) and J.S.D. (1964) degrees from Yale University.

Elisabeth Mann Borgese is professor of political science at Dalhousie University in Halifax, Nova Scotia, Canada, and is head of the International Oceans Institute, based in Malta. She has spent much of the past thirty years working to build an international ocean regime to serve the needs of all the peoples of the world, beginning with her work at the Center for the Study of Democratic Institutions in Santa Barbara, California, in the 1960s and 1970s, and has organized many *Pacem in Maribus* conferences to help achieve this end. She participated in the Third United Nations Conference on the Law of the Sea as a member of the Austrian delegation. Among her many publications are *The Drama of the Oceans* (1976); *Seafarms: The Story of Aquaculture* (1980); and *The Ocean Regime* (1981). She is also coeditor of *The Ocean Yearbook*. She received her diploma from the Conservatory of Music in Zurich, Switzerland, in 1937.

William T. Burke is professor at the School of Law and at the Institute for Marine Studies at the University of Washington in Seattle. His extensive list of publications includes *The Public Order of the Oceans* (coauthor, 1962); *Ocean*

Sciences, Technology and the Future of the Law of the Sea (1966); *Contemporary Legal Problems in Ocean Development* (1969); *National and International Law Enforcement in the Ocean* (coauthor, 1975); and *International Law of the Sea: Documents and Notes* (1992).

Professor Burke received his B.S. degree from Indiana State University in 1949, his J.D. from Indiana University in 1953, and his J.S.D. from Yale Law School in 1959.

Claudia J. Carr is professor of conservation and research studies at the University of California, Berkeley, where she teaches courses in international rural development, sustainable development, and natural resource policy and indigenous peoples. She was previously a professor of environmental studies at the University of California, Santa Cruz. Her publications include *Pastoralism in Crisis* and *Socioeconomic Transformation in Northeast Africa*, as well as numerous articles and reports.

She received a Ph.D. in geography at the University of Chicago.

James Carr is an international coordinator for the Greenpeace Fisheries Campaign, based in Amsterdam. He has participated in meetings of the International Whaling Commission and consulted on environmental issues. His undergraduate degree is in anthropology, and he completed his graduate work on comparative systems of fisheries management at the University of California, Santa Cruz.

Melinda P. Chandler is an attorney-adviser at the Office of the Legal Adviser in the U.S. Department of State, currently working with the Office of Oceans, International Environmental, and Scientific Affairs. Before joining the Legal Adviser's Office in 1986, she worked in private practice in Washington, DC, specializing in international trade law.

She earned her A.B. degree at Indiana University in 1978 and her J.D. at Northwestern University in 1981.

Clifton E. Curtis is the U.S. adviser for Greenpeace International's Political Division, based in the Greenpeace-USA office in Washington, DC. He was president of the Oceanic Society from 1985 to 1990. He has represented nongovernmental organizations in many international negotiations and conferences, including the consultative meetings of the members of the London Dumping Convention, and has written actively on issues of ocean law and policy.

He earned his J.D. degree from George Washington University's National Law Center in 1971.

W. Jackson Davis is professor of biology and affiliate professor of environmental studies at the University of California, Santa Cruz, and is also executive director of the Environmental Studies Institute. In recent years, he has represented the governments of Kiribati and Nauru at the meetings of the International Atomic Energy Agency, the London Dumping Convention, the Nuclear Non-Proliferation Treaty, the Framework Convention on Climate Change, and the United Nations Conference on Environment and Development. He has written widely on scientific and environmental problems; his publications include *The Seventh Year: Industrial Civilization in Transition* (1979).

Professor Davis earned his B.A. degree at the University of California, Berkeley, in 1965, and his Ph.D. in neural sciences at the University of Oregon in 1967.

Catherine L. Floit is an attorney in solo practice in Seattle, Washington. She earned her J.D. degree from the University of Puget Sound in 1984 and her LL.M. from the University of Washington in 1985.

Gracie Fong has served as attorney for the South Pacific Forum Fisheries Agency in the Solomon Islands and in the Crown Law Office of the government of Fiji, where she participated in the negotiations leading to the 1986 Convention for the Protection of the Natural Resources and Environment of the South Pacific Region and the 1985 South Pacific Nuclear Free Zone Treaty.

Ms. Fong has received B.A. and LL.B. degrees from the University of Auckland and an LL.M. from Harvard Law School.

Matthew Gianni coordinates the international efforts of Greenpeace on issues related to the high seas drift net fisheries. He worked on commercial fishing vessels for ten years and represented Greenpeace on fisheries issues at the United Nations Conference on Environment and Development.

Florian Gubon now teaches at the law school of the University of Papua New Guinea. Previously, he taught in the Ocean Resources Management Programme, which is run by the South Pacific Forum Fisheries Agency, at the University of the South Pacific in Suva, Fiji. In addition to his studies in Papua New Guinea, Professor Gubon earned an LL.M. degree at the University of Washington in Seattle.

Joshua Handler is research coordinator for the Nuclear Free Seas Campaign at Greenpeace-USA in Washington, DC. He has written extensively on issues related to military and environmental matters and recently served as coauthor of the *Encyclopedia of the U.S. Military* (1990).

He earned his B.A. degree at the University of Illinois, Champaign-Urbana, and his M.A. in international relations at the University of Chicago.

Grant Hewison currently serves as legal counsel to Greenpeace New Zealand, Inc. Previously, he was both a barrister and a solicitor for Martelli, McKegg, Wells and Cormack in Auckland. He is the author of *The Global Factor: Issues and Images in International Law* and several journal articles focusing on ocean law.

He earned his LL.M. degree at Auckland University, where he also received his LL.B.

Moana Jackson has been director of Maori legal services (*Nga Kaiwhakamarama I Nga Ture*) in Wellington, New Zealand (Aotearoa), since 1987. He is a member of the tribal nations of Ngati Kahungunu and Ngati Porou.

He earned his LL.B. degree in law and criminology at Victoria University in Wellington in 1972 and his LL.M. at Columbia Law School in New York in 1975.

Douglas M. Johnston is professor of law at the University of Victoria in British Columbia, and he holds the Chair in Asia-Pacific Legal Relations at the Centre for Asia-Pacific Initiatives. His publications include *The International Law of Fisheries* (1965, 1987); *The Environmental Law of the Sea* (coauthor, 1981); *Canada and the New International Law of the Sea* (1985); and *The Theory and the History of Ocean Boundary-Making* (1989).

Professor Johnston has an M.A. degree (1952) and an LL.B. degree (1955) from St. Andrews University, an M.C.L. degree (1958) from McGill University, and an LL.M. (1959) and a J.S.D. (1962) from Yale University.

Poka Laenui (Hayden Burgess) is director of the Institute for the Advancement of Hawaiian Affairs and is an active leader of the Pacific–Asia Council of Indigenous Peoples and the World Council of Indigenous Peoples. He has recently participated in the negotiations leading to the drafting of the International Labor Organization's Convention 169 on Indigenous and Tribal Peoples in Independent Countries (1989), and earlier in the 1980s he served as a trustee for the Office of Hawaiian Affairs.

He earned his J.D. degree from the William S. Richardson School of Law at the University of Hawaii in 1976 and also earned his B.A. degree at the University of Hawaii.

Andrew Mack is professor of international relations at the Australian National University in Canberra. Before assuming this chair, he was director of the Peace Research Centre and senior research fellow in the Strategic and Defence Studies Centre at ANU. He has written widely on security matters; his most recent publication is *Security and Arms Control in the North Pacific*.

Professor Mack attended Essex University in the United Kingdom and spent six years in the Royal Air Force.

Joseph R. Morgan is associate professor of geography at the University of Hawaii and is a research associate in the Program on International Economics and Politics at the East-West Center in Honolulu. Before beginning his academic career, he served in the U.S. Navy for twenty-five years. His many publications include the *Atlas for Marine Policy in Southeast Asian Seas* (coeditor, 1983); *International Navigation: Rocks and Shoals Ahead?* (coeditor, 1988);

and the *Atlas for Marine Policy in East Asian Seas* (coeditor, 1992). He is also coeditor of *The Ocean Yearbook*.

Professor Morgan received his B.A. degree in chemistry from the University of Pennsylvania and his B.S. in oceanography from the University of Washington. He earned his M.A. (1971) and Ph.D. (1978) in geography from the University of Hawaii.

Arvid Pardo provided the major impetus for the negotiations that led to the 1982 United Nations Convention on the Law of the Sea when, in 1967, he spoke to the United Nations General Assembly to outline the opportunities and challenges facing the world community with regard to ocean resources. At that time, he was serving as the ambassador of Malta to the United Nations. Subsequently, he taught at the University of Southern California, and he has written extensively on issues of ocean policy. Currently, he lives in Texas and is semiretired but remains active on issues of ocean governance.

Mere Pulea is a legal consultant for the Pacific region, focusing on environmental law, family and population law, and women's issues. She has served as a consultant for the South Pacific Regional Environment Programme (SPREP) since 1982 and participated in the drafting of the 1986 Convention for the Protection of the Natural Resources and Environment of the South Pacific Region. Her many publications include *The Family, Law and Population in the Pacific Islands* (1986) and *Pacific Courts and Legal Systems* (coeditor, 1988).

She is a native of Fiji and earned her bachelor's degree in social studies from the University of Queensland, Australia, and her master's degree in philosophy from the University of the South Pacific in Fiji. She was admitted as a barrister through Lincoln's Inn in London in 1975.

Artemy A. Saguirian is a senior researcher at the Department of International Problems of World Ocean and Environment at the Institute of World Economy and International Relations in Moscow. He has written actively on questions related to the international law of the sea and the Arctic; recently, he has focused on theories of international regime formation.

He graduated from the Moscow State Institute of International Relations in 1976 and completed his Ph.D. in 1980, writing his thesis on U.S. ocean policy and the Third United Nations Conference on the Law of the Sea.

Michael D. Smith is currently assistant director of the South China Sea Informal Working Group at the University of British Columbia in Vancouver. His research interests include aspects of North-South relations, the political economy of development, and environmentalism, as well as feminist and critical social theory.

He has degrees in commerce and law from the University of British Columbia in Vancouver.

Margaret Spring is an attorney with Sidley & Austin in Washington, DC. In 1991, she was a John A. Knauss Fellow on the National Ocean Policy Study of the U.S. Senate Committee on Commerce, Science, and Transportation, through the National Oceanic and Atmospheric Administration's Sea Grant Program. In 1992, she was a legal researcher with the Center for International Environmental Law in Washington, DC.

She earned her J.D. degree at the Duke University School of Law in 1992.

Christopher D. Stone is Roy P. Crocker Professor at the University of Southern California, where he teaches international law, jurisprudence, and corporate law. He is currently rapporteur for the American Bar Association's Intersectional Committee to Draft Intergenerational Accords on the International Law of the Environment. He has written widely on environmental ethics; his books include *Law, Language and Ethics* (coauthor, 1972); *Should Trees Have Standing?* (1972); *Where the Law Ends* (1975); *Earth and Other Ethics* (1987); and *The Gnat Is Older Than Man* (1993).

Professor Stone received his A.B. degree from Harvard University in 1959 and his J.D. from Yale University in 1962.

Kazuo Sumi is professor of international law at Yokohama City University in Japan. His books include *Protection of the Marine Environment in East Asian Waters* (in English, 1977); *Theory and Practice of the 200-Mile Zone: Problems in Japan and the World* (in Japanese, 1977); and *Deep Sea-bed Mineral Resources and International Law* (in Japanese, 1979).

He received his B.A. degree from Yokohama City University in 1965 and his LL.M. from Hitotsubashi University in Tokyo in 1967.

Ian Townsend-Gault is currently associate professor of law and director of the Centre for Asian Legal Studies at the University of British Columbia in Vancouver as well as regional director (West Coast) of the Oceans Institute of Canada, of which he was a founding director. He taught at universities in the United Kingdom and Norway before moving to Canada in 1980. He has written in the field of natural resource law, especially as it applies to the oceans, the marine environment, joint development, and maritime boundary making.

He graduated from the Faculty of Law of the University of Dundee, Scotland, in 1976.

Alaelua V. Saleimoa Va'ai is an attorney in Apia, Western Samoa. He served as consultant during the drafting and negotiations of the 1986 Convention for the Protection of the Natural Resources and Environment of the South Pacific Region. In 1992, he began working toward a Ph.D. on Samoan customary law at the Australian National University in Canberra.

Jon M. Van Dyke is professor of law at the William S. Richardson School of Law at the University of Hawaii at Manoa. His previous books include *North Vietnam's Strategy for Survival* (1972); *Jury Selection Procedures: Our Uncertain Commitment to Representative Panels* (1977); *Consensus and Confrontation: The United States and the Law of the Sea Convention* (editor, 1984); *International Navigation: Rocks and Shoals Ahead?* (coeditor, 1988); and *Checklists for Searches and Seizures in Public Schools* (coauthor, 1992 and 1993).

Professor Van Dyke received his B.A. degree from Yale University in 1964 and his J.D. from Harvard Law School in 1967.

Miranda Wecker is principal staff person for the Willapa Alliance, a grass-roots group undertaking planning to coordinate multiuse activities in the Willapa Bay area in the state of Washington. From 1985 to 1991, she was associate director of the Council on Ocean Law in Washington, DC. She also serves as oceans counsel for the Center for International Environmental Law in Washington, DC.

She earned her J.D. and LL.M. degrees from the University of Washington in Seattle in 1984 and 1985.

Dolores M. Wesson is currently assistant director of the California Sea Grant College at the University of California, San Diego. Previously, she was acting associate director of the Council on Ocean Law in Washington, DC, and an independent consultant to the Organization of American States' Department of Regional Development. She has published several journal articles in the field of ocean law.

She received her A.B. degree in history from the University of California, San Diego, and her master's degree in marine affairs from the University of Washington, Seattle.

Durwood Zaelke is president of the Center for International Environmental Law (CIEL) and adjunct professor of law and scholar-in-residence at the American University's Washington College of Law in Washington, DC. He also is codirector of the joint CIEL–American University Research Program in International and Comparative Environmental Law. Previously, he was a senior research fellow at King's College in London, director of the Sierra Club Legal Defense Fund's Washington, DC, and Alaska offices, and special litigation attorney for the U.S. Department of Justice. He has published widely in the fields of trade and the environment, democratic institutions, multilateral development banks, and international environmental law.

He received his B.A. degree from the University of California, Los Angeles, in 1969 and his J.D. from the Duke University School of Law in 1973.

Index

Allott, Philip, 9–11, 49
Amoco Cadiz, 198
Anand, R.P., 9–10, 72
Anglo-Norwegian Fisheries case (*Great Britain v. Norway*), 313
Antarctic Ocean, 150–163, 238, 300
Aotearoa. *See* New Zealand
Apia Convention. *See* South Pacific
Archipelagic waters, 13. *See also* Law of the sea
Arctic Ocean, 36, 150
Artificial islands. *See also* Atolls
    Johnston Atoll, 17
    Okino-Torishima, 18
Atlantic Ocean, 16, 150–163, 193, 233, 280
Atolls. *See also* Artificial islands
    Bikini, 17, 214, 221
    Christmas Island, 17, 221
    coral, 215
    Enewetak, 17, 214, 221
    Erikub, 214
    Fangataufa, 221
    Johnston, 17, 215–216, 221
    Kwajalein, 17, 214
    low-lying, 17, 214–227
    Marshall Islands, 104, 214, 221, 455
    Moruroa, 124, 215, 221
    pollution, 214
Australia, 117, 126, 137, 215, 292, 415

Baltic Sea, 150–163, 202, 422
Basel Convention on the Control of Transboundary Movements of Hazardous Wastes and their Disposal. *See* Pollution
Beesley, J. Allen, 180
Bering Sea, 16, 235. *See also* Fisheries, donut hole

Biodiversity
    compartmentalized approach to protecting, 167
    protection in new regime, 19, 286
Black Sea, 202
Borgese, Elisabeth Mann, 9–11, 23, 179–180, 377, 388
Brundtland Commission, *Our Common Future*, 27, 36, 173, 187–188, 395
Burke, William T., 234, 235

Canada, 16, 254
    Canada International Development Agency, 358–360
    drift net fishing, 136, 138, 236, 292, 300
    military activities, 415
Carbon tax. *See* Economics
Caribbean
    Conference of Plenipotentiaries Concerning Specially Protected Areas and Wildlife in
        the Wider Caribbean Region, 218, 335
    Convention for the Protection and Development of the Marine Environment of the
        Wider Caribbean Region, 207–208, 328, 335
    generally, 210
    Protocol Concerning Specially Protected Areas and Wildlife (SPAW), 218–219, 328,
        335
Carr, Claudia, 231, 340
Carr, James, 232, 272
Center for the Study of Democratic Institutions, 9, 27–28
    The Ocean Regime (Santa Barbara Draft), 26–29
Chandler, Melinda, 233, 327
China
    drift net fishing, 138
    membership in SPNFZ Treaty, 126
    military activities, 421, 436
China Sea, 150–163
Clean production principle, 191
Climate change, 167
Coastal zones, 149–165, 195, 203, 293, 363–365. *See also* Fisheries
Codes of conduct, 194
Common Heritage Fund, 180
Common heritage of humankind, 38–39, 82, 189, 377, 388, 392
Commons. *See* Global Commons
Conference on the Human Environment in the South Pacific. *See* South Pacific
Convention Concerning the Protection and Integration of Indigenous and other Tribal
    and Semi-tribal Populations in Independent Countries, 95
Convention on Conservation of Nature in the South Pacific (the Apia Convention). *See*
    South Pacific
Convention on the Control of Transboundary Movements of Hazardous Wastes and their
    Disposal. *See* Pollution

Convention on Fishing and Conservation of the Living Resources of the High Seas. *See* Fisheries

Convention on the High Seas (1958), 242, 319. *See also* Fisheries

Convention on International Trade in Endangered Species of Wild Fauna and Flora, 175, 233, 276, 277, 328, 334

Convention on the Law of the Sea. *See* Law of the sea, United Nations Convention on the Law of the Sea

Convention on Liability and Compensation in Connection with the Carriage of Noxious and Hazardous Substances by Sea, 204

Convention for the Prohibition of Fishing with Long Driftnets in the South Pacific. *See* Fisheries; South Pacific

Convention for the Protection and Development of the Marine Environment of the Wider Caribbean Region. *See* Caribbean

Convention for the Protection of the Natural Resources and Environment of the South Pacific Region (SPREP Convention). *See* South Pacific

Corfu Channel case (*United Kingdom v. Albania*), 458

Curtis, Clifton, 145, 187

Davis, Jackson, 145, 147

DDT. *See* Pollution

Deep sea bed. *See* Nonliving resources

Democracy, 61–64

Developing nations, 166, 175, 272, 380, 413, 436. *See also* Fisheries, assistance; Military activities

  assistance to, 191

Development banks, 354–368

Distant Water Fishing Nations. *See* Fisheries

Dolphins. *See* Fisheries, marine mammals

Draft Article on the Protection of the Marine Environment Against Pollution, 303

Draft Convention on Liability and Compensation in Connection with the Carriage of Noxious and Hazardous Substances by Sea, 204–205

Drift nets. *See* Fisheries

Dumping. *See* Pollution

East-West Center, 93

Economics

  carbon tax, 171, 178

  cost-benefit, 352

  free markets, 409

  free-rider, 182, 409

  laissez-faire, 34, 77, 80–82, 304, 314, 320, 400

  natural resources, 33, 94, 132, 172, 239, 284

  scarcity versus abundance, 93

  tax, 175–181

Economics (*continued*)
  tourism, 123
  trade, 73–76, 233–234, 327–336
  wealth redistribution, 179
Endangered species. *See also* Fisheries
  Endangered Species Act, 329
  trade in, 327–330, 334
Environmental impacts assessment (EIA), 193, 285, 303–304
European community, 207, 233, 400
Exclusive Economic Zone (EEZ). *See* Fisheries; Law of the sea; Military activities;
    Nonliving resources;
Exxon Valdez, 81, 198, 208

Fiji, 13, 137
Fisheries, 231–372
  anadromous species, 249, 250–252, 258
    salmon, 15, 236, 237, 251–252, 294–297, 306
  Antarctic ocean, 238
  assistance and aid, 340–368
    aquaculture, 361–362
    Asian Development Bank 354–355, 362
    Canada International Development Agency (CIDA), 358–360, 362
    Fisheries Project Information System (FIPIS), 343, 355, 366
    Inter-American Development Bank, 357
    International Monetary Fund, 357
    Japan International Cooperation Agency (JICA), 357
    Norwegian bilateral agency (NORAD), 360, 362
    Organization for Economic Development (OECD), 355
    overfishing, 345–346
    Overseas Economic Cooperation Fund (OECF), 357
    southern nations, 340, 343–347
    sustainable fisheries development, 367
    United Nations Food and Agriculture Organization, 345, 355, 362
    World Bank, 345, 350
  Bering Sea, 235, 255
  Bering Sea Fishery Authority, 253–254
  by-catch, 250–251, 257, 264, 273, 277, 278, 281, 285, 293–294, 301, 311, 330
  distant water fishing nations (DWFNs), 19, 104, 132–134, 138, 280, 346, 362
  donut hole, 236, 254, 258
  drift nets, 127, 133–140, 233, 237, 248, 252–261, 272–287, 292–309, 401
    high seas, 136, 231–232, 250, 277, 279, 280, 284–286, 292–309, 310–311, 313–
        316, 321
    North Pacific, 238, 250–251, 255, 273, 276, 280, 282–283, 296–298, 316
    South Pacific, 131–135, 255–256, 273, 276, 282–283, 315
    Western Pacific, 135

exclusive economic zone (EEZ), 235, 236, 240, 246, 257–261, 263, 283, 286, 304, 333
ghost fishing, 275, 301–302
high seas, 235, 244–245, 252–261, 318–320
Lacey Act, 329
law regarding
    Anglo-Norwegian Fisheries case (*Great Britain v. Norway*), 313
    Convention on Fishing and Conservation of the Living Resources of the High Seas,
        242–243, 314
    Convention on the High Seas (1958), 242, 319
    Convention for the Prohibition of Fishing with Long Driftnets in the South Pacific
        (Wellington Convention), 126, 127, 138, 256, 316, 331
        convention area, 138, 331
    Convention for the Protection of the Natural Resources and Environment of the
        South Pacific Region (SPREP Convention), 17, 90, 103–130, 161–163, 222
    Driftnet Impact Monitoring, Assessment, and Control Act of 1987, 297, 316, 330
    Fisheries Jurisdiction Case (*United Kingdom v. Iceland*), 242, 245, 314
    Fisherman's Protective Act, 262, 297
        Pelly Amendment, 298, 329–330
    generally, 246–248
    Geneva Fishing Convention. *See* Convention on Fishing and Conservation of the
        Living Resources of the High Seas (this heading)
    International Commission for the Conservation of Atlantic Tunas (ICCAT), 334
    International North Pacific Fisheries Convention, 250, 295, 304
    Magnuson Act, 236, 262, 295, 297–298, 330, 332–333
    Pelly Amendment, 298, 332
    Treaty on Fisheries Between the Governments of Certain Pacific Island States and
        the United States, 133–134
    United Nations Convention on the Law of the Sea (1982)
        provisions
            article 61, 257
            article 63, 245, 246, 318–319
            article 64, 245, 249, 318–319
            article 65, 249, 318–319
            article 66, 246–247, 250–252, 318–319
            article 67, 319
            article 87, 242, 282, 314, 318
            article 116, 243, 246–249, 282, 318–319
            article 117, 282, 304, 314, 318, 320
            article 118, 282, 318, 320
            article 119, 259, 282, 284–285, 318, 320
            article 120, 249, 282, 318
            article 123, 252, 254
            article 192, 303
            article 197, 282
            article 204, 303
            article 205, 303

Fisheries
  law regarding
    United Nations Convention on the Law of the Sea (1982)
      provisions (*continued*)
        article 206, 303
        article 300, 313
        part V, 246–247
        part VII, 318, 320
        part XII, 303
    United Nations Resolution 44/215, 273, 281–282, 286
    United Nations Resolution 44/225, 232–234, 255–261, 273, 281, 283–284, 286,
      292–294, 302–303, 306, 316–317
    United States Fishery Conservation and Management Act of 1976. *See* Magnuson
      Act (this heading)
    United States Public Law, 101–162, 332
  marine mammals, 15–16, 236–237, 243, 249–250, 258, 264, 273–277, 294–295, 299–
    300, 311
    dolphins, 16, 237, 273, 299–300, 331–332
    Marine Mammal Protection Act, 255, 264, 295, 329, 331–332
    whales, 16, 236, 274, 276, 330
  new regime regarding, 239–241, 244–245, 252–265, 281, 284, 471–476
  nontarget species, 3
  Northwest Atlantic, 236
  Northwest Atlantic Fisheries Organization (NAFO), 248, 254
  South Pacific Albacore Research Group (SPAR), 133, 136, 138
  South Pacific Forum Fisheries Agency (FFA), 121, 131–140, 140 (note 2), 141. *See also*
    Law of the sea; South Pacific
    access fees, 132
    FFA Secretariat, 132
    Forum Fisheries Committee, 132
      harmonized minimum terms and conditions of access, 133
      member nations, 126–127, 131–139, 140 (note 3), 141
    straddling stocks, 16, 235–236, 239–240, 241, 244, 248–249, 252–256, 400
      cod, 236
      pollock, 235, 258
    total allowable catch, 236, 251
  tuna, 14, 15, 127, 132–140, 236–237, 249, 255, 256, 262, 278, 349
Floit, Catherine, 232, 310
Fong, Gracie, 90, 131
Food and Agriculture Organization (FAO). *See* United Nations
Forum Fisheries Agency (FFA). *See* South Pacific
Fowler, S.W., 150–159
France. *See also* Military activities
  drift net fishing, 233
Freedom for the seas
  concept of, 4, 9–10, 41–44

evolution of, 14
fisheries, 231, 235, 243
Freedom of the seas
    challenges to, 78–83, 310–321
    concept of, 3–4, 9, 41–44, 47, 145, 211, 393, 449
    evolution of, 41–44, 72–76, 83 (note 1),
    inclusion of military activities, 4
    revision of, 14
French Polynesia. *See* Military activities

General Agreement on Tariffs and Trade (GATT), 233–234, 330, 332–333, 335
GESAMP. *See* Pollution
Gianni, Matthew, 232, 272
Ginsburg, Norton, 9
Global Commons, 19, 32, 92–93, 171–186, 231, 238
    international governance of, 165–166
    military uses of, 409–418
    nonliving resources, 386
    ocean guardians, 171–174
    Tragedy of the Commons, 409–411
    trust fund, 145, 171, 174–186
Goldberg, E.D., 150
Great Britain, 126, 137, 421–422, 431, 436
Greenpeace, 47, 187, 210, 295, 299–300, 448
Grotius, Hugo, 14, 16, 38, 41, 44, 72–75, 81, 445
Gubon, Florian, 90, 121

Haiti, 207
Handler, Joshua, 407, 420
Hawaii
    hymn of creation (Kumulipo), 88–101
    indigenous perspectives, 91–102
Hazardous materials. *See* Pollution
Heavy metals. *See* Pollution
High seas. *See* Fisheries; Freedom of the seas; International law; Law of the sea; Oceans

Incidents at Sea Agreement (1972). *See* Military activities
India, 175, 178
Indian Ocean, 73, 74, 150–163, 207, 280
Indigenous peoples. *See also* Convention Concerning the Protection and Integration of
        Indigenous and other Tribal and Semi-tribal Populations in Independent Countries;
        South Pacific
    colonization of, 41–48, 94–97
    divine cause concept, 91

Indigenous peoples (*continued*)
    generally, 4, 168
    Hawaiian perspective, 91–102
    International Year for the World's Indigenous Peoples, 95
    Pacific Islanders, 4, 89, 91–102, 123–124
    Polynesian custom, 92
    technology transfer (fisheries), 352
    view of stewardship, 89
Indonesia, 13
Intergovernmental organizations, 67, 285
International Commission for the Conservation of Atlantic Tunas (ICCAT). *See* Fisheries
International Convention on Civil Liability for Oil Pollution Damage. *See* Pollution
International Convention on the Establishment of an International Fund for
    Compensation for Oil Pollution Damage. *See* Pollution
International Convention on Oil Pollution Preparedness and Response. *See* Pollution
International Convention for the Prevention of Pollution from Ships. *See* Pollution
International Convention for the Regulation of Whaling, 330
International Convention Relating to Intervention on the High Seas in Cases of Oil
    Pollution Casualties. *See* Pollution
International Convention on Safety of Life at Sea (SOLAS), 200–201
International Convention on Salvage, 203
International Convention on Standards for Training, Certification, and Watchkeeping, 202
International law
    abuse of right, 312–313
    customary, 13, 80, 172, 182 (note 1), 207, 252, 282, 310, 311–317, 320–321, 454–457
    democracy, 61–64
    early treaties regulating the high seas, 23
    ecocentric perspective, 190
    Eurocentric perspective, 41–45, 73, 77
    generally, 25–28, 41–44, 211, 297–299, 384
    hazardous materials, 198–213
    indigenous law, 41–48
    jus cogens, 298
    lack of input from Third World in formulation of, 77–78, 83, 208
    origins of, 49–71
    pacta sunt servanda, 385
    philosophy of, 49–71
    piracy, 296–297
    rebus sic stantibus, 385, 388–391
    sovereignty, 38–40, 42, 44, 78–80, 83, 173, 231, 246–248, 253, 259
    structural inadequacy of, 4, 31, 81–83, 165–168, 172–174, 190, 196, 260–265, 272
    Truman Proclamations, 78–79
    Vattel's approach to, 45
    Vienna Convention on the Law of Treaties, 298
International Law Commission of the United Nations, role of in new ocean regime, 36
International Maritime Dangerous Goods Code, 201

International Maritime Organization (IMO), 189, 200, 204, 209, 219–220
International North Pacific Fisheries Convention. *See* Fisheries
International Oil Pollution Compensation Fund, 205
International Sea-Bed Authority. *See* Nonliving resources
International Whaling Convention, 15, 236, 283, 285. *See also* Fisheries, marine mammals

Jackson, Moana, xv, 9–10, 41, 168
Japan
     atolls, 221
     as distant water fishing nation, 132
     drift net fishing, 104, 127, 136–139, 232–233, 236, 250–251, 276, 280, 292–309, 316
     generally, 90, 221, 436
     Japan International Cooperation Agency (JICA), 357
     ocean dumping, 124
Johnston, Douglas M., 471

Korea
     as distant water fishing nation, 132
     drift net fishing, 136, 138, 232, 236, 250, 279–280, 296, 298, 306, 316

Laenui, Poka (Hayden Burgess), 89, 91
Langkawi declaration, 137
Lauterpacht, Hersch, 81
Law of the sea
     anthropogenic perspective, 190
     early application, 76–77, 310,
     early treaties, 23, 312, 319
     ecocentric perspective, 190
     enforcement, 191, 241
     new principles, 50–64
     new regime, 14, 18, 36–39, 49–71, 147–170, 171–182, 190, 194–196, 237–241, 306, 471–476. *See also* Fisheries; Military activities
          clean production principle, 191
          enforcement, 191, 232, 279–282
          global approach, 163–167
          nongovernmental organizations, 192
          polluter pays principle in, 18
          pollution addressed in, 194
          precautionary principle in, 19, 191, 234, 279, 316–317
          regional approach, 160–163, 190
     United Nations Memorandum on the Regime of the High Seas, 79
          Anglo-Norwegian Fisheries case (*Great Britain v. Norway*), 79
     United Nations Conference on the Law of the Sea, 1958 (First UN Conference), 312
          fragmentation and unworkability of, 23

Law of the sea (*continued*)
  United Nations Conference on the Law of the Sea, 1960 (Second UN Conference), 23
    attempt to codify two hundred mile limit, 80
    struggle between developed and developing countries, 80
  United Nations Conference on the Law of the Sea, 1974 (Third UN Conference), 9,
    25, 27, 34–35, 40, 83, 180, 231, 386, 389
  United Nations Convention on the Law of the Sea (1982)
    anadromous species, 250–252, 258
      salmon, 15, 236, 237, 251–252, 294–297
    archipelagic waters, 13, 441
    artificial islands, 18
    atolls, 17, 18
    as basis for a preconceived law of the sea, 11, 49–50, 58, 64–68
    dispute resolution, 34
    exclusive economic zone, 3, 13, 83, 104, 122, 123, 126, 132–135, 180, 200, 231,
      232, 257–263, 297, 346–347
    generally, 3, 9–10, 28, 33–35, 38–40, 64–68, 131, 164, 181, 188–189, 195, 207,
      211, 231, 244, 272–287, 310, 335, 379, 397
    hazardous materials, 200
    innocent passage, 441
    interdisciplinary considerations in formulation of, 29, 167, 168
    International Sea-Bed Authority, 25
    low-lying atolls, 215, 218
    marine mammals, 15–16, 236–237, 243, 249–250, 255
    member nations, 20 (note 4)
    military activities, 17, 440–444. *See also* Military activities
      peaceful purposes clause, 452–454
    nonliving resources. *See* Nonliving resources
    provisions
      article 19, 453
      article 51, 453
      article 55, 65
      article 56, 65
      article 59, 66
      article 61, 257
      article 63, 245, 255, 318–319
      article 64, 245, 249, 255, 318–319
      article 65, 249, 318–319
      article 66, 246–247, 250–252, 319
      article 67, 319
      article 87, 242, 255, 282, 312, 314
      article 116, 243, 246–249, 255, 261, 282, 319
      article 117, 255, 282, 304, 314, 320
      article 118, 255, 282, 320
      article 119, 255, 259, 282–283, 285, 320
      article 120, 249, 282

article 123, 66, 252, 254
article 135, 387
article 136, 66, 386
article 137, 387
article 140, 387
article 141, 386
article 145, 389
article 150, 66
article 151, 391
article 162, 389
article 190, 66
article 192, 303
article 204, 303
article 205, 303
article 206, 303
article 209, 389
article 301, 453
article 309, 384
part V, 65, 246–247
part VII, 261, 320
part XI, 65, 68, 179, 189, 377, 380–384, 386–387, 388
part XII, 66, 188, 195, 303, 441
part XV, 64
ratification, 20 (note 4)
as reflection of customary international law, 311–317, 320
as reflection of property rights ethos, 64–68
straddling stocks, 16, 234, 244, 248, 252–253, 256, 260
treaties preceding, 13
tuna, 15, 249, 255, 349
United Nations Law of the Sea Office, 189
London Dumping Convention (LDC). See Pollution
Low-lying atolls. See Atolls

Mack, Andrew, 407, 409
Magnuson Act. See Fisheries
Maori peoples, 41–48
Marine mammals. See Fisheries; Law of the sea
MARPOL. See Pollution
Marshall Islands, 104, 214, 221, 455
Mauritania, 137
Mediterranean Sea, 36, 150–163
Mexico, 233, 236, 331–332, 347
Military activities, 17, 407–470
    Association of Southeast Asian Nations (ASEAN), 413
    confidence building measures, 415

Military activities (*continued*)
  developing nations, 413
  exclusionary and warning zones, 407, 445–458
  exclusive economic zone (EEZ), 436, 443
  France, 4, 104, 407, 421, 436, 455
  freedom of the seas, 449
  French Polynesia, suspension of nuclear testing in, 5, 407
  high seas, 407, 411, 445–458
  law regarding
    abuse of right, 451
    Corfu Channel case (*United Kingdom v. Albania*), 458
    Helsinki/Stockholm Conference on Security and Cooperation in Europe,
      416–418
    Incidents at Sea Agreement (1972), 417
    North Atlantic Treaty Organization (NATO), 417
    Nuclear Test cases (*Australia v. France; New Zealand v. France*), 455
    Prevention of Dangerous Military Activities Agreement (1989), 417
    Seabed Treaty (1971), 454
    South Pacific Nuclear Free Zone Treaty, 106, 125–126, 207, 416
    Strategic Arms Reduction Talks, 424–425
    United Nations Convention on the Law of the Sea (1982)
      archipelagic waters, 441
      innocent passage, 441, 443
      peaceful purposes clause, 452–454
      provisions
        article 19, 443, 453
        article 51, 453
        article 236, 442
        article 301, 454
        part XII, 441
    Warsaw Pact, 417
  navies, 407, 413, 420–432, 435–444
  new regime regarding, 411–418, 444, 471–476
  nuclear weapons and vessels, 407, 420–432
  relationship to indigenous law, 41–48
  sea lines of communication (SLOC), 413, 443
  South Pacific Nuclear Free Zone Treaty (SPNFZ Treaty), 106, 125, 126
    member nations, 126
  Soviet Union (former), 104, 207, 216, 414–416, 420–432, 456
  United States
    generally, 407, 414–416, 420–432, 450–456
    Indian Ocean, 207
    Johnston Atoll, 104
    Marshall Islands, 104, 455
    nuclear weapons, 420–432
    South Pacific, 207

Morgan, Joseph, 407, 435
Moruroa, 124, 215, 221

National Oceanic and Atmospheric Administration (NOAA), 173
Nauru Agreement Concerning Cooperation in the Management of Fisheries of Common
    Interest, 126
New Zealand, 41–48, 117–118, 137, 292, 300, 455
    drift net fishing, 135
Nongovernmental organizations (NGOs), 25, 47, 48, 91, 95, 96, 124, 172, 175, 192, 195,
    232, 299, 340, 361, 475–476
Nonliving resources, 16–17, 26, 82, 179, 377–403
    ad hoc sea bed committee, 82
    as common heritage, 377, 388, 392
    deep sea bed, 26, 65, 82, 172
    exclusive economic zone (EEZ), 390, 394, 474
    law regarding
        International Sea-Bed Authority, 16, 25, 40, 383, 388, 390, 474
            preparatory commission for, 36, 380
        pacta sunt servanda, 385
        rebus sic stantibus, 385, 388–391
        Seabed Treaty (1971), 454
        Truman proclamation regarding, 20 (note 1), 394
        United Nations Conference on the Law of the Sea, 1974
        (Third UN Conference), 386, 389
        United Nations Convention on the Law of the Sea (1982), 377–387
            "freezing" articles, 387–391
            provisions
                article 135, 387
                article 136, 386
                article 137, 387
                article 140, 387
                article 141, 386
                article 145, 389
                article 151, 391
                article 162, 389
                article 209, 389
                article 309, 384
                part XI, 377, 380–384, 386–387, 388
    new regime regarding, 380, 385, 392–401, 471–476
North Pacific Fur Seal Convention, 249. *See also* Fisheries, marine mammals
North Sea, 150–163, 195
Norway. *See also* Fisheries, assistance and aid
    Norwegian bilateral agency (NORAD), 358, 360, 362
Nuclear Test cases (*Australia v. France; New Zealand v. France*), 455
Nuclear weapons. *See* Military activities

Oceans. *See also* Fisheries; Pollution
  assimilative capacity of, 148
  biology of, 148–149, 150–163, 199, 216, 286, 304
  phytoplankton bloom, 148–149
Ocean space, 38–39, 200
Ocean Space treaty, 30–33
Ocean wilderness, 4
Okino-Torishima, 18
Our Common Future. See Brundtland Commission
Ozone depletion, 167, 175, 182, 194, 317

Pacem in Maribus, 26, 30
Pacific Islanders. *See* Indigenous peoples
Pacific Ocean, 41–48, 91–102, 103–112, 150–163, 220, 236. *See also* South Pacific
Papua New Guinea, 13, 137
Pardo, Arvid, 9, 26–28, 30–33, 38, 81–82, 189. *See also* Ocean Space Treaty
Pell, Claiborne
  model treaty, 24, 25
  The Challenge of the Seven Seas, 24
Pelly Amendment. *See* Fisheries, law
Philippines, 13
  as distant water fishing nation, 132
Pinto, Chris, 28
Planet Protection Fund, 178
Polluter pays principle, 18
Pollution. *See also* Law of the sea
  assimilative capacity of oceans, 148
    chlorinated hydrocarbons, 153, 162
  DDT, 150–154
  generally, 16, 145, 147, 150–172, 191, 194, 389, 410, 442, 476
  Group of Experts on the Scientific Aspects of Marine Pollution (GESAMP), 145, 150–160, 173, 202
  hazardous materials, 198–213, 221–222
  heavy metals, 151–160, 162
  land-based sources, 17, 145, 164, 165, 193
  law regarding
    Basel Convention on the Control of Transboundary Movements of Hazardous Wastes and their Disposal, 206–207, 210
    Convention on the Control of Transboundary Movements of Hazardous Wastes and their Disposal, 205
    Convention for the Protection of Natural Resources and Environment of the South Pacific Region, 17, 114, 161–163
    Draft Article on the Protection of the Marine Environment Against Pollution, 303
    Draft Convention on Liability and Compensation in Connection with the Carriage of Noxious and Hazardous Substances by Sea, 204–205

generally, 16, 160–167, 171–175
International Convention on Civil Liability for Oil Pollution Damage, 205
International Convention on the Establishment of an International Fund for
    Compensation for Oil Pollution Damage, 205
International Convention on Oil Pollution Preparedness and Response, 204
International Convention for the Prevention of Pollution from Ships (MARPOL
    73/78), 189, 201–202, 206, 209, 219, 301–302
International Convention Relating to Intervention on the High Seas in Cases of Oil
    Pollution Casualties, 203
International Convention on Safety of Life at Sea, 200–201
International Convention on Salvage, 203
International Oil Pollution Compensation Fund, 205
London Dumping Convention (LDC), 17, 111, 161–163, 167, 189, 194–195, 206,
    209, 301
Oil Pollution Act of 1990, 209
Paris Commission for the Prevention of Marine Pollution from Land-based Sources
    in the Northeast Atlantic, 193
polychlorinated biphenyls (PCBs), 150–154, 162
prevention
    environmental impact assessment, 193
    environmental quality standards, 192
    hierarchical waste management, 193
    industrial sectors approach, 193
    uniform emissions standards, 192
Portugal, role in early use of the high seas, 74
Precautionary Principle, 19, 147, 164, 166, 191, 234, 316–317, 349
Prevention of Dangerous Military Activities Agreement (1989). See Military activities
Public Trust Doctrine, 19
    international aspect of, 57–61
Pulea, Mere, 90, 103

Regional seas programs. See United Nations Environment Programme

Saguirian, Artemy, 377, 379, 388
Salmon. See Fisheries
Santa Barbara. See Center for the Study of Democratic Institutions
Seabed Treaty (1971). See Military activities; Nonliving resources
Selden, John, the writing of Mare Clausum, 75–76
Smith, Michael, 377, 392
SOLAS Convention. See International Convention on Safety of Life at Sea
Solomon Islands, 13, 137
South Pacific Ocean, 91–102, 103–142, 210, 214–218
    drift net fishing in. See Fisheries

South Pacific Ocean (*continued*)

    international law regarding

        Committee for Coordination of Joint Prospecting for Mineral Resources in the South Pacific Offshore Areas (CCOP/SOPAC), 122

        Conference on the Human Environment in the South Pacific, 105, 124–125

        Convention on Conservation of Nature in the South Pacific (the Apia Convention), 106, 124, 125

        Convention for the Prohibition of Fishing with Long Driftnets in the South Pacific (Wellington Convention), 126, 127, 138, 256, 316, 331

            convention area, 138, 331

        Convention for the Protection of the Natural Resources and Environment of the South Pacific Region (SPREP Convention), 17, 90, 103–130, 161–163, 222

            convention area, 116–117

            environmental impact assessment under, 107

            island states, 118, 121, 122–130

            liability, 119

            pacific rim, 117

            pollution, 114–115, 119

            protocols under, 106, 113, 117

            provisions, 108–110, 215–218

        Langkawi declaration, 137

        Nauru Agreement Concerning Cooperation in the Management of Fisheries of Common Interest, 126

        South Pacific Albacore Research Group (SPAR), 133, 136, 139

        South Pacific Commission (SPC), 104, 122–124, 132–133

            member nations, 133

        South Pacific Conference on Nature Conservation, 137

        South Pacific Declaration on Natural Resources and the Environment, 124

        South Pacific Forum Fisheries Agency (FFA), 121, 131–140, 140 (note 2), 141

            access fees, 132

            FFA Secretariat, 132

            Forum Fisheries Committee, 132

            harmonized minimum terms and conditions of access, 133

            member nations, 126–127, 131–139, 140 (note 3), 141

        South Pacific Forums, 105, 127, 136, 255–256, 279, 281

        South Pacific GeoSciences Commission (SOPAC), 121

        South Pacific Nuclear Free Zone Treaty (SPNFZ Treaty), 106, 125–126, 207

            member nations, 126

        Tarawa Declaration, 136, 137, 255–256

        Treaty on Fisheries Between the Governments of Certain Pacific Island States and the United States, 133–134

    military activities, 207

Soviet Union (former), 16, 17, 77. *See also* Military activities

    fisheries, 235, 296

    marine mammals, 249

    nonliving resources, 379–382

perestroika, 35
SPNFZ Treaty, 126
SPAW Protocol. *See* Caribbean
Stewardship, 89, 92–93, 131, 188, 189, 313. *See also* Indigenous peoples
Stone, Christopher, 145, 171, 234, 397
Sumi, Kazuo, 232, 292
Sustainable development, 4, 24, 27, 188, 231–232, 348, 366–367, 395, 399
Sustainable living. *See* Sustainable development

Taiwan
    as distant water fishing nation, 132
    drift net fishing, 104, 127, 136, 138, 232–233, 236, 250, 279, 296, 298, 306, 316
Technology transfer, 163, 164, 178, 188–190, 350–353
Townsend-Gault, Ian, 377, 392
Treaty on Fisheries Between the Governments of Certain Pacific Island States and the
    United States. *See* Fisheries
Treaty of Tordesillas, 38
Truman Proclamations, 20 (note 1), 78–79, 394
Tuna. *See* Fisheries

United Kingdom. *See* Great Britain
United Nations. *See also* Fisheries
    high seas fisheries, 245–252
    Law of the Sea Office, 189, 195
    role in ocean governance, 167
United Nations Conference on Environment and Development (UNCED), 47, 234, 285–286
United Nations Conference on the Human Environment, 125, 314–315
United Nations Convention on the Law of the Sea (UNCLOS). *See* Law of the sea
United Nations Development Program (UNDP), 360
United Nations Educational, Scientific, and Cultural Organization (UNESCO), 189
United Nations Environment Programme (UNEP), 17, 104, 150, 175–176, 194, 200, 315
    Oceans and Coastal Areas Program Activity Center, 189
    regional seas program, 17, 19, 35, 160–163, 190–191, 209–210,
United Nations Food and Agriculture Organization (FAO), 189, 231, 272, 276, 278, 284,
    305–306, 310, 343, 360, 362
United Nations Industrial Development Organization (UNIDO), 360
United Nations Resolution 44/215 and 44/225. *See* Fisheries
United States. *See also* Fisheries; Military activities
    challenge to "freedom of the seas," 78
    Department of Commerce, 330, 333
    as distant water fishing nation, 132
    drift net fishing, 136–139, 233–234, 236, 253, 256–261, 298–300, 316
    free trade agreement with Canada, 333

United States (*continued*)
    Hawaii, 94–96
    hazardous materials, 199
    membership in SPNFZ Treaty, 126
    military assertion, 104, 207, 216
    nonliving resources, 378
    Oil Pollution Act of 1990, 209
USSR. *See* Soviet Union

Va'ai, A. V. S., 90, 113
Van Dyke, Jon, 9, 13, 145, 214, 407, 445
Vanuatu, 137, 292
Vienna Convention on the Law of Treaties, 298, 317

Wecker, Miranda, 145, 198
Wellington Convention. *See* South Pacific
Wesson, Delores, 145, 198
Western Pacific, purse seine net management, 135
World Commission on Environment and Development. *See* Brundtland Commission
World Meteorological Organization (WMO), 189

Yugoslav Model, assembly of ocean regime under, 29

Zaire, 137

Island Press Board of Directors